MODERN TRANSPORT GEOGRAPHY

MODERN TRANSPORT GEOGRAPHY

Edited by
B. S. Hoyle and R. D. Knowles
on behalf of the
Transport Geography Study Group
of the
Institute of British Geographers

JOHN WILEY & SONS
Chichester · New York · Brisbane · Toronto · Singapore

First published in Great Britain in 1992 by
Belhaven Press

Published December 1994 by
John Wiley & Sons Ltd, Baffins Lane, Chichester, West Sussex
PO19 1UD, England

National 01243 779777
International (+44) 1243 779777

Other Wiley Editorial Offices

John Wiley & Sons, Inc., 605 Third Avenue,
New York, NY 10158-0012, USA

Jacaranda Wiley Ltd, 33 Park Road, Milton,
Queensland 4064, Australia

John Wiley & Sons (Canada) Ltd, 22 Worcester Road,
Rexdale, Ontario M9W1L1, Canada

John Wiley & Sons (SEA) Pte Ltd, 37 Jalan Pemimpin #05-04,
Block B, Union Industrial Building, Singapore 2057

ISBN 0 471 94689 3

Typeset by Mayhew Typesetting, Rhayader, Powys
Printed and bound in Great Britain by
Redwood Books, Trowbridge, Wiltshire

Contents

List of Figures

List of Photographs

Notes on Contributors

Michael Browne is a Reader in Logistics in the Transport Studies Group, Polytechnic of Central London, 35 Marylebone Road, London NW1 5LS, UK.

Clive Charlton is a Senior Lecturer in the Department of Geographical Sciences, Polytechnic South West, Drake Circus, Plymouth, Devon PL4 8AA, UK.

Dr John Farrington is a Senior Lecturer in Geography in the Department of Geography, University of Aberdeen, St Mary's High Street, Aberdeen AB9 2UF, UK.

Dr Richard Gibb is a Senior Lecturer in the Department of Geographical Sciences, Polytechnic South West, Drake Circus, Plymouth, Devon PL4 8AA, UK.

Dr Derek Hall is Head of Geography in the School of Social Studies, Sunderland Polytechnic, Forster Building, Chester Road, Sunderland SR1 3SD, UK.

Dr David Halsall is a Senior Lecturer in Geography at Edge Hill College of Higher Education, St Helens Road, Ormskirk, Lancashire L39 4QP, UK.

Dr Yehuda Hayuth is an Associate Professor in the Department of Geography, University of Haifa, Haifa 31999, Israel.

Dr David Hilling is a Senior Lecturer in Geography in the Department of Geography, Royal Holloway and Bedford New College, University of London, Egham Hill, Egham, Surrey TW21 0EX, UK.

Dr Brian Hoyle is a Reader in Geography in the Department of Geography, University of Southampton, Southampton SO9 5NH, UK.

Dr Richard Knowles is a Lecturer in Geography in the Department of Geography, University of Salford, Salford M5 4WT, UK.

Dr Stephen Nutley is a Lecturer in Geography in the Department of Environmental Studies, University of Ulster, Coleraine, Co. Londonderry, Northern Ireland, BT52 1SA, UK.

Dr Kenneth Sealy is Reader Emeritus in the Department of Geography at the London School of Economics and Political Science, Houghton Street, London WC2A 2AE, UK.

José Smith is a Lecturer in Geography in the School of Geography, Kingston Polytechnic, Kingston-upon-Thames, Surrey KT1 2EE, UK.

Dr Brian Turton is a Senior Lecturer in Geography in the Department of Geography, University of Keele, Keele, Staffordshire ST5 5BG, UK.

Dr Alan Williams is Reader in Economic Geography and Head of the Department of American and Canadian Studies in the University of Birmingham, PO Box 363, Birmingham B15 2TT, UK.

Preface and Acknowledgements

Transport is a subject of universal interest and importance. The investigation and analysis of transport is therefore of concern to a wide variety of students, researchers and planners as well as to those directly involved in the operation of transport systems. The variety and complexity of issues involved, however, is such that no single author can realistically be expected authoritatively to cover the subject as a whole.

From its inception in the early 1970s the Transport Geography Study Group of the Institute of British Geographers has developed and maintained a close interest in the spatial aspects of transport phenomena, and its members have published extensively within the field of transport geography. The basic purpose of this book, which represents the outcome of a long-held ambition on the part of TGSG members, is to draw on the collective expertise of the Group and to present a wide-ranging contribution to the study of modern transport geography which we hope will be of use and value to students in a variety of inter-related disciplines.

Our policy as editors has been to encourage contributors to draw upon their own experience in the context of a particular problem or specific area of reference within modern transport geography. We have attempted to cover the major components of the subject, at a level appropriate to undergraduates following courses in transport geography and other cognate subjects, in an open-ended manner indicating where more detailed information may be located and where further research is needed.

It is a pleasure to record our gratitude to our fellow contributors for their cooperation in the preparation of this book, and to Iain Stevenson and his colleagues at Belhaven Press for their support and encouragement. Some of the maps and diagrams were prepared or amended in the Cartographic Unit of the University of Southampton, under the direction of Mr Alan S. Burn, and some were drawn in the Department of Geography at the University of Salford; others were submitted by authors in a finished condition, thus accounting for some variation in cartographic style. We also record our appreciation of secretarial help provided by Mrs Judy Rhodes at Southampton and by Mrs Marie Partington and Mrs Moira Armitt at Salford.

B. S. Hoyle
University of Southampton
R. D. Knowles
University of Salford

1

Transport Geography: an Introduction

Brian Hoyle and Richard Knowles

Transport is part of the daily rhythm of life. Mobility is a fundamental human activity and need, but is restricted by the friction of distance. As a complex industry in terms of land use, employment and functions, transport is a major factor interlinked with the environment and with the spatial distribution and development of all other forms of economic and social activity. Geographical theories, methods and perspectives contribute significantly towards an understanding of transport problems and their eventual solution.

Introduction

Transport is a topic of universal interest and importance. Most people wish to travel from one place to another, regularly or occasionally. Goods collected, extracted or manufactured, almost without exception, are distributed from place to place before consumption. If people need or wish to use services, which are generally provided at a limited number of places, they must travel in order to do so. Transport industries exist to provide for the movement of people and goods, and for the provision and distribution of services; and transport thereby fulfils one of the most important functions and is one of the most pervasive activities in any society or economy. 'There is no escape from transport . . .' (Munby, 1968, 7). In advanced countries, modern transport systems involving road, rail and air networks are generally well-developed; global economic integration relies upon efficient maritime transport; and the development of the less-developed parts of the world is substantially dependent upon transport. 'Even in the most remote and least developed of inhabited regions, transport in some form is a fundamental part of the daily rhythm of life' (Hoyle, 1973, 9).

The general importance of transport is not in doubt, but its very ordinariness leads to its general acceptance and to a widespread assumption that transport is not a particularly interesting subject of study, except when things go wrong. Transport is a focus of media attention when disasters occur, when strikes paralyse services, or when exciting innovations capture public interest or become the subject of controversy.

Transport services cannot operate perfectly all the time, and to a greater or lesser degree the travelling public always has a ready-made target for complaints, justified or otherwise. Yet most rational people would agree that the vast majority of transport services operate at a reasonable level of efficiency for most of the time: were this not so, economic and social systems would grind to a halt.

All transport systems, however, are capable of improvement: more extensive, faster and above all more efficient services are constantly in demand. The level of efficiency and customer satisfaction achieved by any transport service is fundamentally a political issue, for it rests largely on the level and pattern of public-sector resource allocation to transport as opposed to other economic sectors and on the conditions under which private-sector investment in transport is permitted or encouraged. It follows that 'almost every transport decision is a public issue' (Munby, 1968, 7); and that transport is 'an enormously varied, exciting and controversial area of study' (Whitelegg, 1981, 4).

The study of transport is not the sole prerogative of any one academic discipline, and transport is too important to be left entirely in the hands of its practitioners. Transport, by its very nature, lends itself to multidisciplinary study and to interaction between those who operate or use transport systems and those who control or seek to analyse them. Geographers have much to contribute to the study of transport, and transport geography is increasingly widely recognized as a useful and important component in the broad field of transport analysis.

Why transport geography?

Why is transport *geography* important? There are two main reasons. First, transport industries, facilities, infrastructures and networks occupy substantial areas of geographical space, constitute complex spatial systems and provide substantial numbers of widely spread jobs. Second, geography is concerned with interrelationships between phenomena in a spatial setting and with the explanation of spatial patterns; and transport is frequently one of the most potent explanatory factors. Transport is a measure of the interactions between areas; it also enables a division of labour to occur. Spatial differentiation, wider market areas and economies of scale in production are partly a product of transport availability and use; and the demand for transport, in turn, is partly a product of these factors (Gauthier, 1970; Hay, 1973; Button and Gillingwater, 1983).

Transport geography is thus concerned with the explanation, from a spatial perspective, of the socioeconomic, industrial and settlement frameworks within which transport networks develop and transport systems operate. The subject therefore centres upon dynamic interrelationships within transport itself and in transport-related contexts. A substantial and growing literature and an increasing interdisciplinary involvement on the part of transport geographers have led to an enhanced awareness of the importance of the spatial dimension in transport studies, and of the contributions transport geographers are making, individually and severally, to the further understanding and eventual solution of transport problems.

Some of these issues have been highlighted at a series of conferences organized by the Transport Geography Study Group of the Institute of British Geographers. For example, in the early 1980s, transport geographers and planners examined a wide range of transport and recreation issues at a meeting in 1982 (Halsall, 1982). As the Conservative government's policies of privatization, deregulation and competition were beginning to change the framework in which British transport operates, a range of public issues in transport were discussed by transport geographers (Turton, 1983). Transport geographers joined with planners, Transport 2000 and British Rail to identify problems of and prospects for rail-based rapid transit in Britain's increasingly congested and decentralized conurbations (Williams, 1985). Implications of the then imminent deregulation of local bus services were examined in depth at a symposium of transport geographers and economists (Knowles, 1985).

More recently, the Channel Tunnel and its estimated impacts on short Channel crossings and ferry terminals were evaluated by transport geographers, economists and planners together with representatives of Eurotunnel, British Rail, ferry companies and Dover Harbour Board (Tolley and Turton, 1987). The effects of technological innovation in transport on spatial change were examined in relation to high-speed railways, minibuses, unconventional modes, small ports, retailing and telecommunications (Tolley, 1988). Green modes of transport and traffic-calming measures were analysed on a European scale at a symposium of transport geographers, planners and transport pressure groups (Tolley, 1990). A joint British-Italian seminar of geographers, economists and mathematicians evaluated aspects of transport policy and urban development (Knowles, 1989). The revitalization of derelict docklands and waterfronts was examined on an international scale at a symposium of transport geographers, planners and political scientists (Hoyle, 1990).

Mobility

The study of transport rests essentially on two cardinal principles. The first is that *mobility is a fundamental human activity and need*. In all societies, environments and economies the movement of goods and people is a necessary element in functional and developmental terms. The word 'transport' describes this activity, whether in terms of a relatively straightforward transfer of people or goods from one location to another, over a short distance, or in terms of the infinitely more complex systems involving many different directions, modes and locations on an international scale. The transport industries constitute, basically, a response to these activities and needs; transport facilities are normally provided in response to, rather than in anticipation of, demand. As in all demand-led industries, there can rarely if ever be a perfect match between the transport facilities and services required or desired by a population or economy and the available infrastructure at a particular time.

It follows that, although transport in one form or another is part of the daily rhythm of human life in all societies and economies, most places and people suffer from restrictions on mobility. Such restrictions may be temporary or long-lasting – even, in personal terms, permanent; they may be very seriously disruptive or only marginally inconvenient. Most commonly these restrictions arise from economic factors, especially the cost of transport: most individuals and families cannot afford to make all the journeys they would ideally wish to undertake. In developed countries the mobility gap is widening, especially in rural areas, between the growing majority with regular access to a private car and the minority who are entirely dependent on declining public transport for access to shops, medical services, families and friends (Moseley, 1979). Industries and businesses naturally attempt to reduce transport costs by limiting movements and few governments can afford to provide modern transport facilities and services to satisfy existing demand, let alone to cater for anticipated future requirements. Of course, demand for transport is in a sense a function of available facilities and services: people always want more than they can have. Similarly, the services and facilities provided are clearly a function of demand, for unless demand exists there is no point in providing them.

Restricted mobility is inevitably a brake on development, in every sense. In modern Western cities, particularly, people are increasingly used to seeing, expecting and perhaps using transport facilities for mobility-deprived individuals and groups: those confined to wheelchairs, for example, or who have difficulty in using public transport and need specially adapted vehicles and access points to buildings and public spaces. Political factors, too, directly restrict the movement of individuals – refugees, hostages, guest workers, would-be immigrants – and in some countries it is still difficult or not permitted for most people to travel beyond national boundaries. Indirectly, political decisions underpin resource allocation to the transport sector, so that governments can restrict or enhance mobility by withholding or advancing investment in transport facilities and services. More generally, however, it is usually the broad level of economic development, together with the technological level of transport provision, that create a restrictive environment in transport terms.

Factors affecting the relative restriction of mobility are clearly interrelated, but we should also consider *the friction or restrictive impact of distance* itself. Everyone is aware that some places are more expensive than others to live and work in, or to trade from; and a major factor contributing to these spatial variations is the cost of transporting people to work and goods to market. Inland countries in West Africa, such as Mali and Niger, suffer economically because of the cost of transporting trade goods to or from coastal ports in neighbouring countries. The downward economic gradient which runs northwards from the coasts of Ghana, Côte d'Ivoire or Nigeria reflects ever-rising transport costs, as well as deteriorating environmental conditions, as distance from the sea increases.

In Australia, where the phrase 'the tyranny of distance' has gained a certain currency, 'the distance of one part of the Australian coast from another, or the distance of the dry interior from the coast, was and is a problem as obstinate as Australia's isolation from Europe' (Blainey, 1966, viii). The peripheral distribution of population, urbanization and economic development is an expression of the high cost of transport between coastal ports and distant hinterlands, as well as of the attractions of the coastal zones as compared with the frequently less-favourable environments of interior areas. The Norwegian maxim 'the forests divide, the mountain plateaux unite' reflects the fact that before mechanical transport became available, forests constituted a greater barrier to movement (Steen, 1942, translated in Knowles, 1976). The mountains also linked economically complementary areas thereby generating a demand for movement, unlike the valleys which linked areas with similar economies. Distance – and its chief enemy, efficient transport – are potent factors in any explanation of the geography of economic and social activity.

Multidisciplinarity

The second cardinal principle on which the study of transport rests is that *transport studies are essentially multidisciplinary*. Well-developed components within transport studies include *transport engineering*, concerned with the design and development of transport infrastructures and facilities; *transport economics*, dealing with the analysis of transport demand and the costs of meeting that demand in relation to other forms of economic activity; and *transport history*, concerned with the evolution of transport facilities, partly in terms of their intrinsic interest, in relation to past societies and economies, and partly as an explanation of the origins of modern transport systems. Politics and law are other major fields of study, as well as activity, in which transport issues loom large; for in all societies and economies, in various ways, transport is necessarily subject to some forms of political control and legal regulation, yet is itself a factor in the modification of political and legal systems (Banister and Hall, 1981).

Transport studies are therefore *multidisciplinary* in character, and are sometimes *interdisciplinary* as well. Fields of enquiry and activity such as transport economics and transport law are necessarily discrete, up to a point, reserving to themselves a specific body of information and a specific range of methods or techniques derived from the wider experience of their particular disciplines. Yet the evolution of transport law, for example, is conditioned by transport economics; the history of transport is an expression of transport technology; and transport technology, in turn, is intimately connected with transport engineering, which is dependent upon transport economics.

In all these ways, no specific academic

discipline, no subject-defined body of theory or methodology, can ever be totally self-contained; each must work with others, to a greater or lesser degree, drawing as required on a common fund of knowledge, in the pursuit of objectivity and truth. As in some other fields of study where interdependence is potentially a key to success, transport analysis requires that the scope and nature of an enquiry must be defined by the problem set, not by any preconceived notions of what is or is not relevant to a particular discipline.

The role of transport geography

Transport geography lies at the heart of this interlocking web of relationships; for, as an integrative science, geography draws some of its materials from related subjects and focuses upon the analysis of interrelationships, especially those expressed in spatial dimensions. It is obvious that so vast, exciting and varied a field as the study of transport geography encompasses a great variety of approaches. 'The skills which geographers have to offer are by definition useful ones and need no sterile efforts at carving out some indefensible space of disciplinary exclusivity' (Whitelegg, 1981, 4).

Transport geography, like the study of transport as a whole, rests on two essential ideas. The first is that *transport is itself a major complex industry in terms of land use, employment and functions.* Transport infrastructures and facilities occupy large areas of land and water space, and transport services provide substantial employment. In both these dimensions, transport is highly significant geographically. The second idea is that *transport facilities and services, taken as a whole or in terms of their component parts, are a major factor affecting the environment and the spatial distribution and development of all other forms of economic and social activity.* In this sense, transport is a major influence on virtually all other phenomena capable of analysis in terms of spatial variations and structures (Taaffe and Gauthier, 1973).

In this context, it is possible to approach the study of transport geography from several different directions. Perhaps the most common method is the *modal* approach which looks separately at road, rail, air and maritime transport systems and problems. This method is exemplified by Bird's studies of seaports (1963, 1971) and Sealy's work on air transport (1966). Such an approach is necessary for some purposes, especially the examination of specific issues, but it disregards the basic underlying intermodal interdependence of transport systems serving an area, however limited their degree of integration might appear to be. Another common approach to transport geography is through the study of factors – environment, economy, etc. – affecting the demand for, and the development of, transport networks (White and Senior, 1983). While this is a necessary element in an appreciation of the potentialities and limitations of transport systems, emphasis upon specific factors must avoid losing sight of the interdependent web of relationships of which they form a part. In this book traditional approaches based on modes and factors are largely rejected in favour of integrated analysis based on issues, problems, principles and examples. This approach, transcending factors and modes *per se*, shows how geographical dimensions can contribute significantly towards an understanding of transport problems and towards their eventual solution.

Explanation and assessment

It is, however, useful at the outset to suggest a broad, conceptual factor-based framework within which most if not all of the diverse elements of modern transport geography can be said to find a place. Figure 1.1 describes such a framework. The present-day transport system of any country or area cannot normally be explained by one factor alone. Explanations can be found, however, in a series of interrelated factors. Some of the more important factors are indicated in Figure 1.1 which shows how transport

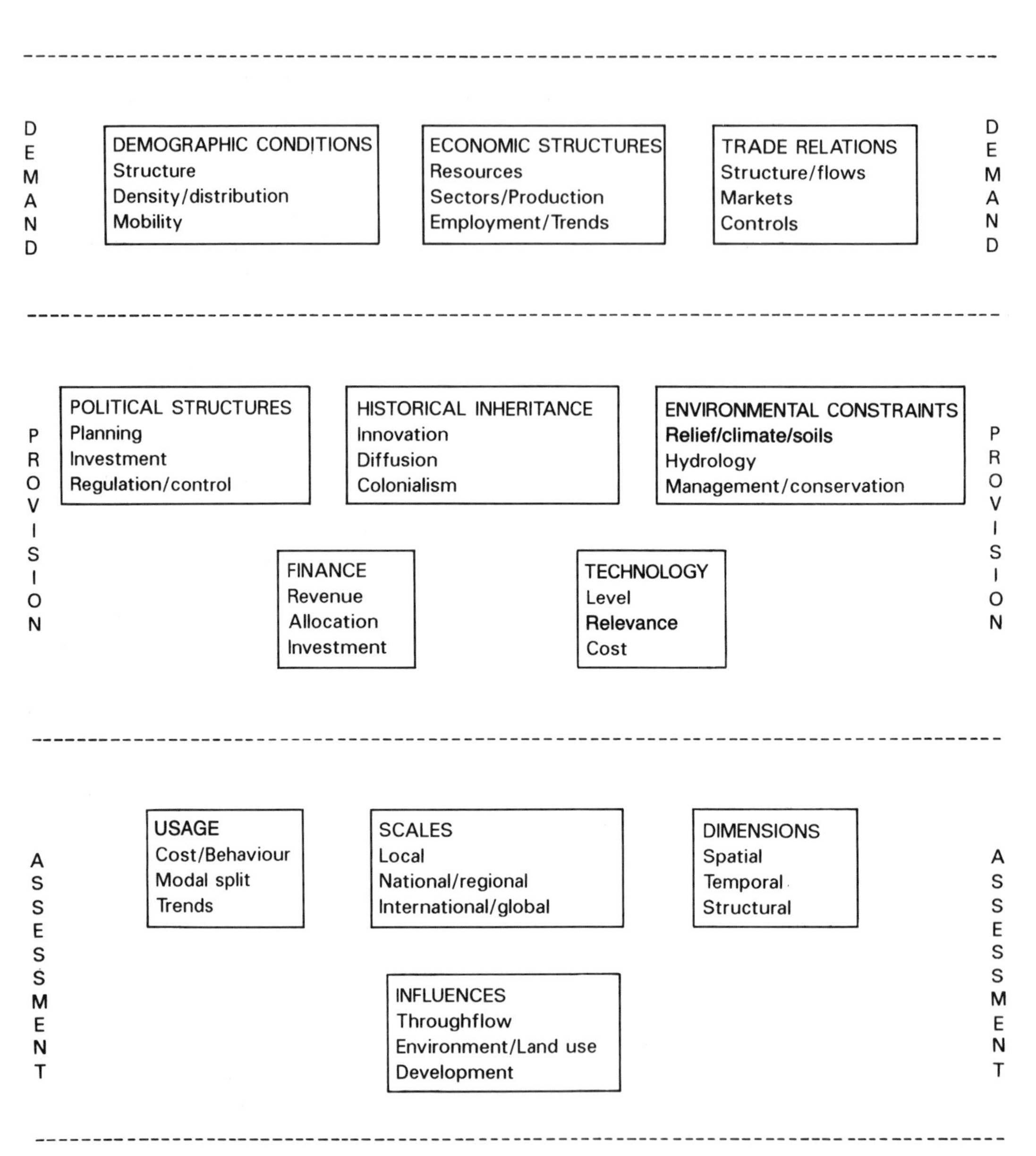

Figure 1.1 Some factors involved in transport demand, provision and assessment.

demand is influenced by economic and demographic circumstances and by trading conditions; how transport provision is constrained by political structures, inherited transport networks, environmental factors, available technology and finance; what affects transport usage and modal split; how the effects of transport are assessed at different scales and in different dimensions; and what influence transport has on land use, volume of activity, economic development and the environment (Fullerton, 1975; Blunden and Black, 1984; Barke, 1986).

These factors affect transport in different ways, influencing each other as well as affecting transport systems directly and indirectly.

Transport systems themselves also influence all these different areas of human activity and the physical environment within which they are set. Each factor may operate in positive, negative or neutral ways; each may affect transport on different scales, from the local to the global; and two basic dimensions – time and space – are involved. Such factors are not only useful as sources of explanation and understanding of transport systems and patterns, past and present; they are also prominent areas of influence and concern in terms of assessment, traffic forecasting and planning for the future.

Changes in transport demand usually originate with changes in the pattern of resource exploitation and are often stimulated by changes in population structure, density, distribution or mobility as well as by people's desire to improve living standards. Resource exploitation involves the extraction, processing and marketing of resources, requires an increasingly specialized division of labour to generate higher living standards and although stimulated by trading opportunities is limited by trade constraints.

The form of the transport network provided and the nature of the transport system which operates upon it is a product of competing constraints. Historical factors are essential to a proper understanding of modern transport systems, for all existing transport networks have been inherited from the recent or more distant past and from decision-making processes now modified or superseded. In Britain or France, this principle applies equally to Roman road networks and to motorways currently under construction. Although it is true that, as L. P. Hartley put it, 'the past is a foreign country – they do things differently there' (Hartley, 1953, 1), the importance of inherited transport systems and of superseded decision-making processes is that they provide part of the framework within which present-day decisions are taken and future developments planned. Historical legacies, in other words, provide one set of constraints which condition, positively or negatively, the ways and methods in which future transport systems can be designed and implemented.

The physical environment influences the development of transport infrastructures – roads, railways, seaports and airports – both directly and through the comparative costs of construction. The morphology of any specific component of a transport network – a railway station, an air terminal, a motorway, a container terminal – is set within a specific environmental context and its development raises particular environmental questions, problems and perhaps controversies.

All the factors discussed above are underpinned in many respects by technological factors. The technological characteristics of individual transport modes – pipelines, railways, canals, roads – impose limitations with regard to usage and maintenance costs, for example. Similarly, vehicles offer potentialities and impose limitations by reason of their individual or collective technological characteristics: bicycles, cars, ships, trains, aeroplanes, lorries and hovercraft all have appropriate physical, social and economic environments within which they operate and without which they either cannot operate or are unsuitable. Advanced technology is expensive and transport costs are therefore frequently a reflection of technological inputs. Together, technology and cost factors are closely related to environmental issues, for the adaptation of a transport system to physical conditions or to environmental concerns is dependent upon technological capacity and available financial resources.

There is a sense in which political factors transcend the logic of other factors discussed above and their interrelationships. Political decisions involving transport investment, like those in other spheres, hinge upon issues both broader and more specific than those outlined here. There is often a conflict between the demand for transport and the political will to provide it, or between the political objective of a transport innovation and its economic purpose or value. For example, the trans-Siberian railway completed to consolidate Russian rule over Siberia, and the widespread introduction of railways in Africa during the early European colonial period, underscore the significance of political motivations for transport innovation. The political entity of Canada was created by the British North America Act (1867) which required the building of the Intercolonial Railway to link the four colonies of Ontario, Québec, Nova Scotia and New Brunswick so as to enable the new country to function as a political and economic unit (Leggett, 1973). The Canadian Pacific Railway was a similar but much larger-scale legal requirement. British Columbia entered the Canadian Confederation in 1871 under an agreement which guaranteed the construction of a railway connecting the Pacific coast with Ontario within ten years. The controversies surrounding the Channel Tunnel and its connecting railways in Britain and France provide a contemporary example (Tolley and Turton, 1987).

Political considerations are significant in another sense in relation to transport. Governments are a major source of capital for investment in transport infrastructure, although private investment is also very important in some countries. In addition, governments are involved in the regulation of the supply of transport services, in the control of intermodal competition (to varying degrees), in safety control, in the coordination of investment allocation between modes and areas, and in decisions concerning pay and working conditions. In all these ways, governments are often in a position to decide what happens in transport terms, but decisions can only be taken in the context of consideration, evaluation, acceptance or rejection of all the relevant factors involved.

Economic factors involve a different set of perspectives. Traditionally, economic approaches to transport have involved the assessment and analysis of traffic flows – the collection, dissection and discussion of movements along a line, through a node or within a network, in relation to demand and costs. The objective of such approaches underlines the essential economic perspective, based on demand/cost relationships and on the comparative claims of other forms of investment or activity for available finance. These perspectives have led transport economists and planners to develop sophisticated quantitative transportation models to attempt to forecast future traffic trends and to identify interrelationships between different transport modes, expressed as the *intermodal split*. Implications have been assessed for investment and planning.

Transport can be defined not only as an economic facility but also as a social enabler, so it is therefore impossible to disregard social factors in transport analysis and planning. Social activities and characteristics constitute a basis for transport planning, and they may be developed or modified by available transport facilities. The analysis of journey-to-work patterns provides an important spatial link between economic and social factors, as does the wider question of accessibility to modern transport services, especially in rural areas. Recreational travel and shopping patterns are two other areas where social characteristics form an important part of transport analysis. In rural and urban areas, each socially distinctive locality generates its own type and pattern of demand for transport, and responds in its own way to the available services and facilities.

Transport and land use therefore influence each other. The development of tram and railway systems, for example, enabled a separation to occur of workplace from place of residence (Kellett, 1969; Ward, 1964). The later developments of motor buses, lorries and above all of mass-produced cheap private motor cars enabled a much greater suburbanization and deconcentration of urban and economic activity to occur. As a consequence it is now increasingly expensive to provide a decentralized urban area with adequate public transport services. City centres have become less accessible due to road congestion and parking difficulties, while bypasses and ring roads have enhanced the accessibility of the suburban fringe. However, where land use and transport planning have been coordinated and the decentralization of urban activity has been tightly channelled over several decades (as, for example, in Greater Copenhagen, Greater Oslo and Greater Stockholm) the city retains its accessibility (Fullerton and Knowles, 1991).

Transport investment can positively influence economic development, and often does so. Improving the speed, capacity and reliability of transport or reducing its price provides opportunities for widening market areas and increasing market share. This will only occur if finance, productive capacity, entrepreneurial skill and trading conditions permit. The frictional effect of distance has steadily reduced over time. This has helped, together with lower tariff and technical trade barriers, to produce a worldwide area for the supply of many raw materials and a world market for many finished products. Lower-cost transport, however, can lead to economic decline where inefficient high-cost industries or areas of production were previously sheltered from lower-cost competitors by the cost barrier of distance.

The influence and effects of transport on the environment is widely regarded as a critical issue in terms of air pollution, noise pollution, the overcrowding of urban places and the deterioration of rural areas. Modern efficient transport can substantially improve the quality of life in many areas and ways, but any improvement inevitably brings costs of various kinds. Environmental impact assessment is therefore a critical element in transport planning.

In considering the relative importance of these and other factors affecting transport in a particular country or area, geographers not only use general models but also emphasize the diversity of place and the specific combination of factors which helps to explain an individual transport problem or situation. In some cases, one particular factor may be overwhelmingly important. In most cases, however, we can only understand how a transport system operates, how it has grown and developed and how it is related to socioeconomic progress, if we examine all the relevant factors involved and the relationships between them. This principle applies equally to countries and areas at any particular level of development, from the most advanced to the least developed.

Summary and conclusion

Geography is concerned with environmental and human interrelationships in a spatial context, and transport geography is the study of transport systems and their spatial impacts. Just as population distribution is a delicate parameter of the interaction of a wide range of environmental and human factors affecting people and their activities, so also transport is an epitome of physical and socioeconomic interrelationships between individuals and groups in society, an ultimate yet dynamic expression of the demands and constraints that condition human mobility. In this wide-ranging context, transport geography plays an important role. It rests on the principles of universal mobility and multidisciplinarity; and is concerned with transport as an industry, a source of employment and a user of land and water space. Above all, however, transport geography investigates the effects of transport demand and availability on almost all other aspects of human society. Whether through modal, factor-based, demand-led or problem-based approaches, there is truly no escape from transport, and the study of transport geography helps us to understand why this is so. To a greater or lesser extent, the tyranny of distance affects us all.

References

Banister, D. and Hall, P. (ed.) (1981), *Transport and public policy planning* (London: Mansell).

Barke, M. (1986), *Transport and trade* (Edinburgh: Oliver and Boyd).

Bird, J. H. (1963), *The major seaports of the United Kingdom* (London: Hutchinson).

Bird, J. H. (1971), *Seaports and seaport terminals* (London: Hutchinson).

Blainey, G. (1966), *The tyranny of distance: how distance shaped Australia's history* (Melbourne: Sun Books).

Blunden, W. R. and Black, J. A. (1984), *The land-use/transport system* (Oxford: Pergamon, 2nd edn.).
Button, K. J. and Gillingwater, D. (eds) (1983), *Transport, location and spatial policy* (Aldershot: Gower).
Fullerton, B. (1975), *The development of British transport networks* (Oxford: Oxford University Press).
Fullerton, B. and Knowles, R. D. (1991), *Scandinavia* (London: Paul Chapman).
Gauthier, H. L. (1970), 'Geography, transportation and regional development', *Economic Geography* 46, 612–19.
Halsall, D. (ed.) (1982), *Transport for recreation* (Transport Geography Study Group, Institute of British Geographers).
Hartley, L. P. (1953), *The go-between* (London: Hamish Hamilton).
Hay, A. (1973), *Transport for the space economy* (London: Macmillan).
Hoyle, B. S. (ed.) (1973), *Transport and development* (London: Macmillan).
Hoyle, B. S. (ed.) (1990), *Port cities in context: the impact of waterfront regeneration* (Transport Geography Study Group, Institute of British Geographers).
Kellett, J. R. (1969), *The impact of railways on Victorian cities* (London: Routledge & Kegan Paul).
Knowles, R. D. (1976), *An analysis of transport networks in selected marginal areas with special reference to Norway*, unpublished Ph.D. thesis, University of Newcastle-upon-Tyne.
Knowles, R. D. (ed.) (1985), *Implications of the 1985 Transport Bill* (Transport Geography Study Group, Institute of British Geographers).
Knowles, R. D. (ed.) (1989), *Transport policy and urban development: methodology and evaluation* (Transport Geography Study Group, Institute of British Geographers).
Leggett, R. F. (1973), *Railways of Canada* (Vancouver: Douglas & McIntyre).
Moseley, M. J. (1979), *Accessibility: the rural challenge* (London: Methuen).
Munby, D. L. (ed.) (1968), *Transport: selected readings* (Harmondsworth: Penguin Books).
Sealy, K. R. (1966), *The geography of air transport* (London: Hutchinson).
Steen, S. (1942), *Ferd og fest: reiseliv i norsk sagatid og middelalder* (Oslo: Aschehoug).
Taaffe, E. J. and Gauthier, H. L. (1973), *Geography of transportation* (Englewood Cliffs: Prentice Hall).
Tolley, R. S. (ed.) (1988), *Transport technology and spatial change* (Transport Geography Study Group, Institute of British Geographers).
Tolley, R. S. (ed.) (1990), *The greening of urban transport* (London: Belhaven).
Tolley, R. S. and Turton, B. J. (eds) (1987), *Short sea crossings and the Channel Tunnel* (Transport Geography Study Group, Institute of British Geographers).
Turton, B. J. (ed.) (1983), *Public issues in transport* (Transport Geography Study Group, Institute of British Geographers).
Ward, D. (1964), 'A comparative historical geography of streetcar suburbs in Boston, Massachusetts and Leeds, England: 1850–1920', *Annals of the Association of American Geographers* 54, 477–89.
White, H. P. and Senior, M. L. (1983), *Transport geography* (Harlow: Longman).
Whitelegg, J. (ed.) (1981), *The spirit and purpose of transport geography* (Transport Geography Study Group, Institute of British Geographers).
Williams, A. F. (ed.) (1985), *Rapid transit systems in the UK: problems and prospects* (Transport Geography Study Group, Institute of British Geographers).

2

Transport and Development

Brian Hoyle and José Smith

Transport provides a key to the understanding and operations of many other systems at many different scales, and is an epitome of the complex relationships that exist between the physical environment, patterns of social and political activity, and levels of economic development. Viewpoints on relationships between transport and development continue to evolve. This chapter focuses on transport as a social facility and as an economic enabler at various levels within the transport spectrum. Through the medium of case-studies it interprets the concept of transport as a permissive factor rather than as a direct stimulus to economic development or spatial change.

Introduction

Transport is an epitome of the complex relationships that exist between the physical environment, patterns of social and political activity and levels of economic development. In advanced and developing countries, investment in transport is a matter for political negotiation, economic calculation and environmental consideration (Adams, 1981). Modern economies require, and assume, relatively sophisticated transport systems; yet changes are frequently contentious. A majority of the world's peoples, however, inhabit areas underprovided with even rudimentary forms of transport, and the need to facilitate economic development by providing improved transport services is often overwhelming.

Transport systems provide a key to the understanding and operation of many other systems at many different scales. At one extreme, intercontinental transport provides essential communication between the advanced and the developing worlds. At the other extreme, local transport to rural markets in many parts of the Third World is a vital component in changing dynamic socioeconomic structures (Barke and O'Hare, 1984; Mabogunje, 1989; Todaro, 1989). Viewpoints on relationships between transport and development, at various levels in the development spectrum, continue to evolve (Hoyle, 1973 and 1988; Leinbach et al., 1989). Theoretical approaches to transport network development provide useful perspectives, and historical evidence is often relevant, but present-day problems require a multifaceted approach.

This chapter examines the role of transport in the process of spatial change, in a variety of contexts, at various levels of development and in terms of a range of modes and intermodal systems. The focus is on transport as a social facility and as an economic enabler, at various levels within the development spectrum, and the interpretation rests on the concept of transport as a permissive factor rather than as a direct stimulus to economic development or spatial change. Two case-studies – Québec and Zimbabwe – allow varying interpretations of the balance between factors affecting the transport/development relationship; and a discussion on telecommunications transfer of information focuses upon a relatively new and rapidly developing link between less-developed societies.

Theoretical frameworks

Relationships between transport and development are underpinned, in one sense, by five essential ideas, whatever the specific modes of transport or level of development under discussion might be. The first is *the continuing relevance of the historical dimension*, in the sense that all existing transport networks have been inherited from the recent or more distant past, and not infrequently were designed to serve purposes rather different from those they are now expected to fulfil. Critics of the inadequacies of transport systems should bear in mind the costly process of adaptation to present-day and foreseeable requirements.

A second basic idea concerns *the degree of intermodal choice* available to an individual, group or society. The selection of a specific transport mode for a particular purpose depends upon a range of factors including the range of modes available, their relative cost, safety factors and convenience. In some areas, however, relatively few modes are available and therefore choices are severely restricted. Very broadly, a relative lack of intermodal choice characterizes the developing countries, while people in advanced countries usually have a much wider choice available. The restrictive impact of limited transport availability is a major factor affecting the development process in poor countries, while conversely the availability of a wide range of modes facilitates socio-economic progress in advanced areas of the world.

The *relative significance of different transport modes* is a third essential idea, and focuses particularly on road/rail competition and the declining importance of railways. Historically, railways provided the pioneer transport arteries in many world areas, but over time roads have proved more flexible and more competitive as well as providing more convenient door-to-door transport. Although rail-track mileage is increasing worldwide, specialized networks such as urban rapid-transit systems are increasingly popular and rail modernization schemes are frequently seen as a good investment, railways have nevertheless lost their arterial role.

A fourth essential idea centres on *the critical role of the seaport* in the context of regional, national and international transport systems. As a node within a multimodal surface transport system, linking water-borne and land-transport systems, a seaport is well placed to act either as a generative focus of development diffusion or, alternatively, as a parasitic node drawing off resources from its hinterlands and restricting economic growth. In transport studies, seaports are often ignored in favour of roads, railways and other land-based modes; yet, like airports, their pivotal position within an intermodal transport system is indispensable.

Fifth, it is important to emphasize that *transport systems are dynamic wholes*, and that their evolution and operation should be perceived and analysed in this context. In order to understand how any system operates today, what its problems are and how it might be improved, it is usually helpful to know how it has originated, grown and developed. Transport history, like history in general, is not bunk; it is part of mankind's inheritance and helps us to appreciate the potentialities and the limitations of the transport systems we have today or are planning for the future.

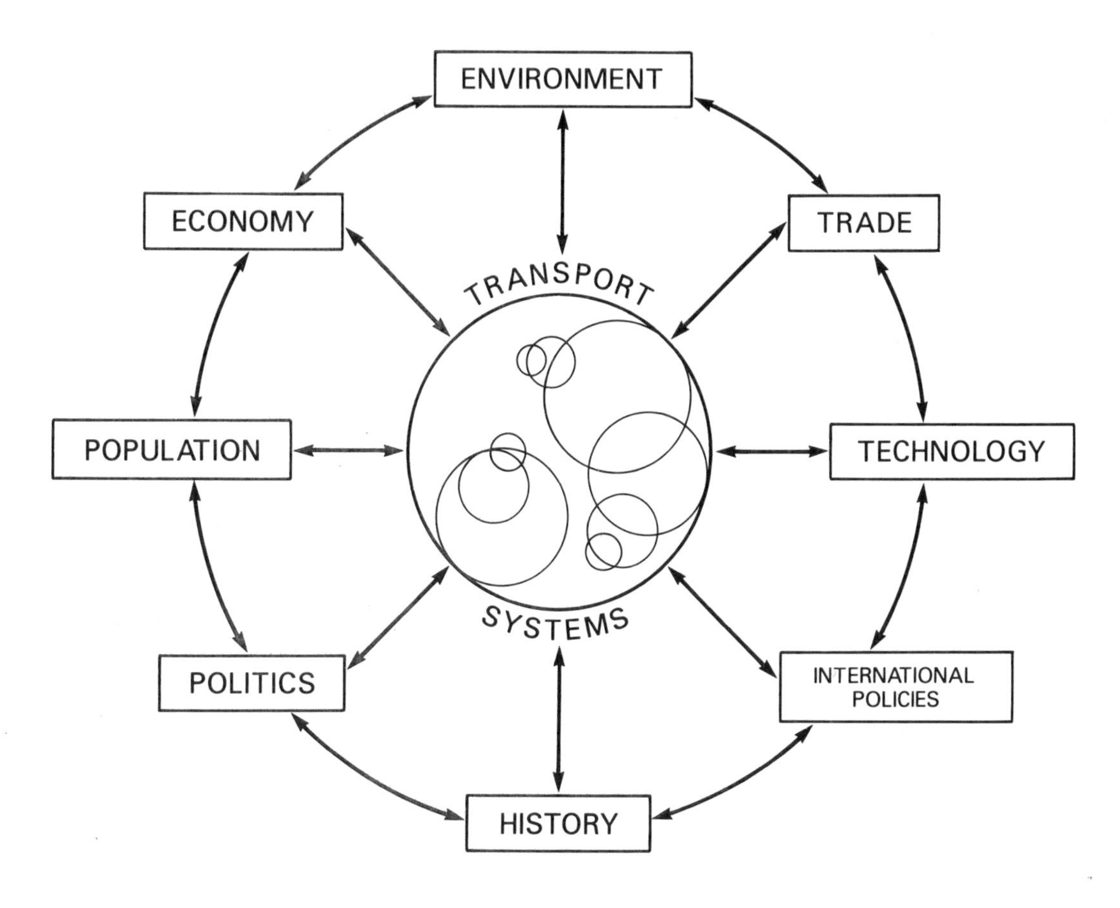

Figure 2.1 Some factors influencing the development of transport systems and transport/development relationships.

Factors involved in transport and development

Many factors are involved in the complex relationships between transport and development, and the present-day transport system of a country or area cannot normally be explained by one factor alone. Explanations can usually be found, however, in a series of interrelated factors. Some of the more important factors are indicated in Figure 2.1 which shows how transport systems of various types are influenced by environmental characteristics and constraints; by historical trends and conditions; by economic, political and demographic circumstances; by technological changes and by trading conditions.

Figure 2.1 emphasizes that these factors affect transport in different ways, influencing each other as well as affecting transport systems directly and indirectly. Transport systems themselves, together with the physical environment within which they are set, also influence all these different areas of human activity. Each factor may operate in a positive, negative or neutral way; each may affect transport on different scales, from the local to the global; and two basic dimensions – time and space – are involved. Examples of some of these factors at different scales of activity are given in Table 2.1.

In considering the relative importance of factors affecting transport in a particular country or area, geographers not only use general models but also emphasize the diversity of place, and the specific combination of

Table 2.1 Some examples of factors involved in the development of transport systems

Scale	*Environmental*	*Historical*	*Technological*	*Political*	*Economic*
Local	Soils/drainage Geomorphology	Settlement Culture	Roads	Enterprise Administration	Employment Core zones
Regional	Altitude Crop environments	Colonies	Railways	Trade	Road/rail competition
Continental	Distance	Colonialism	Sea routes	Independence	Markets
Global	Oceans	Isolation	Energy Air transport Telecommunications	Neocolonialism	Interdependence Prices/demand levels

factors, which help to explain an individual transport problem or situation. In some cases, one particular factor may be overwhelmingly important. In most cases, however, we can only understand how a transport system operates, how it has grown and developed and how it is related to socioeconomic progress, if we examine all the relevant factors involved and the relationships between them. This principle applies equally to advanced and less-developed countries.

The Taaffe, Morrill and Gould model

The interrelationships between transport and development in a developing area of the world have been drawn together in a very well-known model (Figure 2.2a) (Taaffe, Morrill and Gould, in 1963). The ideas it contains were originally derived from research in Ghana and Nigeria. (Figure 2.2b shows an adaptation to East Africa.) A series of six diagrams suggests how, in a hypothetical developing country, a transport network may gradually evolve from a pre-colonial situation of underdevelopment, through a period of external political intervention to the period of political independence.

The model represents the parallel evolution of political, economic and transport systems in a developing country. It is based on the usually valid assumption that, in such countries, transport networks are rooted, both physically and historically, in seaports. The first of the six small diagrams in Figure 2.2 suggests how a scatter of small, unconnected coastal trading ports represents the initial points of political contact and economic exploitation, and forms a basis for the introduction of inland transport modes. Arterial railways and roads are introduced, and in time extended, and feeder lines are added. Inland transport routes pass through or focus upon places of established political significance or economic opportunity, so the changing transport network in part controls, and is in part controlled by, the changing political and economic geography of the port hinterlands. As the sequence of diagrams indicates, some places grow and prosper, others decline, and the connectivity of the network is gradually increased. Eventually, in the final diagram of the series, the model suggests a fully integrated transport network closely attuned to a mature economy, a diversified development surface and a mature political system.

This model, like all models, is an over-simplification of reality. It provides a useful point of view but is open to discussion and reinterpretation. In terms of the relationship between transport and development, a number of questions may be asked. How, for example, does a developing country make the transition from one stage of the model to the next? What is the nature of the development process involved? How far do the processes implied in the model represent the

(a)

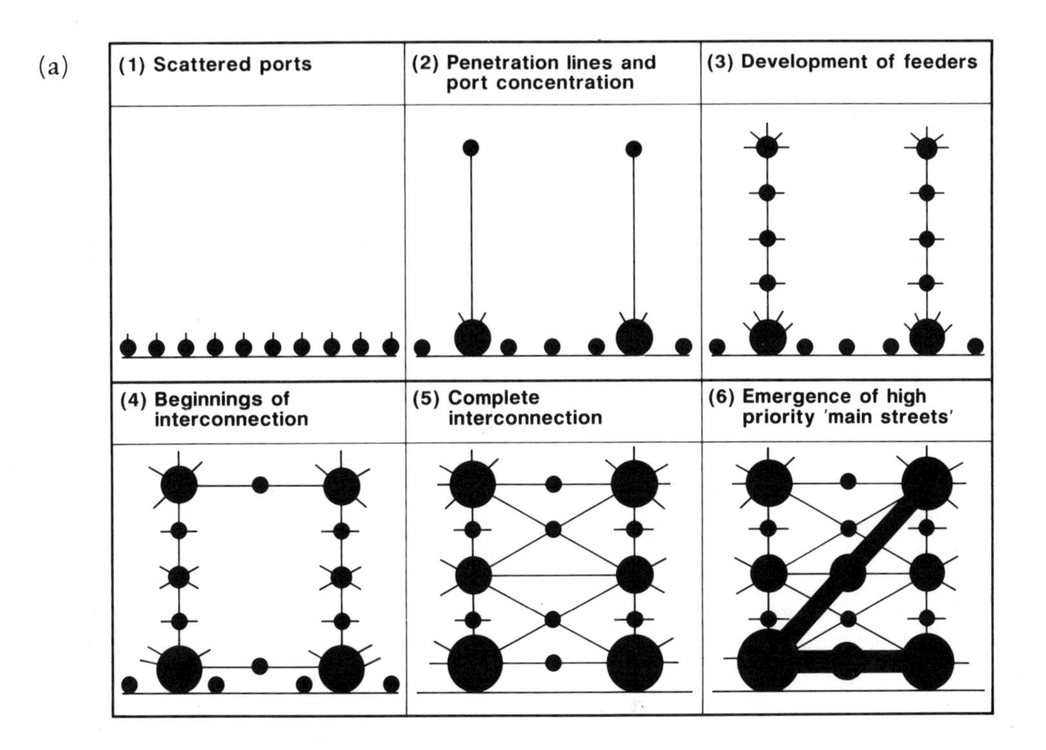

(b)

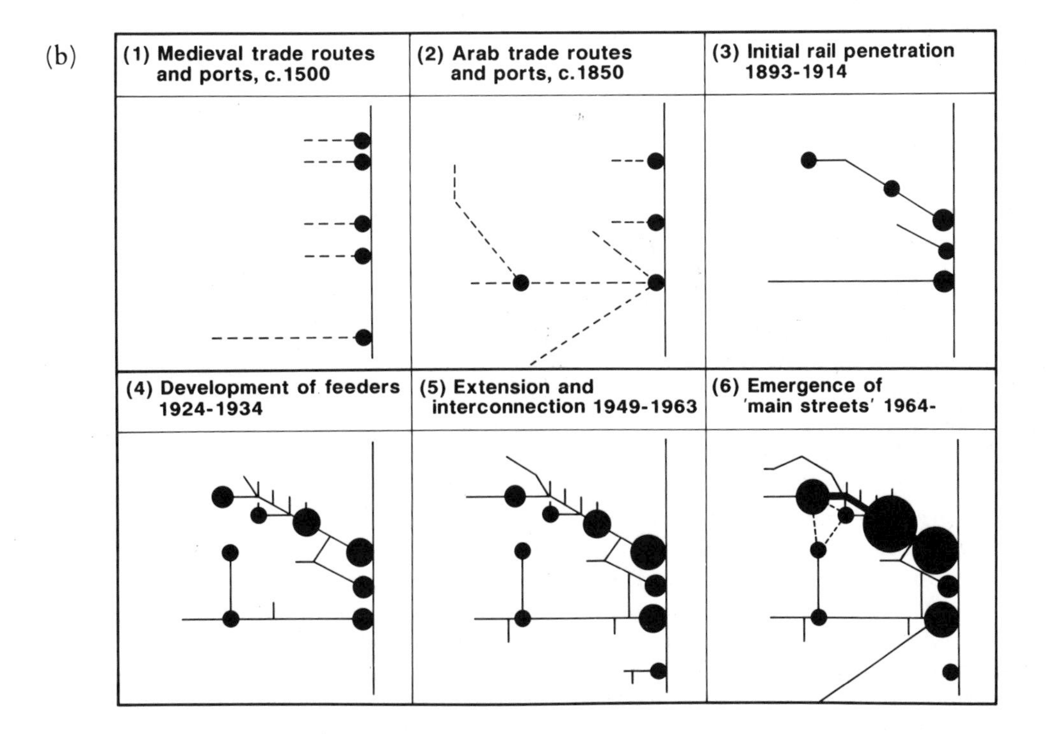

Figure 2.2 The Taaffe, Morrill and Gould model (a) (1963), and an adaptation to East Africa (b) (Hoyle, 1973).

real needs of developing countries rather than the aspirations of external, colonial powers? How relevant is the model to an understanding of precolonial roots of modern transport systems, or to present-day demands and objectives? Does the model bear any relation to the processes by which transport systems in advanced countries have evolved? All such questions admit a variety of answers, but whatever the nuances of interpretation, the intimate interrelationships between transport and development provide an essential underpinning.

Hybrid transport systems

An alternative and complementary perspective is provided by Rimmer (1977) who outlined the development of a hybrid transport system in less-developed countries, derived from the colonization process by which metropolitan powers used revolutionary modes of transport to penetrate indigenous systems and to gain both political control and cultural and economic dominance. The resultant restructuring of resource use, patterns of circulation, organization and outlook transformed the indigenous system, and instituted an interdependent relationship in which the colonizing power to a substantial extent controlled a two-way exchange of goods and services. This process eventually yielded a hybrid transport system in developing countries containing both indigenous and imported elements, often inadequately integrated.

Using terminology derived from Brookfield (1972 and 1975), Rimmer identified four phases in the evolving interrelationships between metropolitan and Third World countries in transport terms (Figure 2.3).

(a) A *pre-contact phase* involved no links between a Third World country and a distant power in the advanced world. Within the Third World country, a limited network of tracks, together with navigable waterways, supported a relatively restricted socioeconomic and political system.

(b) An *early colonial phase*, second, involved the establishment of direct contacts by sea between advanced and developing countries but did not produce radical changes in Third World societies: Europeans were largely content to dominate sea-transport routes and to establish foothold settlements such as trading posts and garrisons.

(c) A third phase of *high colonialism* involved more fundamental changes including the introduction of roads and railways, port facilities and inland transport nodes, and the diversification of economic activity (including industrialization and commercial agriculture) and settlement patterns (including rapid urbanization).

(d) A fourth *neocolonial* phase involves a substantial further diversification of the economic development surface of the Third World country and continuing (if modified) trade links with the former metropolitan power. The modernization of the transport system in the Third World country involves, at this stage, elements of rationalization, adaptation and selective investment in response to changing demands. There is, however, no radical adjustment to the systems inherited from earlier phases.

Just as the Taaffe, Morrill and Gould model was conceived in West Africa, the Rimmer model is essentially associated with South-east Asian experience. Both models are useful, as external and generalized viewpoints, in helping us to understand some of the processes of transport change. But because both models are broadly confined to the period of European colonialism and its immediate aftermath, they do not tell us much about how such transport systems may be expected to change during a more extended period of political independence. In this regard, the experience of the Americas and Australia is relevant, and suggests a fifth *mature independence* phase characterized by

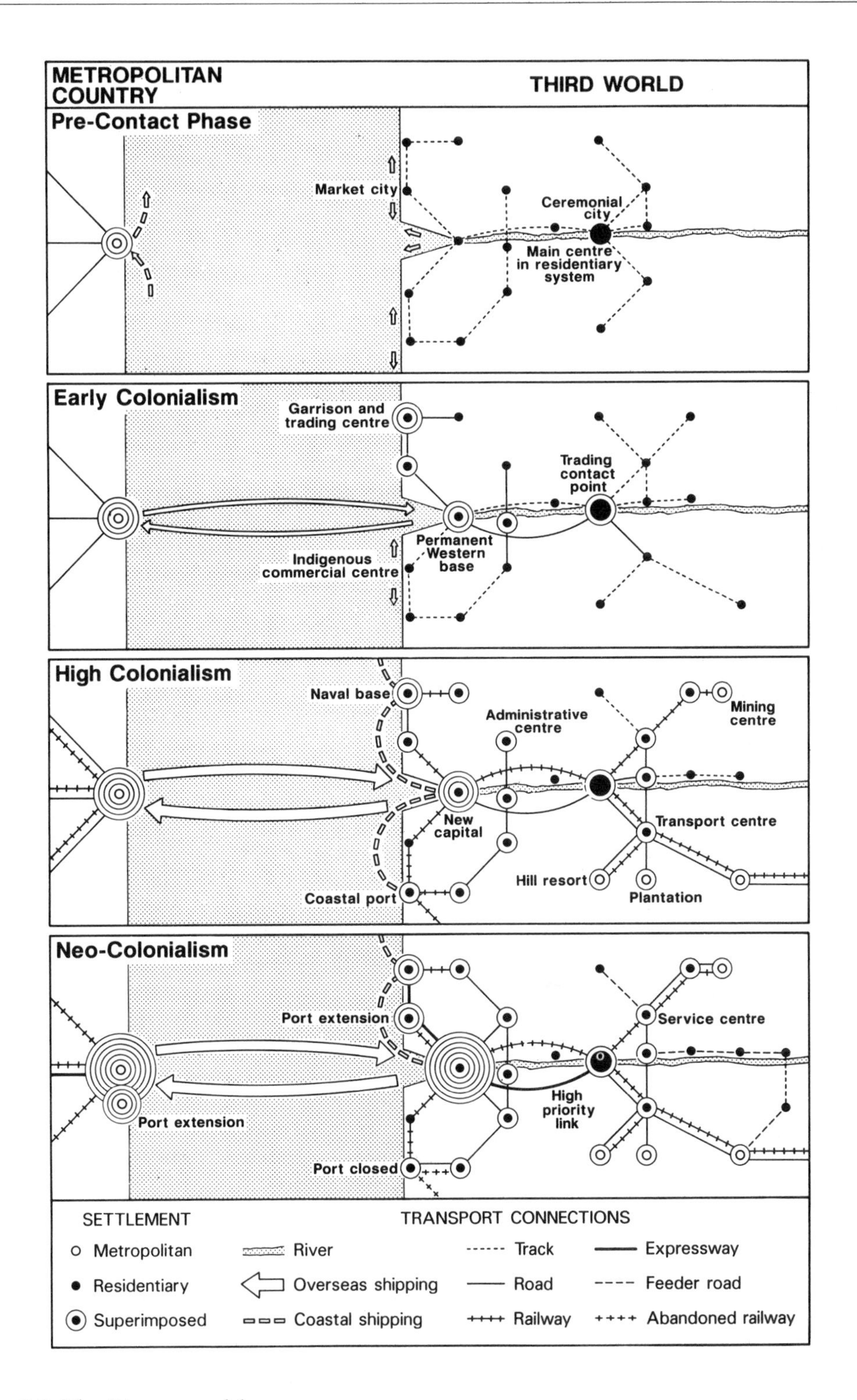

Figure 2.3 The Rimmer model.

continuing transport links along established lines (both within the former dependent territory and with the former colonizing power) together with substantial diversification and reorientation of external links and continuing intensification of internal networks.

Transport and development in Québec

The Canadian province of Québec (Figure 2.4) provides an illustration of relationships between transport and development in a variety of contexts, past and present. Québec is Canada's largest province in area (1.7 million sq. km.), but contains only about 6.5 million people, most of whom live in the St Lawrence lowlands, along the transport axis linking Montreal and Québec City. Transport is undoubtedly a fundamental factor affecting the development and character of the modern Québec economy and society (Langman, 1985; Allison and Bradshaw, 1990).

Overall the Québec environment is not a particularly promising one for settlement and development, in comparison with other North American areas into which European migrants moved. Yet the combination of the beaver, the offshore fisheries, the agricultural potential, the timber and above all the magnificent transport artery provided by the river made the St Lawrence lowlands a zone of attraction from an early date, and man's use of this environment and in particular its arterial routeway has had significant repercussions throughout much of North America (Patterson, 1982).

Heartland and hinterland: Québec as a focus of in-migration

Québec is the historic heartland of Canada, the location of the first substantial European settlements in North America, and the geographical core area of the confrontation between Britain and France in this part of their colonial empires. In terms of heartland and hinterland, Québec is (historically at least) central to the Canadian core (Waddell, 1987). Today, Québec remains a highly distinctive entity geographically, politically and socioeconomically.

European intervention, based on sea transport, took place in the context of earlier, overland migrations of so-called Indian peoples from another direction. Before the coming of the Europeans across the North Atlantic, much of Québec was inhabited by groups of Indians. Following the retreat of the last ice age over 10,000 years ago, migrants from Asia crossed via Alaska into Canada and gradually moved south and east, on foot or by canoe. For the Indians, the St Lawrence lowlands were an attractive but terminal zone in their long-continued migrations from the far north west. For the Europeans coming westwards across the North Atlantic, in contrast, the St Lawrence artery provided ready-made access, a gateway into new North American territories ripe for exploitation.

New France

The beginnings of European settlement and economic exploitation in Québec, initially dependent upon sea transport, date back many centuries. European explorers came across the Atlantic in search of new resources and new ocean-transport routes to the Orient and its riches. The St Lawrence became known as a convenient navigational entry route into new lands of economic opportunity, and trading posts were set up along the river. Between 1570 and 1650 many French fur traders came to the shores of Québec, bartering with the Indians who brought furs down to the St Lawrence from the north and west, overland and by canoe, discouraging maritime traders from penetrating inland.

New France, as areas coming under French influence in North America were known, centred on Québec City, founded in 1608 as a gateway in settlement and transport terms. The fur traders who set up their extensive

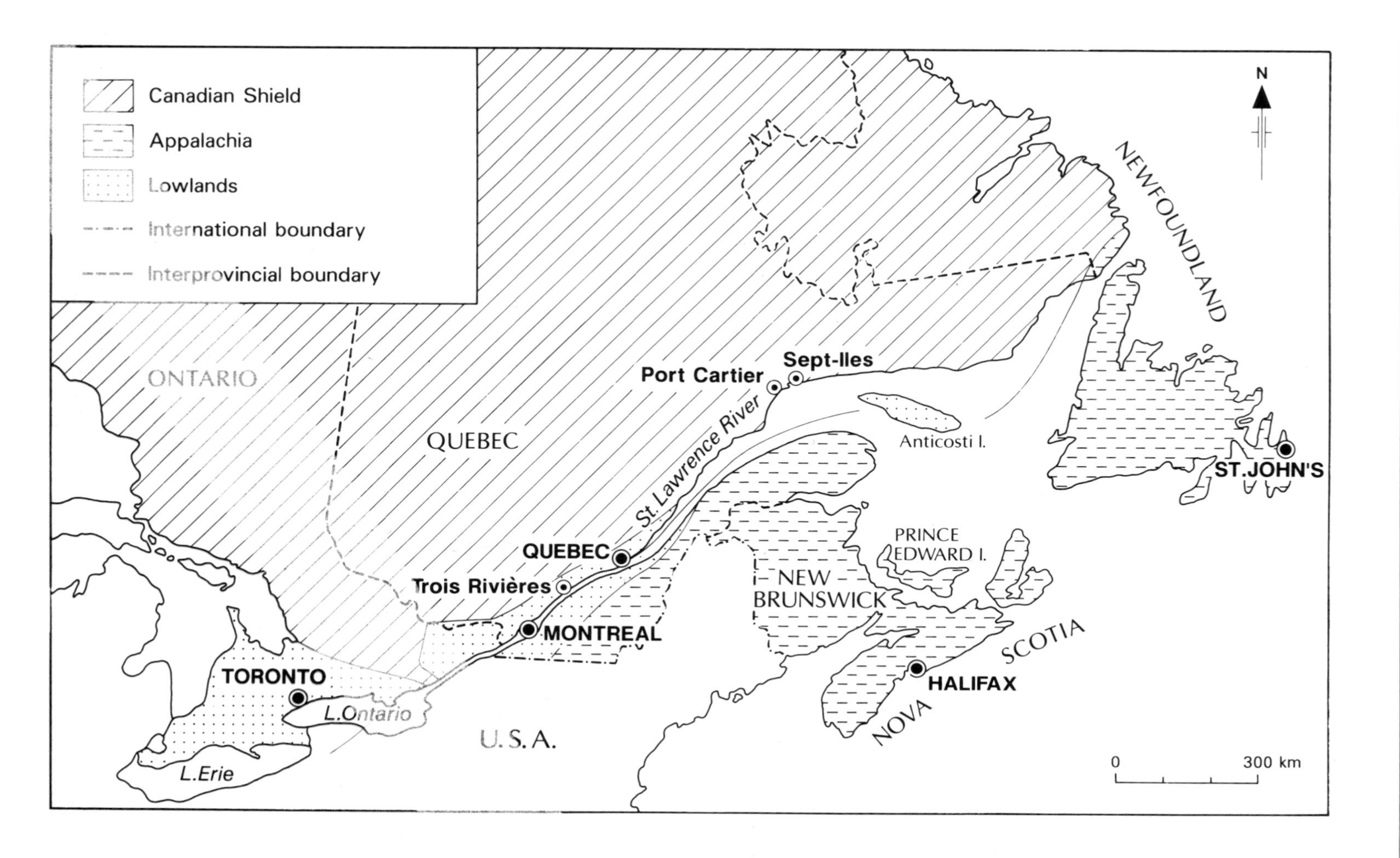

Figure 2.4 The St Lawrence artery.

supply and trading networks were known as the *coureurs de bois* – literally, the runners of the woods – an important if elementary form of transport at the time. When New France was declared a crown colony in the 1660s this encouraged immigration and settlement, the development of transport facilities and the diversification of the economy. Explorers travelled west to the Rockies and south to Louisiana. At its greatest extent New France covered a vast area; but it was no more than a trading empire, dependent upon slow, primitive forms of transport, especially rivers.

French pioneer settlers in seventeenth-century Québec lived off the land and off trade as far as possible. But they remained dependent on maritime transport for the expanding fur trade with Europe, for imported textiles and clothing, as well as wines and brandy to help them survive the rigours of winter. The farm unit and the family were the keynotes of this peasant society, and river transport was the essential network tying the communities together. The later seventeenth and early eighteenth centuries were punctuated by Anglo-French wars in north-eastern North America as settlements, transport routes and territories were fought over. Eventually, Britain clearly gained the upper hand: Québec City, the strategic gateway, was captured by the English in 1759 – an encounter in which the river played a crucial part; and all of New France was formally ceded to Britain in 1763. In spite of the advantages of the St Lawrence transport artery, therefore, the French endeavour in North America was relatively unsuccessful, partly because the English colonies along the eastern seaboard of the continent were far more accessible and more attractive to settlers.

British rule

At first, the change from French to British rule did not fundamentally alter the Québec economic and transport systems, but as the nineteenth century progressed a society undergoing rapid transformation began to generate new forms of economic activity and new transport demands. The creation of an English Upper Canada (Ontario) and a French Lower Canada (Québec) in 1791 reflected Anglo-French rivalry and also reactions to the early politico-geographical evolution of the USA. The American Civil War (1861–5) convinced Canadians of their need to remain politically independent, and produced a political climate favouring union between the various British colonies in Canada. The British North America Act of 1867 created the Canadian Confederation, signalling the end of the British colonial regime and the birth of modern Canada. These political changes were in part a reflection of transport factors, and they gave rise to new demands for transport and to transport innovations.

Political independence, economic expansion and transport development

The expansion of transport networks, especially railways, has played an important part in the process of economic development in Québec. The first great period of railway construction began in the 1850s, and by 1885 the Canadian Pacific Railway linked the Atlantic and Pacific coasts with Montreal as a focal point within the system. By the 1870s industrialization, urbanization and the influx of French urban employees meant that Montreal had become eastern Canada's major economic and transportation focus, where entrepreneurs were investing heavily in industries, railways, and banks.

The early twentieth century brought accelerated economic development in Québec, and Montreal became increasingly a nodal service centre for the development of the Canadian west – exporting wheat from the prairies through its growing port and sending its own industrial products along the railways to the pioneer fringe. During the First World War the trans-Canadian Pacific Railway was duplicated by the consolidation of other privately owned railways such as the

Canadian National Railway. In Québec, new industries based on hydro-electric power exploitation began to appear – aluminium, chemicals – as well as a modernized forestry and paper industry. This industrial progress, partly based on US investment, clearly led to increased transport demands met by port development, road construction and rail-network improvements. While transport improvements clearly helped to bind the emergent Canadian nation together in political as well as economic terms, by the 1930s they also facilitated the beginnings of the migration of economic power south-westwards through the Canadian heartland away from Montreal and towards Toronto.

Recent decades have seen a scientifically-based modernization drive, as Québec tried to keep its place in the forefront of twentieth-century North American development, although increasing economic prosperity has been paralleled by increased dissatisfaction with Québec's place within a federal Canada. Rapid postwar economic growth – in mining, manufacturing and service industries – involved rapid urbanization, improved transport and the further marginalization of the farming community. Today, the Québec economy depends substantially upon natural resource exploitation, including vast hydro-electric power resources, iron-ore deposits, abundant timber and numerous other minerals including asbestos, copper, mica and lead. Growing industries include pulp and paper, chemicals, aircraft, and food processing.

The Québec port system

Ports and port cities have played, and continue to play, an important part in the economic life of Québec. The development of the province, as in a sense the development of Canada as a whole, is rooted in maritime exploration and trade, and in the foundation of coastal settlements which provided an initial basis for movement into the interior. Today, several of Canada's major cities – including Montreal and Québec – continue to function as seaports, although their urban economies are now greatly diversified.

Québec possesses a great variety of ports, paralleled by a complex management system involving federal, provincial and local government. Four of the top ten Canadian ports, measured by total cargo throughput, are located in Québec province (Table 2.2). Two of these, Montreal and Québec City, are major multifunctional general cargo ports; the other two, Port Cartier and Sept-Iles, located on the north shore of the St Lawrence estuary, serve primarily as exporting ports for iron ores and other minerals extracted from the northern interior of Québec province. The province is also served by a variety of smaller ports such as Trois-Rivières which fulfil important local and regional functions.

Table 2.2 The top ten Canadian ports by total throughput, 1986 to 1989 (million tonnes)

	1986	*1987*	*1988*	*1989*
1 Vancouver (BC)	57.2	64.6	70.3	63.8
2 Sept-Iles (Qué)	20.7	19.6	23.0	23.3
3 Port Cartier (Qué)	19.3	23.1	22.5	21.3
4 Montreal (Qué)	20.8	21.4	21.8	20.3
5 Halifax (NS)	13.5	15.1	14.8	16.3
6 Québec (Qué)	11.3	18.3	17.9	15.4
7 Saint John (NB)	11.9	13.0	14.7	14.6
8 Thunder Bay (Ont)	17.7	19.4	17.3	13.4
9 Hamilton (Ont)	10.4	10.9	12.9	12.5
10 Prince Rupert (NB)	10.6	13.8	12.7	11.6
Total top ten ports	193.5	219.3	227.9	212.4
All Canadian ports	327.6	362.2	389.9	363.4

Source: Statistics Canada, *Shipping in Canada, 1989* (Ottawa, 1991).

Table 2.2 lists all Canadian ports handling over 10 million tonnes in 1987 and shows that no Québec port approaches Vancouver, the leading Canadian port, in this respect (Forward, 1982). It is interesting to note, however, that Thunder Bay (Ontario), the interior terminal on Lake Superior of the route plied by ocean-going vessels using the St Lawrence Seaway, handles a similar total throughput to the four Québécois ports on the list.

Québec today is an important component in the Canadian overseas trading economy, exporting vast quantities of grains, ores and forest products to a wide range of destinations, as well as relying on international container trades for a wide variety of imported and exported goods. It follows that the efficiency of Québec's ports is a matter of critical provincial and national importance in developmental terms. Any port system is by definition dynamic, and the changing character of port activities, together with the broad sweep of late twentieth-century urban economic and social change, have radically altered the historic relationships between ports and cities, and between ports and hinterlands in recent decades.

This has affected the character of individual Québec port-cities, modified the provincial port system, and produced some striking examples of the now widespread phenomenon of waterfront revitalization. While continuing to serve as maritime terminals, older zones within port cities such as Québec and Montreal have acquired new functions and characteristics in their new role as tourist-historic cities. In this sense the sites of some of the earliest European settlements in North America, doorsteps from which development spread far and wide, have themselves experienced further changes as their functions are once more transformed.

Transport and rural development in Zimbabwe

Over the past 100 years the relationship between transport and development in Zimbabwe has reflected the experience of many developing countries. On achieving independence in 1980, Zimbabwe inherited a transport network adapted to the needs and aspirations of external powers and typified the colonial stages of the spatial models proposed by Taaffe, Morrill and Gould (1963) and Rimmer (1977). Post-independence policies have sought to maintain this network and its external linkages which are vital to the nation's economy (Griffiths, 1990; Smith, 1988). However, development policies under the socialist government have given higher priority to the needs of the African rural areas (Drakakis-Smith, 1987). As in many other developing countries, rural accessibility and mobility have become key features in rural development planning (Barwell, et al., 1985).

The colonial transport network

The emergence of Zimbabwe's modern transport system dates back to the 1890s. Following the 1885 Treaty of Berlin which initiated the 'Scramble for Africa', Cecil Rhodes expanded his interests northwards into the Limpopo and Zambezi valleys. Extending the South African railway network through Bechuanaland (now Botswana), the British South Africa Company reached Bulawayo in 1897 and crossed the Zambezi at Victoria Falls in 1902. Meanwhile, a second major penetration route was being established from the east, linking the Mozambique port of Beira with Umtali (now Mutare) in 1898 and Salisbury (now Harare) in 1899. Finally the two routes were connected in 1902 along the ridge of the high veld by the Bulawayo to Salisbury line (Kay, 1970).

The early railway routes were to exert considerable influence over subsequent network development and settlement patterns. The policy of settling Africans into reserves and selling land rights to European immigrants began in the 1890s and the patterns of land ownership were well established by the time of the 1930 Land Apportionment Act. European settlement, attracted by minerals or agricultural potential in climatically favourable areas, dominated the high veld and eastern highlands. African reserves were confined to the less productive middle and low veld areas, isolated from the major centres of economic activity (Figure 2.5).

Such contrasts were to become more marked once road building began in

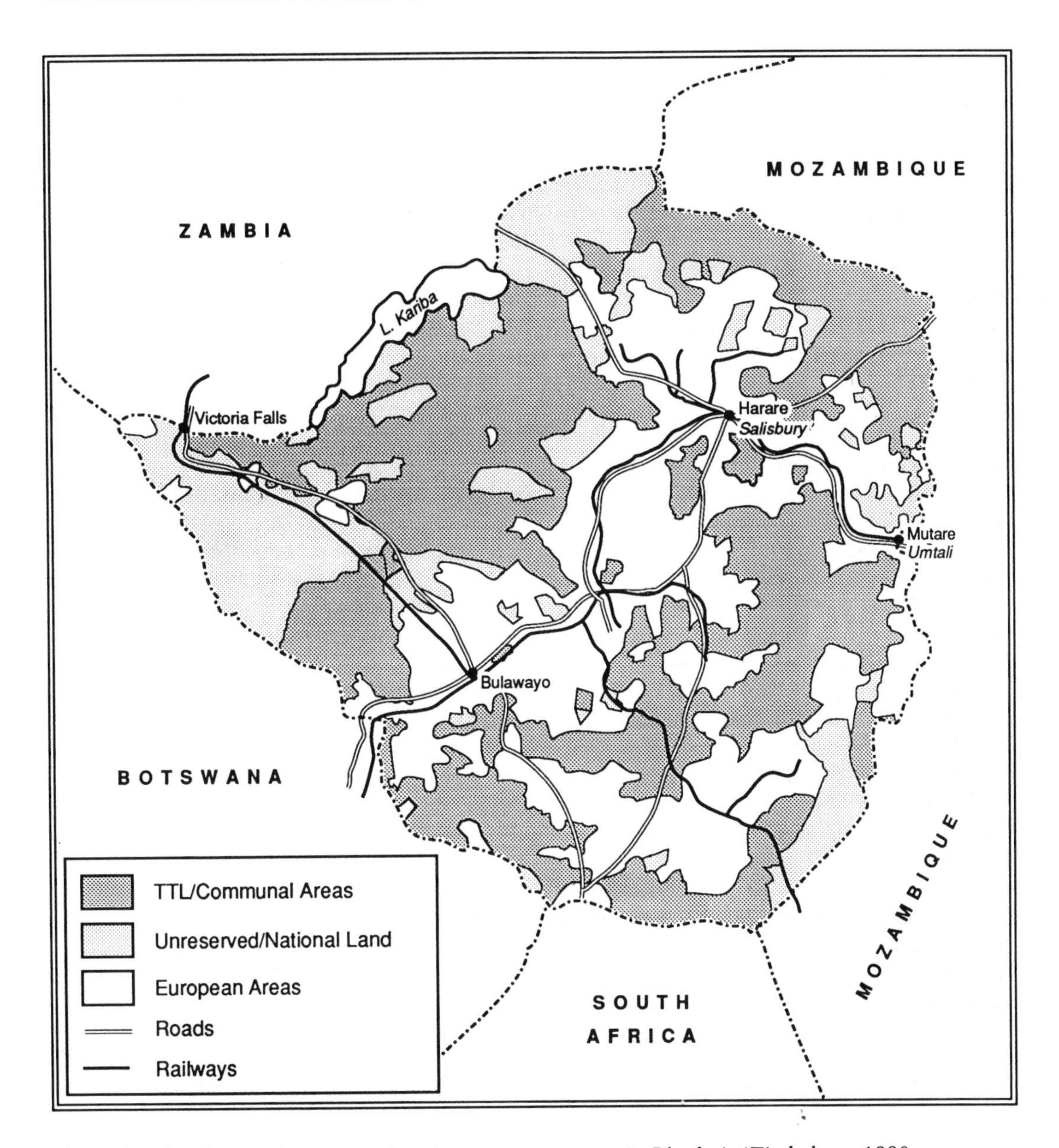

Figure 2.5 Land apportionment and major transport routes in Rhodesia/Zimbabwe, 1980.

Rhodesia during the 1930s. Main roads paralleled the railways linking major centres of European commercial development and were maintained in good condition, especially during the years of the Central African Federation (1953–63) and, for security reasons, during the period of Southern Rhodesia's illegal unilateral declaration of independence (1965–80). Moreover, throughout the European commercial farming areas, rural councils invested local taxes in a substantial feeder road network. In contrast, lack of resources severely limited expenditure on the road network in the Tribal Trust Lands, which were heavily dependent on subsistence agriculture. The liberation struggle which preceded independence in 1980 prohibited even limited maintenance and many African rural areas were left without a basic transport infrastructure.

Post-independence rural transport

Since independence in 1980 Zimbabwe has focused much of its national development effort on the reconstruction of the Communal Areas (the former Tribal Trust Lands). Fifty-seven per cent of the country's population live in these areas, which are characterized by dependence on agriculture, land degradation and outward migration. Most are located in ecologically marginal zones and drought is a major problem. Upgrading the rural road network has been regarded as a major prerequisite for improving the social and economic conditions in such areas.

Immediately after independence, road transport was re-established in the Communal Areas through an emergency road construction programme. In 1984, a Rural Roads Programme was announced which is scheduled for completion in 1992. The objectives of the programme include: facilitating the marketing of surplus agricultural produce; equitable access for all households to basic facilities and service centres; and the provision of all-weather roads suitable for bus transport. In total the programme aims to provide 16,600 km. of primary rural roads at a 1986 cost of 60 million dollars. Foreign aid and expertise, mainly from West Germany, have been of major importance to the programme which is managed by the District Development Fund (DDF) under the guidance of the Ministry of Local Government, Rural and Urban Development.

Household access to primary roads has been a key element in the design of the feeder road network. 'Scotch carts' pulled by animals are the major form of transport in the Communal Areas and the network has been designed so that ideally each household is within ten km. of a primary road (this distance represents a day's return journey using a scotch cart). Primary roads have all-weather surfaces suitable for bus transport and link with rural service centres and the national road network. They are supplemented by a secondary road system (not all-weather) which aims to reduce the homestead-to-road distance to five km. In hilly terrain, where scotch carts are impracticable and loads are carried by human porterage, only primary roads are provided but these aim to limit the homestead-to-road distance to three km. Road construction is preceded by a standard appraisal scheme devised by the DDF and implemented at regional level throughout the country.

The provision of primary roads has encouraged the expansion of the long-distance rural bus networks. Since Communal Areas are characterized by a lack of urban development, most services link these areas with major urban centres located in the former European commercial areas, especially Harare and Bulawayo. By 1985 nearly 200 firms operated throughout Zimbabwe: each tended to serve a specific locality, all were privately owned and most were family concerns. However, post-independence expansion has not been without its problems. The industry is highly dependent on imported equipment and has faced rapid increases in costs and shortages of spare parts consequent upon the growth of Zimbabwe's foreign debt. Fare increases have been limited by the government and have not kept pace with costs. Moreover, the network exhibits major inequalities; marginal areas have low levels of service provision while some major routes, for example those linking Communal Areas of the north-east with Harare, offer up to ten services a day.

Rural transport and agricultural marketing

Combined with schemes to promote smallholder cultivation, the rural roads programme has facilitated the expansion of agricultural marketing throughout many of the Communal Areas. This trend is most apparent for staple food crops, especially

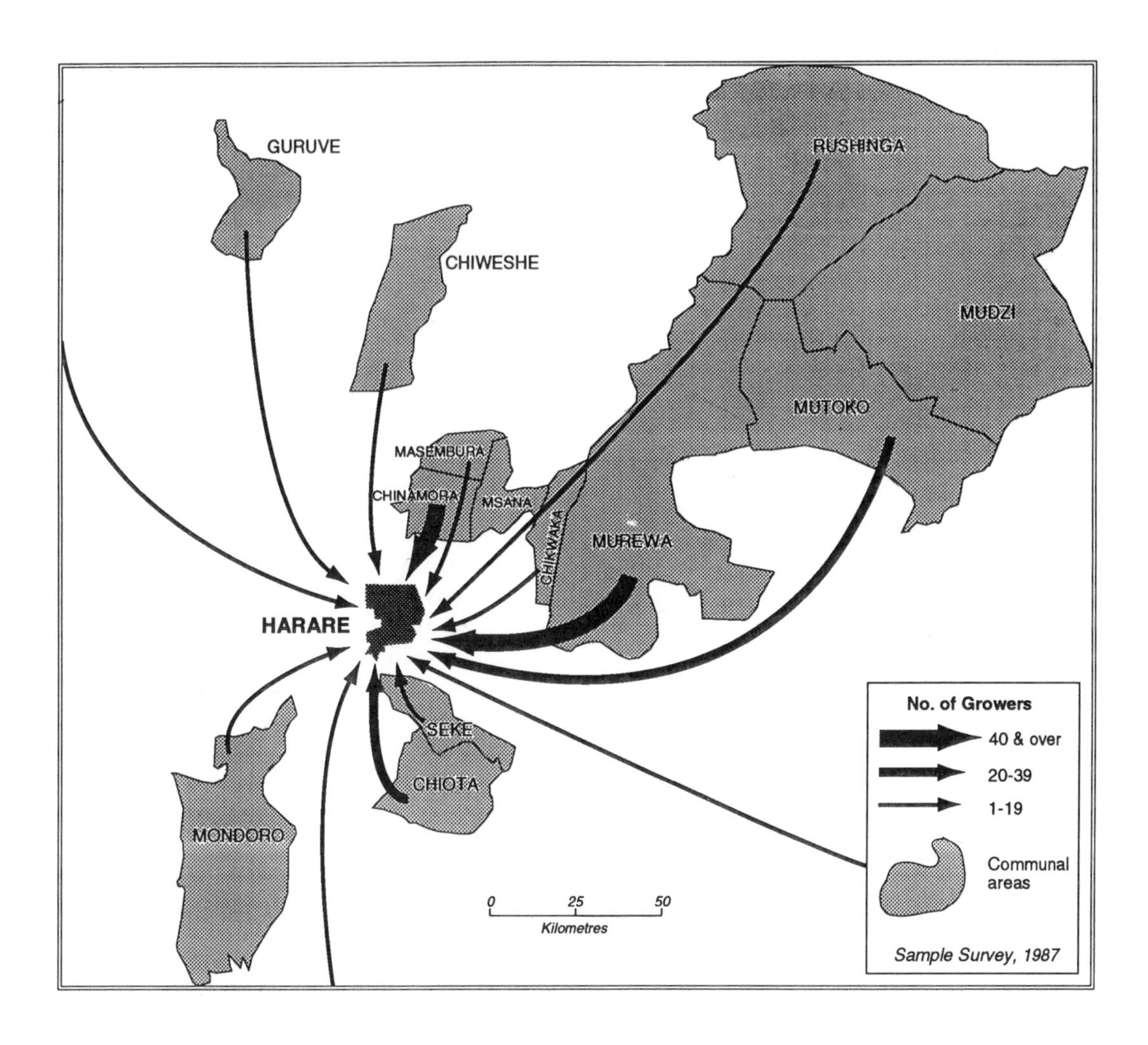

Figure 2.6 Horticultural marketing into Harare, Zimbabwe.

maize, and export crops such as cotton which are collected and sold by the state-marketing boards. The expansion in maize output, for example, has not only allowed Zimbabwe to become self-sufficient in its major food crop, but also to export to neighbouring food-deficit countries. In addition, the expansion of the rural road network and the operation of long-distance buses has enabled many communal farmers personally to participate in the private-marketing sector through the sale of horticultural crops. These are widely grown for home consumption on small plots of irrigated land throughout the Communal Areas. In addition, it is estimated that 65 per cent of farmers in northern Zimbabwe also sell horticultural crops, providing families with an important source of regular income.

The Communal Areas and middle-order towns throughout northern Zimbabwe offer only limited marketing opportunities for horticultural crops and many farmers travel to Harare to sell their produce. The spatial impact of the Harare market is therefore considerable and extends into the Communal Areas located in the north-east of the country (Figure 2.6). Of necessity, trips are frequent

Table 2.3 Farm to main road journeys in Zimbabwe

Distance		*Transport mode*	
Distance (km)	*% growers*	*Mode*	*% growers*
1.0 or less	32	Porterage	13
1.1–3.0	33	Wheelbarrow	10
3.1–5.0	14	Scotch cart	58
5.1–10.0	7	Private car	2
10.1 and over	2	Lorry	16
Not given	12	Other	1
Total (N = 182)	100		100

Table 2.4 Journeys into Harare

Distance		*Transport mode*	
Distance (km)	*% growers*	*Mode*	*% growers*
1–25	3	Grower and produce by bus	70
26–50	21	Grower by bus/ produce by lorry	6
51–75	24	Grower by bus and lorry/produce by lorry	5
76–100	30	Grower and produce by lorry	16
101–125	10		
126–150	6	Grower and produce by private car	3
151 and over	6		
Total (N = 182)	100		100

since fruit and vegetables are produced in small quantities and most are highly perishable. Following a recent field survey, detailed analysis of the travel patterns of a sample of almost 200 growers selling in Harare illustrated the complexity of the journeys.

Most journeys were made in several stages, each of which might involve a different transport mode and a separate cost. In the absence of direct access to motorized transport, over 80 per cent of growers had to carry their produce to the main road. The majority owned either a wheelbarrow or an ox-drawn scotch cart but 16 per cent had to hire transport (Table 2.3). Sixty-eight per cent of growers had to travel over one kilometre to a primary road but few recorded any problems of access. In this respect, communal farmers have greatly benefited from the post-independence provision of rural access roads.

Less than four per cent of growers had their own transport and 81 per cent used long-distance buses at least once a week to transport their produce and/or themselves into Harare (Table 2.4). Substantial distances were a feature of the journey patterns: only 24 per cent travelled less than 50 km. while 22 per cent of all journeys were over 100 km. Few growers reported any problems with the frequency of bus services since nearly all had access to two or more buses into Harare each weekday. However, the competitive nature and the low returns of the rural bus industry during the 1980s have made operators more reluctant to carry high-bulk, low-value horticultural crops. Over half of those dependent on the bus services experienced problems with drivers refusing to take their produce. Nor were private haulage firms able to offer a more reliable service since many were operating with inadequate or insubstantial fleets. For the growers, this unreliable aspect of their transport resulted in crops not reaching the market before they had perished.

On arrival in Harare most growers hired a cart to transport their produce from the bus terminus to the city market. This added a standard charge to the total transport costs in which the main element was the grower's bus fare. Given the small volume of produce marketed by individual farmers, transport costs varied mainly with distance, rather than with the amount of produce being sold. Consequently, for 66 per cent of all growers, such costs represented at least 20 per cent of their sales and most growers experienced some trips during which they did not recoup their transport costs. Although 86 per cent of growers undertook other activities in Harare, such as purchasing household goods and agricultural inputs, the marketing of agricultural crops represented a considerable time and cost input for communal farmers.

In general, therefore, the post-independence road-building programme has improved the accessibility of Communal Areas to major centres of economic activity. The provision of feeder roads and the subsequent development of rural bus services have facilitated the growth of agricultural marketing in areas formerly dominated by subsistence agriculture. In this respect, the programme has begun to change the spatial patterns inherited from the pre-independence period. However, the development programme must be matched by greater investment in both road maintenance and in expanding bus and lorry fleets if communal farmers are to continue to take advantage of increased accessibility. Unfortunately, road maintenance has difficulty in attracting funding while the transport sector is dependent on imports, both of new vehicles and spare parts. Overcoming these problems is essential if the post-independence momentum of rural development is to be maintained in the Communal Areas.

Telecommunications, information transfer and development

Technological innovation has played a major role in the evolving relationship between transport and development. As Table 2.1 illustrates, modal change was a major factor in the growth of world trade and the emergence of the global economy. The current telecommunications revolution, with its emphasis on the exchange of information and expertise, represents the latest stage in the development of a dynamic system. As with earlier modes, these 'highways of the future' are a permissive factor. Initially concentrated in advanced economies, they have exacerbated rather than reduced both regional and global disparities (Gillespie and Goddard, 1986). However, networks are evolving which illustrate the potential for telecommunications to reduce the mobility gap both within and between the less-developed countries and thus play a central role in the development process (Owen, 1987).

An example of such a network is the Technological Information Pilot System (TIPS) which was established in 1984 to link sources and users of technology and trade information in developing countries. The scheme recognized that over the previous 30 years, developing countries had created significant industrial research and manufacturing capabilities based on either indigenous or adapted technologies. However, due to Western orientation of trade and transport links, developing countries had failed to expand technology exchanges and trade relations. For example, during the 1980s less than ten per cent of Africa's trade occurred within the continent. Thus the main objective of the scheme was to promote technology transfer and economic cooperation among developing countries by creating greater awareness of each other's capabilities and needs.

Under the auspices of the United Nations Development Programme (UNDP) and the United Nations Fund for Science and Technology for Development (UNFSTD) and funded by the Italian government, the scheme established a network of national bureaux linked to each other through an International Operations Centre (IOC) based in Rome (Figure 2.7). The bureaux receive information via dedicated teleprinter circuits and satellite links. At the country level the bureaux disseminate the information to users through daily and weekly bulletins such as *South Tech* which are despatched by facsimile, teleprinter, messenger or mail. National bureaux are also responsible for the collection of trade and technology data which are fed into the TIPS network.

Ten countries were selected to participate in the pilot scheme which became operational in 1987: Brazil, China, Egypt, India, Kenya, Mexico, Pakistan, Peru, the Philippines and Zimbabwe. Chosen on the basis of their economy being appropriate to the scheme, each country received donor funding until 1990 when national bureaux were expected to become self-financing. TIPS has estimated that establishing technology and trade-information transfer is at least five

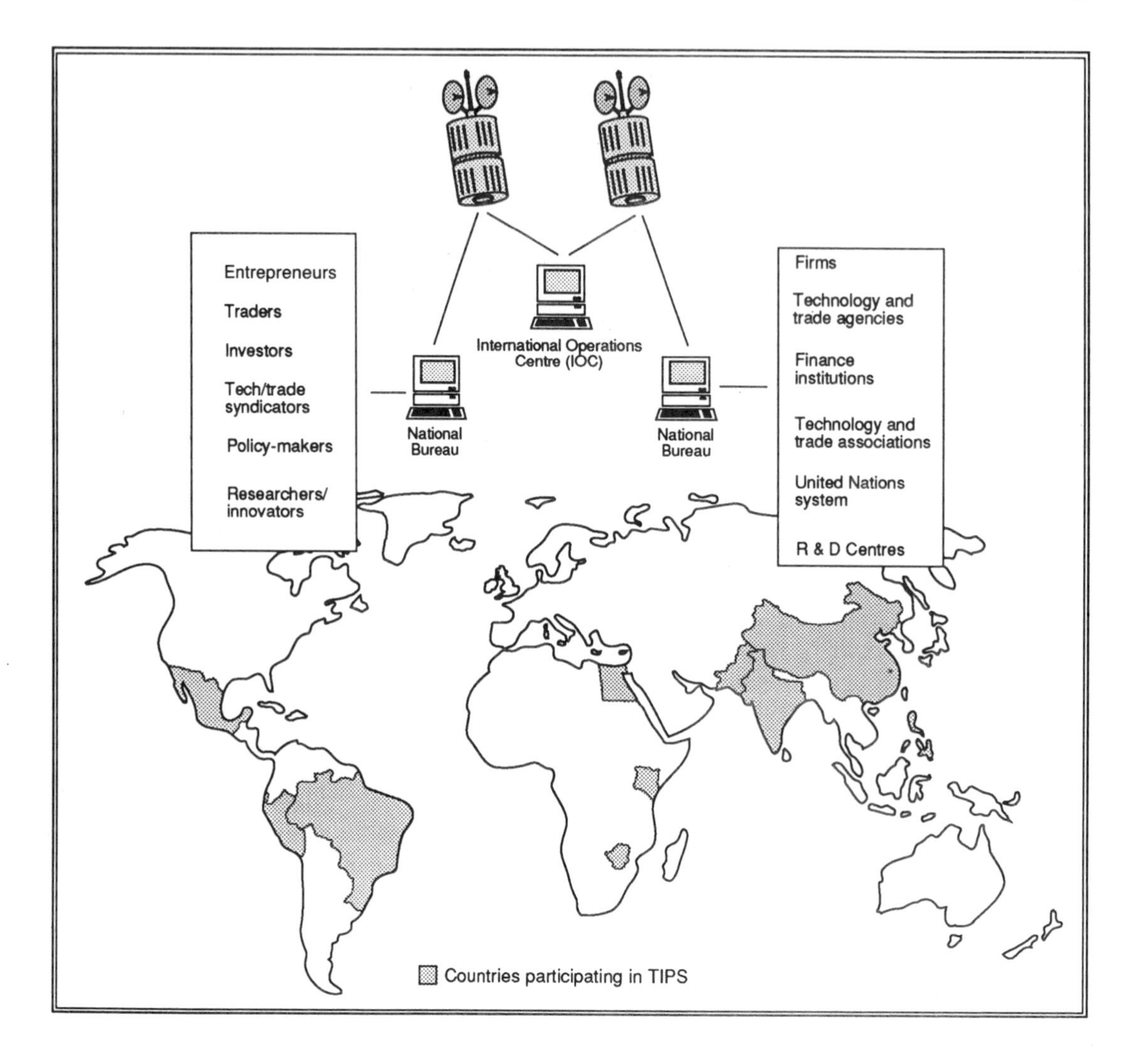

Figure 2.7 The Technological Information Pilot System (TIPS).

times more costly for a firm in a developing country than for its counterpart in the industrialized world (Breede, 1990). Growth in TIPS membership will be determined largely by the availability of international aid (Figure 2.7). Providing this is forthcoming, plans exist for 60 participating countries by 1995.

Ultimately, all areas of trade and technology information relevant to the developing countries will be incorporated into TIPS, together with data from UN information systems. During the pilot phase, however, information exchange was restricted to 11 major sectors including agro-industries, biotechnology, energy, food processing, pharmaceuticals and textiles. Eighty per cent of users within the participating countries came from private- and public-sector manufacturing industry. The remainder represented a diversity of institutional types: government industries, parastatal bodies and research, academic and financial institutions. The purposes for which information transfer have been used vary widely and include applied research, production methods and trade and business opportunities. By 1990 the TIPS network was disseminating 700 to 800 items of information per month and its first year of operation had resulted in over 1,000 trade and technology contracts between participating countries. As geographic and sectoral growth continues through the 1990s, information transfer will continue to create trading opportunities between developing countries.

TIPS provides one example of a growing number of networks aimed at improving information transfer between developing countries. Overall, progress has been much slower than in the industrialized world mainly because of a lack of technology and finance (Ono, 1990). However, sufficient networks are being established to illustrate the impact of telecommunications at a variety of scales within the less-developed world. For example, the Food and Agriculture Organization of the United Nations has coordinated a number of regional networks exchanging information on agriculture and fisheries, while the Group of 77 is establishing a Multi-Sectoral Information System for its members. As technology becomes more widely available and relative costs decrease, the opportunities to use telecommunications to overcome the mobility gap will also increase (Codding, 1989). At the global scale, telecommunications offer the potential to increase information flows from the industrialized to the developing countries. At the local scale, they are being used to overcome the isolation of the rural areas of the Third World. In India, for example, telecommunications are providing educational, health and marketing information to villages lacking all-weather roads (Owen, 1987).

Although in their infancy, telecommunications will play an increasing role in the evolving relationship between transport and development. Telecommunications are part of a dynamic whole and as such they will have a profound effect on established transport links. Initially, their rapid rate of technological development and high cost have widened spatial inequalities at a variety of scales and consolidated existing patterns of physical transport provision. However, as the TIPS example has shown, they are also capable of facilitating new patterns of economic and social activity which in turn require changes in existing transport modes. In this respect, their role in the development process and subsequent impact on patterns of human activity could be as crucial as that of the railways a century before.

Conclusion

This chapter has considered specific case-studies chosen to illustrate the relationships between transport and development at scales ranging from sub-national to global. Each demonstrates the permissive nature of transport in the development process and the multifaceted contexts in which transport systems operate in both advanced and developing societies. All confirm the importance of the historical dimension in understanding the spatial form of present-day networks and the processes which have created them. Above all, the case-studies reflect the evolutionary nature of the relationship between transport and development.

The dynamic nature of transport systems has played, and will continue to fulfil, an important role in the process of spatial change. Throughout the greater part of human history, transport has been a slow and laborious business. The nineteenth and twentieth centuries have been characterized by rapid technological advances which have revolutionized the transfer of people, goods and information. Today, the technological transfer of information is virtually instantaneous, air transport has effectively reduced the world to a global village, and efficient multimodal networks have streamlined goods' movement. As the Québec example has shown, such changes both reflect and encourage socioeconomic evolution, but must be assimilated and adapted in order to promote overall regional development. In most advanced societies, this has been best achieved by balancing the growth of strong external links with the continuing intensification of internal networks.

Transport evolution in the less-developed countries has been in some ways more problematic. Most modern networks owe their origins to the European colonial period with its emphasis on external links rather than internal network development. Early models of the transport/development relationship accepted this external orientation as the norm, but more recent studies have questioned the validity of such views and called for a more 'development-specific' approach to transport provision. As the studies of Zimbabwe and telecommunications have shown, this has resulted in a more policy-orientated approach towards the further development of transport systems and towards development processes in the Third World. There are also signs that the relatively new links provided by telecommunications will offer some solutions to the limitations in the transport systems inherited from earlier phases of development.

A focus on the role of transport as a facilitator of social, economic and political change reveals and underpins two contrasting spatial patterns. Throughout the twentieth century, transport modes have shown an increasing ability to overcome physical barriers and to reduce the friction of distance. As a consequence, the world is becoming a smaller place linked by intercontinental and global networks, a trend which will continue into the twenty-first century. On the other hand, global inequalities in development have never been greater. While advanced nations have developed relatively sophisticated transport systems, a majority of the world's peoples are underprovided with even rudimentary forms of transport. Although transport provision by itself will do little to reduce inequalities, appropriate transport strategies as part of multifaceted development programmes will have a major role to play in overcoming the problems of regional and global polarization.

References

Adams, J. (1981), *Transport planning: vision and practice* (London: Routledge & Kegan Paul).

Allison, R. and Bradshaw, M. (1990), *The concept of French Canada in a geographical context* (Plymouth: College of St Mark and St John, Canadian Studies Geography Project for sixth forms and colleges).

Barke, M. and O'Hare, G. (1984), *The Third World* (Edinburgh: Oliver & Boyd).

Barwell, I., Edmonds, G., Howe, J. and De Veen, J. (1985), *Rural transport in developing countries* (London: Intermediate Technology Publications).

Breede, W. (1990), 'The Technological Information Pilot System: a bridge across the South', *Telecommunications Policy* 14(5), 434–41.

Brookfield, H. C. (1972), *Colonialism, development and independence: the case of the Melanesian Islands in the South Pacific* (Cambridge: Cambridge University Press).

Brookfield, H. C. (1975), *Interdependent development: perspectives on development* (London: Methuen).

Codding, G. A. (1989), 'Financing development assistance in the International Telecommunications Union', *Telecommunications Policy* 13(1), 13–24.

Drakakis-Smith, D. (1987), 'Zimbabwe: the slow struggle towards socialism', *Geography* 72(4), 348–52.

Forward, C. N. (1982), 'The development of Canada's five leading national ports', *Urban History Review* 10(3), 25–45.

Gillespie, A. and Goddard, J. B. (1986), 'Advanced telecommunications and regional economic development', *Geographical Journal* 152(3), 383–7.

Griffiths, I. (1990), 'The quest for independent access to the sea in southern Africa', *Geographical Journal* 155(3), 378–91.

Hoyle, B. S. (ed.) (1973), *Transport and development* (London: Macmillan).

Hoyle, B. S. (1988), *Transport and development in tropical Africa* (London: John Murray).

Kay, G. (1970), *Rhodesia: a human geography* (London: University of London Press).

Langman, R. C. (1985), 'Cultural pluralism: the case of Canada', in *Geographical studies of development*, ed P. P. Courtenay (Harlow: Longman), 197–223.

Leinbach, T. R., Chia, L. S., Kissling, C. C., Robinson, R. and Spencer, A. H. (1989), *South-east Asian transport: issues in development* (Singapore: Oxford University Press).

Mabogunje, A. L. (1989), *The development process: a spatial perspective* (London: Unwin Hyman, 2nd edn.).

Mwase, N. R. L. (1989), 'Transport and rural

development in Africa', *Transport Reviews* 9(3), 217–34.

Ono, R. (1990), 'Improving development assistance for telecommunications', *Telecommunications Policy* 14(6), 476–88.

Owen, W. (1987), *Transportation and world development* (London: Hutchinson).

Patterson, J. (1982), 'The Lower St Lawrence valley', *North America*, 220–9.

Rimmer, P. J. (1977), 'A conceptual framework for examining urban and regional transport needs in south-east Asia', *Pacific Viewpoint* 18, 133–47.

Smith, J. A. (1988), 'The Beira Corridor Project', *Geography* 73(3), 258–61.

Smith, J. A. (1989), 'Transport and marketing of horticultural crops by communal farmers in Harare', *Geographical Journal of Zimbabwe* 20, 1–15.

Taaffe, E. J., Morrill, R. L. and Gould, P. R. (1963), 'Transport expansion in underdeveloped countries: a comparative analysis', *Geographical Review* 53, 503–29.

Todaro, M. P. (1989), *Economic development in the third world* (Harlow: Longman, 4th edn.).

Waddell, E. (1987), 'Cultural hearth, continental diaspora: the place of Québec in North America', in *Heartland and hinterland: a geography of Canada*, ed. L. D. McCann (Scarborough, Ontario: Prentice Hall Canada Inc). 149–72.

3

Transport Policy and Control

Richard Knowles and Derek Hall

In this chapter reasons are identified to explain the increasing regulation of public freight and passenger transport services up to the 1970s and for worldwide moves subsequently to deregulate and privatize transport. Four case-studies examine Jamaica's franchised bus system, the deregulation of United States domestic air services, the privatization of British Airways and bus deregulation and privatization in Britain.

Introduction

Regulation and public ownership

The tradition of transport regulation is well established throughout the world (Button, 1991). Most governments have intervened in the transport market for many years to protect customers and employees by introducing quality and safety controls, controlling the quantity of services to ensure a comprehensive transport network, controlling the price of services, regulating the entry of new transport operators and sometimes by public ownership of transport companies. These controls were built up from the nineteenth century through to the 1970s.

In this protected system, operators were expected to provide some services for social rather than for commercial reasons. An early British example was the requirement of the 1844 Railway Act that any new railway should provide working men's trains to spread the benefits of cheap mechanized travel to all sections of the population. Britain's 1930 Road Traffic Act introduced a system of road service licences and fares controls to regulate the often chaotic and unsafe free market for bus services experienced in the 1920s. Bus operators were protected from unregulated competition but were expected to cross-subsidize unprofitable rural services from profits made elsewhere (Knowles, 1989). Button and Gillingwater (1986) identified the 1930s as a particularly active period of regulation, much of it reversed since the 1970s.

Deregulation and privatization

Deregulation measures started in an *ad hoc* way in response to the requirements of individual modes of transport. For example, Swedish Railways were divided into commercial and social networks as early as 1963, while Swedish road haulage was deregulated by 1968 (Fullerton, 1990). British road haulage was deregulated in 1968, as were United States domestic air services from the mid-1970s. In the late 1970s the Theory of Contestable Markets suggested that the free entry of new operators into the transport

Photograph 3.1 One consequence of the 1980 Transport Act was the mushrooming of commuter coach services. Here beside Victoria Railway Station in 1986 are 'long' and standard Leyland Olympian coaches operated by Eastern National and London Country.

market was the key mechanism to ensure efficiency and welfare maximization (Baumol, 1982; Hibbs, 1985). Regulation was held to be responsible for increasing prices by limiting competition.

Contestability theory was used to underpin the ideological moves to deregulate and privatize transport which started in the United States and Great Britain at the end of the 1970s. This process then spread rapidly throughout the world and included many developing countries, often at the instigation of the World Bank and the International Monetary Fund (Bell and Cloke, 1990). In the United States, the 1978 Airline Deregulation Act, 1980 Staggers Rail Act, 1980 Motor Carrier Act and 1982 Bus Regulatory Reform Act liberalized interstate transport with consequential effects on intrastate regulation (Button, 1991). In Great Britain the 1980 Transport Act abolished fare controls on all bus services and deregulated express bus services, while the 1985 Transport Act deregulated local bus services outside Greater London by abolishing road-service licences (Knowles, 1989). Other measures reduced and capped the level of public transport subsidies. A widespread privatization programme from 1982 onwards sold off publicly owned British transport companies including the National Freight Corporation, British Transport Docks Board, Seaspeed Hovercraft, British Transport Hotels, Sealink Ferries, British Airways, British Airports Authority, National Bus Company, Scottish Bus Group and many municipally owned bus companies (Knowles, 1989). Since the mid-1980s Contestability Theory has been challenged increasingly as the outcome of deregulation and privatization has often been oligopolistic control of particular transport markets instead of competition (Button,

1991; Kahn, 1988). Optimal regulatory systems are now seen by many as those systems such as franchising and compulsory competitive tendering which stimulate competition in bidding for service contracts but control services, quality and fares by preventing the misuse of monopolistic or oligopolistic power. This type of competition reduces operating costs and public subsidy without destroying the integrated network of public-transport services (Knowles, 1985). London Regional Transport, for example, put out to competitive tender nearly a quarter of its bus-route network from 1986 to 1989, reducing costs by 15 per cent while increasing bus-service reliability (Gayle and Goodrich, 1990).

In 1989 and 1990 the sudden collapse of the Communist political system and centrally planned economies throughout Eastern Europe, in favour of multiparty democracy and market economies, heralded a new phase in transport deregulation and privatization. It quickly became clear that the highly centralized Soviet Unified Transport System (Mellor, 1975) would be obsolete, as in future sources of raw materials and finished products and modal split would be determined more by the market than by the state.

Deregulation continues to gather momentum in the rest of the world. For example, road-freight deregulation has been completed in New Zealand and the Irish Republic and started in the United States. Interstate bus services have been deregulated in Australia followed by intrastate services in New South Wales. Here the Canberra to Sydney coach services have cut fares and increased frequencies but the previous monopoly operator still dominates the market and train services have been cut from four to one per day (Higginson, 1990). New Zealand deregulated all scheduled passenger services on 1 July 1991, including buses, taxis and rail services, while Denmark began to competitively tender Copenhagen Capital Region bus services in 1990 to reduce soaring costs (Fullerton and Knowles, 1991; Higginson, 1990). In the European Community the introduction of the Single European Market after 1992 will result in the eventual harmonization and liberalization of national transport regulations. Further privatization measures in developed countries include Air Canada, Air New Zealand and Japanese National Railways, all designed to improve operating efficiency, reduce government debt and help balance budgets. In the UK additional privatization proposals will include British Rail, municipal airports and remaining municipal bus companies if the Conservative government is re-elected.

Welfare states such as Sweden and Norway have also been attracted by the ability of the free market to increase efficiency and adapt public-transport services to match demand. Deregulation, however, threatens the viability of minimum transport services in marginal areas and restricts the accessibility of the carless population to jobs and to retail and health services. Some regulation of transport is therefore necessary to protect the most vulnerable. In 1988 Swedish coach services were deregulated and a National Railway Infrastructure Authority took over fixed rail infrastructure. This infrastructure is charged for on the basis of the marginal cost of use and on a fixed charge equivalent to road tax on each wagon (Fullerton, 1990). Swedish Railway services were redivided into four groups – a profitable business railway and subsidized county railways, Northern Inland railway and the iron-ore railway from the inland iron mines to the Baltic and North Atlantic coasts. Norway deregulated its internal freight transport in 1986 to the benefit of road transport with subsidies decreasing by five per cent in one year and new operators increasing by 41 per cent (Higginson, 1990).

In the developing countries deregulation is often hampered by import restrictions, due to a lack of foreign exchange, causing a shortage of vehicles and spare parts. In Sri Lanka, for example, all bus routes were deregulated in 1979 but the impact was blunted by the continuing subsidies to state-owned buses. Communist China has deregulated long-distance coach services and fares are allowed to vary up to 15 per cent.

Demand-responsive and flexible paratransit services are a common feature of many developing countries but often operate illegally providing 'unsafe and unreliable services at unregulated fares' (Higginson, 1990, 61); (Adeniji, 1983; Rimmer, 1986). Chile has had to reintroduce some bus-route controls to counter such illegal services.

Privatization is a lucrative policy for many developing countries. Nigeria, for example, announced in July 1988 that it would privatize Nigeria Airways and the National Shipping line, end the subsidy to the Ports Authority and reduce government support to the Nigeria Railway Corporation (Lewis, 1990). Singapore started to sell off Singapore Airlines in 1985 and decided to privatize the Mass Rapid Transit Corporation, Singapore port and airport (Low, 1990).

Case-studies

Jamaica's franchised bus system

The franchising of Jamaican urban bus services involved the legalization of the mainly illegal minibuses which had creamed off much of the profitable traffic and had put the state-owned Jamaica Omnibus Service deep into debt and then in 1983 out of business (Anderson, 1990). This problem of illegal competition from minibuses and adapted vehicles is widespread in the Third World, causing similar problems, for example, in Chiang Mai (Thailand) and Lagos and Ibadan (Nigeria), with the Ibadan City Bus Service liquidated in 1976 (Adeniji, 1983; Rimmer, 1986; Thomson, 1978). In 1983 the Jamaican government franchised the 65 bus routes in the Kingston Metropolitan Region into ten packages which were then resold at a higher price to minibus owners, most of whom own one vehicle.

Overcrowding occurs in the peak period and illegal competition survives in the peak and off-peak periods on routes abandoned by the franchise holder because of worries about personal safety. Timetabling has broken down because of competition. Short operation sometimes occurs with services turning back before the end of the route. Non-designated stops are frequent and tickets are rarely issued. In the peak period there is discrimination against school children and pensioners carried at concessionary fares.

The objective of franchising is to remove government subsidy and provide conditions for genuine, permanent competition with an effective division of responsibility between private ownership and limited government control. This form of deregulation has produced negative effects on the quality of service, working conditions and increased road-traffic accidents.

Deregulation of United States' domestic air services

The US airline industry was regulated by the Civil Aeronautics Board (CAB) from 1938. The CAB's policy was to preserve the 16 Trunk Airlines which existed in 1938, but these were reduced to 11 by mergers (American, Braniff, Continental, Delta, Eastern, National, North West, Pan Am, TWA, United and Western). Routes were awarded to these airlines and fares specified by the CAB which in return ensured safe airline-operating practices (Pickrell, 1991). Its major service initiative was to allow a group of local-service airlines to start subsidized feeder services during the 1950s to connect small communities with the cities served by the trunk airlines. Subsidy payments to local airlines increased rapidly as new aircraft replaced ex-military planes. The CAB then began to award more profitable, longer-distance routes to local-service airlines to offset feeder-service losses. Many acquired jet planes and evolved into regional airlines.

Strong economic growth increased domestic air passengers tenfold from 1950 to 1970 but fares stabilized because of the lower operating costs of modern aircraft. After 1970 fares rose with the cost of fuel and labour. Demand was suppressed as the mileage formula for fares overpriced

Photograph 3.2 The growth of high standard services and the resurgence of double-deck coaches followed from the 1980 Transport Act deregulating coach services. Here in National Express 'Rapide' livery, complete with hostess, washroom and catering facilities, is a 1986 Midland Red Coaches MCW Metroliner, seen at Marble Arch in the summer of 1987.

profitable long-haul routes and underpriced the often loss-making short-haul routes. The oil crisis of 1973–4 increased pressure for deregulation. A Congressional Committee recommended relaxing the controls over fares and new airlines entering the market which occurred from 1975. American Airlines started to discount tickets by up to 45 per cent in 1977, in response to charter carriers and low-fare airlines, leading to 60 per cent traffic increases on some routes and copycat fares by competitors. The 1978 Airline Deregulation Act phased out the CAB's route licensing powers over three years and its fares controls over five years. Commuter airlines were permitted to receive subsidies for routes to smaller communities. The CAB itself was abolished in 1984.

A. E. Kahn, the US Transportation Secretary who introduced deregulation, advocated that it would:

1 make airlines more competitive;
2 offer large reductions in average fares and distribute the benefits more equitably;
3 provide new lower-fare and quality options;
4 be more efficient;
5 make airlines less likely to take part in cost-inflating competition;
6 continue to serve pre-existing networks;
7 not subject the airline industry to severe financial distress (Kahn, 1988).

Since 1982 the domestic US airline market has been open to any national carrier. The most significant result of route deregulation has been the development of 'hub and spoke' route networks. Services from smaller towns are fed into large city airports, or hubs, and the hubs are then linked by regular direct flights. Hubbing reduces costs, raises load factors and frequencies on hub-to-hub flights and the cheaper seat/mile costs of larger aircraft can lower fares on competitive routes.

Hubbing can raise the entry costs for new operators as an airline needs to develop many spoke routes to achieve sufficient density of feeder traffic to achieve economies of scale. On-line connections by a single airline are a strong competitive advantage. The proportion of passengers boarding flights at major link airports rose from 19.9 per cent in 1977 to 31 per cent in 1984. This has paradoxically caused congestion and delays at some airports.

Competition has brought fares more into line with resource costs. Average fares have fallen on long-haul routes and discount tickets but risen on short-haul routes and standard 'coach' fares. In 1987 90 per cent of passengers used discount fares compared with only 15 per cent in 1976 before deregulation. Airline costs have been cut by higher productivity and lower pay while the leisure market has accepted lower frequencies and higher load factors which yield lower fares. Safety standards have continued to increase despite the sacking of most air-traffic controllers who were on strike in 1981.

Following deregulation, eight former local-service airlines (Frontier, Hughes Airwest, North Central, Ozark, Piedmont, Southern, Texas International and US Air) grew rapidly and began to compete with the smaller trunk airlines. In the early 1980s they were joined by five intrastate airlines (Alaska and four new carriers: Air Cal, Air Florida, PSA and Southwest) and three new commuter airlines (Air Wisconsin, Empire and Horizon). After 1984, however, the number of regional or national competitors were reduced by financial failures, mergers or takeovers to ten airlines by 1988 and seven by 1991.

By the spring of 1991 only six of the pre-deregulation trunk airlines survived. Braniff, Continental and Eastern were bankrupt and had ceased operating, while National and Western had been taken over by Pan Am and Delta respectively. The main US international airlines, Pan Am and TWA, were in financial difficulties and were selling off assets including international route licences to London Heathrow to United and American respectively (*Guardian*, 1991a, 1991b).

The market share of the eight largest airlines which fell from 81 per cent in 1978 to 70 per cent in 1985 rose to 90 per cent by 1988. The four mega-hubs Atlanta, Chicago, Dallas and Denver each serve as a major link for two airlines while the other hubs are each dominated by one airline. This demonstrates the monopolistic nature of the airline market after deregulation. Competition cannot be effective as long as the entry of new airlines is so costly. Other sources of monopoly power are the ownership of computerized reservation systems which bias displays of seats and fares available to customers and joint marketing agreements with commuter and regional airlines including code-sharing by connecting flights.

Pickrell (1991) shows that 137 out of 183 US airports increased their number of flights between 1979 and 1988, often by more than 50 per cent. However, 46 airports showed a decline, mainly of 10 to 50 per cent, because of deregulation. Seat capacity per plane fell at all except the largest city airports as airlines switched to smaller, often turbo-propeller, aircraft for spoke routes.

The privatization of British Airways and international air transport deregulation

British Airways (BA) was privatized in 1987 to make it more efficient and to remove the need for government financial support. BA's poor productivity levels, in comparison with other international airlines, were improved by cutting staffing levels, modernizing its aircraft fleet and improving its terminal facilities (Ashworth and Forsyth, 1984).

The international airline industry remains heavily regulated. Routes are licensed usually on a bilateral basis by pairs of countries who each usually designate their national airline. The frequency of flights is limited and fares are high and government approved (Sealy, 1966). In the last ten years there have been several relaxations, for example between the UK and West Germany in 1984, the UK and Belgium in 1985 and the UK and Canada in 1987. A new 1991 UK–USA agreement allows a second British airline to fly from London Heathrow to North America, lets British airlines fly via Europe to the USA and beyond and permits better marketing deals with US domestic airlines (*Guardian*, 1991b). In return, United's $290 million deal to buy Pan Am's Heathrow routes and American's $445 million deal to buy TWA's Heathrow routes have been allowed to stave off bankruptcy. Whereas United and American will be able to use Heathrow as a hub to serve Europe, this agreement allows BA to revive its plan to launch Brussels as a Europe and North America route hub jointly with the state-owned Belgian airline, Sabena. This replaced the 1977 agreement which allowed cities in the USA to be served direct from the UK by two airlines from each country except for New York, Los Angeles, Boston and Miami which were limited to one airline each from each country.

BA has benefited from regulation as it receives most of the bilateral route licences and its oligopoly powers enable it to charge high airfares and make supernormal profits. However, domestic deregulation in the USA, Canada, Australia and the UK have generated pressure for international deregulation, especially within the European Community. Liberalization within Europe, for example, has resulted in London to Frankfurt flights increasing from 68 to 97 per week between the summers of 1983 and 1987 (Monopolies and Mergers Report, 1987). The growth of mega-airlines and the internationalization of the industry allowed the BA/British Caledonian merger to be approved despite its monopoly power in the UK. Other agreements include a BA/United worldwide marketing partnership in 1987 yielding connecting United flights in the USA and shared terminal facilities at four major US airports.

Since privatization, BA has nearly doubled its profits on airline operation from £183m (6 per cent) in 1986/87 to £340m (8.2 per cent) in 1988/89. While European turnover is increasing, its profits have fallen to one per cent because of deregulation. BA's North American services account for over half of all its profits probably because of tight regulation which limits competition. BA's Middle East, Far East and Australasia routes yield a 10 per cent profit, North and Latin American routes 13 per cent profit and African routes 15 per cent profit, all in tightly regulated markets.

BA faces a much more competitive future with more international deregulation, competition from the Channel Tunnel on short European routes, erosion of its dominant share of frequencies from London Heathrow, privatization of large state airlines and the growth of a few major airlines to rival the United States big three: American, Delta and United.

Bus deregulation and privatization in Britain

UK public-transport deregulation and privatization — UK public-transport policy has long experienced political partisanship. In the 1980s the government set about pursuing the deregulation and privatization of public-transport services, with road-passenger transport foremost in that process. As a consequence, not only have routes and services been deregulated, but the fundamental structure of Britain's bus and coach-operating industry has been dramatically altered at a time when the structure of the bus-manufacturing industry was also undergoing substantial readjustment.

The UK bus-operating industry — The UK's

first licensed urban motor bus service began operation in 1898. Six years later, with growing numbers of motorized vehicles, a registration system was introduced. But it was not until the 1930 Road Traffic Act that the notion of regulation through route licensing was established, to eradicate what was seen as a dangerous vying for passengers by increasing numbers of competing buses on the streets of London and other major cities.

From then until 1986, the local bus-operating industry remained highly regulated. Route licensing through the office of Traffic Commissioners – a field agency of central government – regulated and regularized public bus services. New potential bus operators needed to conform to certain standards of organization, and applications for a route licence could be opposed – and often were, successfully – by incumbent bus and railway operators.

The 1948 Transport Act nationalized companies within the large Thomas Tilling group and the Scottish Motor Traction Company (SMT). Four groups of bus companies emerged: (a) those in full public ownership under the auspices of the British Transport Commission (from 1962 the Transport Holding Company) (London Transport was also established as a directly run central government operation); (b) those of the British Electric Traction group, in which the government had a minority holding; (c) municipal companies, run by local county boroughs and municipal boroughs; and (d) completely independent private companies. Of the four groups, the first two were made up of relatively large organizations, which dominated much of the country's bus operation, particularly inter-urban and urban-rural services. Municipal companies operated within the confines of their administrative boundaries, sometimes under the terms of joint schemes with a BTC or BET company.

Independent operators were mostly small, with a few notable exceptions. They were – and still are – particularly important in rural and remoter areas, holding detailed local knowledge and patronage where the larger companies were unable to derive economies of operating scale. Significant among independents have been operator associations: owner cooperatives formed to pool the resources of smaller owners in the face of a rapid growth of large operators. While the need for such groupings largely disappeared with the 1930 Act, examples remain (most notably in Ayrshire) of associations which later formed themselves into limited companies which could perhaps act as a model for cooperation among contemporary hard-pressed smaller operators.

UK bus-passenger numbers reached their peak in the early 1950s, after which time increasing car ownership and new patterns of spatial mobility saw bus ridership decrease substantially; services were commensurately reduced in rural and suburban areas and at off-peak times. Partly reacting to this situation, the Labour Government's 1968 Transport Act established the National Bus Company (NBC) as a corporate state enterprise encompassing all former BTC and BET companies, as well as some others, and amalgamating units perceived to be too small to derive economies of scale. North of the border, the Scottish Transport Group was established on a similar basis. The Act also facilitated the setting up of comprehensive passenger transport authorities in urban conurbations in advance of local-government reform in the mid-1970s. Passenger Transport Authorities and Executives (PTAs and PTEs) were publicly controlled and owned to provide locally determined integrated public-transport systems. Government subsidies were introduced for certain types of services and vehicles; attempts to standardize bus design into just a few models through the provision of bus grants possessed parallels in the structural concentration taking place within the bus-manufacturing industry itself.

By the 1970s, over 95 per cent of local bus services were being operated by either NBC or municipally-owned companies. Meanwhile, UK bus patronage continued to decline: from just over 8,000 million journeys in 1975 to just over 6,000 million in 1985, a fall of almost a quarter. Passenger receipts, at 1986 prices, fell from £2,600 million to £2,300 million over the same period, a decline of 12 per cent. Numbers of staff employed in the bus- and coach-operating industry fell from 224,400 in 1975 to 174,300 in 1985, a reduction of some 22 per cent (Department of Transport, 1987, 106–7, 111). The decline in bus-passenger traffic and revenue required an operating subsidy to maintain a comprehensive local service. Nationally, subsidies rose by 1,300 per cent between 1972 and 1982, from £10 million to £520 million. Fares rose 35 per cent in real terms, while services declined by 12 per cent (Knowles, 1989, 82).

The UK bus-manufacturing industry — As in the car industry, structural rationalization through mergers and takeovers saw bus and coach manufacturing increasingly concentrated into fewer hands. The industry had traditionally entailed two essential elements – chassis and body manufacture – two groups of activities which were usually undertaken by separate companies. Famous chassis manufacturers such as AEC, Bristol, Daimler and Guy, and body-builders Eastern Coach Works (ECW), Park Royal and C. H. Roe were absorbed into a larger and seemingly more cumbersome and inefficient Leyland Group, such that customer choice noticeably dwindled as both UK chassis and bodies, encouraged by 50 per cent government grants on new buses, became more standardized and concentrated in the hands of one major manufacturing group.

While in the short term this may have been appropriate for a dwindling home market, it did little for export opportunities. The Leyland Group, with assured sales to home companies based on standard bus types, began to lose overseas markets. Meanwhile, foreign manufacturers began penetrating the British market. In 1988, after the government had written off debts of £55 million, Leyland Bus was bought out by a management consortium. Within months, following the enforced redundancies of 750 workers, the company was resold to Volvo Bus of Sweden: this appeared to sound the death-knell for the wholly UK-owned bus-manufacturing industry.

Structural reorganization of operating units — Although several mergers and absorptions among NBC subsidiaries had taken place in the early 1980s in advance of privatization and deregulation, moves towards decentralization and company individuality began with larger operating units being (re-) subdivided. The most comprehensive approach to this task was taken north of the border. In June 1985 the seven nationalized Scottish Bus Group companies were broken down into 11 smaller units. All but one of the existing companies' operating area boundaries were redrawn, partly providing a closer geographical relationship with Scottish administrative regions.

The Transport Act, 1985 — The Conservative government argued that the existing regulatory system had three inherent shortcomings:

(a) it prevented competition which would keep costs, subsidies and fares down;
(b) it inhibited innovation such as the use of minibuses and midibuses; and
(c) it suppressed passenger demand by overcharging profitable traffic in order to be able to cross-subsidize unprofitable routes.

Early responses to this situation were enshrined in the Transport Act, 1980, which abolished fare controls on all bus services and deregulated express bus provision. Local 'stage-carriage' services remained subject to the provisions of the 1930 legislation on route-licensing, but the entry of new operators was eased by shifting the onus of proof in licence application and objection from the applicant to the objector, in a particular attempt to encourage the better provision of rural bus services (Knowles, 1989, 82–3).

The perceived success of the 1980 legislation encouraged the government to press further ahead with the implementation of 'free-market' measures in the public-transport arena. The Buses White Paper (Department of Transport, 1984), and the subsequent Transport Act, 1985, were the outcome. Their overriding aims were to:

(a) deregulate – introduce far greater competition into – British bus services (initially to exclude Greater London);
(b) considerably reduce public subsidy to the bus operating industry; and
(c) shift the bulk of bus operation from the public to the private sector, in line with other privatization policies being pursued by the government.

The introduction of route competition — The Transport Act, 1980, deregulated long-distance coaching operations and stimulated some competition, higher-quality service and cheaper fares on such trunk routes as between London and Newcastle. Overall, traffic increased by 60 per cent from 1980 to 1982, although direct coach services on secondary routes between smaller, less popular centres were cut back. The Act also allowed for the setting up of trial areas for bus service deregulation; three were subsequently established – in Norfolk, Devon and Hereford (Evans, 1988). The demonstration effect of this legislation was critical for the subsequent 1985 Act, which abolished road-service licensing. Provided that operators had a valid PSV Operator's ('O') licence, they could now introduce, change or withdraw a service simply by registering details with the appropriate Traffic Commissioner's Office, notifying the relevant local authority (which had the role of coordinating and publicizing public transport in its area), and waiting for a 42-day period to elapse. The government argued that greater competition would provide more choice for the travelling public: more buses, greater frequencies and lower fares. While this may have been the case on trunk routes in urban areas, where lucrative pickings were to be had, in Scotland, in the first year and a half of deregulation, passenger journeys dropped by 6.5 per cent, twice the average decline since 1977, and only one completely new operator survived.

The Act explicitly demolished area coordination and integration schemes, some of which dated back to the 1940s. All metropolitan areas with PTEs had developed such schemes. Perhaps the most notable was that in Tyne and Wear, which integrated bus, metro, rail and ferry services, and was subscribed to by three major bus companies and several smaller ones, to the extent that all major companies used the same liveries within the PTE area, had common through-ticketing schemes between modes and shared interchange and other facilities. This was now abandoned: the bus companies had to compete with each other, and with Tyne and Wear PTE's bus operations which were now run by an 'arm's length' company. They were all also in competition with the PTE's own light rail Metro system which had been expressly designed to be integrated with other forms of public transport.

While arguing that fares appeared to have been largely unaffected by deregulation, an early government sponsored report (Rickard et al., 1988) pointed to the significantly increased fares and inconvenience brought about by the loss of through-ticketing and pre-paid tickets, as a result of the abandonment of coordinated schemes. In subsequent years, however, some of the pre-deregulation schemes have been clawed back through renewed cooperation among operators, albeit under veiled threat of investigation by the Office of Fair Trading. In Tyne and Wear by 1988, for example, a separate company, jointly owned by the PTE and the area's bus operators, was established to administer the pre-payment travelcard scheme. The company – Network Ticketing Limited – is responsible for printing, distributing and selling the tickets and for revenue allocation in accordance with agreement among its members. This encouraged all pre-deregulation participants and newcomers to join fully in the travelcard scheme.

Reducing public subsidy — Within the Act, two types of service were recognized:

Photograph 3.3 Despite the disruption of deregulation, most PTEs have maintained integrated networks in their area. Perhaps the best example is Tyne and Wear, which runs the light railway Metro system and encourages the maintenance of bus links at major interchanges. Here at Bank Foot, north of Newcastle in the spring of 1988, are PTE supported minibus links to Newcastle Airport (to which the Metro was being extended in 1991) and to local residential areas, operated by an integral Metrorider of Moor-Dale and a van chassis Mercedes L709D of Busways, respectively.

commercial and tendered. Bus companies were now to operate services as a business and not as a public service, registering as commercial routes being considered to be economically viable. The legislation forbade the cross-subsidy of routes, as had often happened in the past. The appropriate local authority or PTE was charged with assessing the transport needs not met by registered commercial networks. With a restricted amount of central-government subsidy and its own limited resources, the authority was empowered to put out to competitive tender routes which it considered essential for social or other reasons. However, given the ever-changing nature of the registered commercial services, particularly in areas of strong competition, local authorities and PTEs have faced an extremely difficult task in keeping ahead of new developments and in assessing future need for tendering purposes: one of the most significant long-term public concerns about the consequences of deregulation has been the uncertainty brought by continually changing patterns of services and operators. Within the metropolitan areas, for example, in 1988 Greater Manchester experienced over 1,900 new and changed registrations, affecting 90 per cent of all services, some changing three or four times during the year.

Even after deregulation, however, the concept of a free market-service industry is more apparent than real. By the very nature of their statutory role, local authorities subsidize services through the formal tendering procedures, and more indirectly subsidize the operating companies by providing publicity and information on all transport services, both tendered and commercial. Bus companies also continue to receive central government subsidy through the fuel tax rebate system.

Looking at the financial results of the first year of deregulation, a government sponsored report (Balcombe et al., 1988) pointed to substantial savings made in direct bus-service subsidies, but also argued that these were partially offset by increases in administrative and publicity costs to local authorities and by the costs of establishing the new 'arm's length' companies. A report undertaken for the Association of Metropolitan Authorities (Tyson, 1988) found a reduction in subsidy levels of about ten per cent for the metropolitan areas, but also recognized that part, at least, of this saving had been obtained by deflating wages rather than by any real increase in efficiency. Uncertainty among passengers had led to significant lost patronage - between 7.5 and 12.5 per cent for most PTE areas, but a hefty 32.5 per cent in Merseyside. These losses partly reflected the substantial fare increases experienced because of previous low-fare policies and the loss of such facilities as through-ticketing. After two years of deregulation, fares in metropolitan areas had risen 27.5 per cent above inflation (Tyson, 1989). In non-metropolitan areas, a government-sponsored study (Rickard et al., 1988) also suggested that early cost savings might not be sustainable for local authorities and that operators had cut costs largely by wage-level cuts and redundancies.

In this way, while the Department of Transport suggested on more than one occasion that the savings to rate-payers of bus deregulation amounted to £40 million in 1987, such calculations neglected to take into account the £17 million fuel-duty rebate, greatly increased administrative costs and the unquantifiable loss of amenity to large numbers of non-car owners.

The case of London — The 1984 London Regional Transport Act removed control of London Transport from the elected Greater London Council and vested it in central government in order to reduce subsidy levels. Within the overall coordinating ambit of London Regional Transport (LRT: later to take on the old name of London Transport), London Buses Limited (LBL) and London Underground Limited (LUL) were established for the day-to-day running of their respective modes and given company status in April 1985. In preparation for the likelihood of deregulation, LBL was restructured into 11 (later 12) operating units during 1988.

Although the LRT operating area was excluded from the provisions of the 1985 Transport Act for full deregulation, under the 1984 Act LRT was empowered to put out groups of routes to competitive tender with the stated aim of providing LBL with the discipline of competition. This process began in 1985 and the first service to be lost by LBL to a competitor commenced operation with its new private operator in July of that year.

After the timetable for full deregulation in Greater London had apparently been put back several times, in March 1991 a consultation paper was published (Department of Transport, 1991) indicating the government's intention to introduce legislation in the 1991/2 parliament to deregulate bus services fully in London and to privatize the now 12 operating units of LBL. Acknowledging the already severe traffic-congestion problems of the capital, the Department of Transport would be accompanying deregulation with a review of bus lanes and other bus priority measures.

Privatization of the bus-operating industry — The 1985 Act provided for the sale of the National Bus Company (NBC) in whole or in part to the private sector within three years. Its constituent companies were subsequently

Photograph 3.4 Dedicated route liveries include Go-Ahead Northern's Gateshead MetroCentre Supershuttle, seen here on a then new DAF/Optare Delta actually servicing the last day of the Gateshead National Garden Festival in October 1990.

divided into three groups of non-adjacent operating areas, in order that sales of adjacent companies should not be sold to the same purchaser and thereby establish a geographical monopoly. On the other hand, to ensure that no NBC company was too big to dissuade competitors from taking advantage of deregulation, in February 1986 the government insisted that the four largest NBC companies be broken down into smaller units before deregulation day. The main short-term aims were therefore to produce medium-sized companies without monopolistic advantage but with commercial appeal.

The NBC had been the world's largest bus and coach operator, at its height employing 80,000 staff and running 20,000 vehicles. By the time of privatization these numbers had been reduced to 48,000 and 14,000 respectively. By mid-1988 all NBC subsidiaries and assets had been sold, well within the three-year period stipulated by the 1985 Act. Of the 62 NBC subsidiary companies disposed of, 18 were sold to a single bidder for less than their asset value, with, in some cases, discounts of more than 65 per cent. By the end of 1990, some £6 million had been clawed back from property sales for the benefit of the taxpayer, although such arrangements were only incorporated into the sale of 14 of the companies privatized (Department of Transport, 1990a).

Increasingly, inter-company takeovers

helped to establish a limited number of large private groupings such as Badgerline and Drawlane, each buying up several ex-NBC companies. For example, by 1991 Stagecoach Holdings, based in Perth, dominated the English south coast, and had become in just a few years one of the largest private bus companies in Europe, with additional bus interests in Malawi and Canada. Some groups diversified into other fields or entered the bus industry having established themselves elsewhere. In the former case, Proudmutual Holdings, owners of Kentish Bus, Northumbria and three smaller bus companies in Northumberland, diversified into operating car transporters and Honda motorcycle sales and Renault car-sales outlets.

The 1985 Act also required local authorities and PTEs to transfer their bus operations to separate public-transport companies (which may have private capital invested in them). This became known as an 'arm's length' relationship (Department of Transport, 1987, 54), and the government soon made it clear that this was just one step towards full privatization of municipal and PTE bus operations. As a consequence, a number of municipal companies and PTEs subsequently negotiated the selling-off of their transport companies, usually as management or management/worker buyouts. Examples include Yorkshire Rider (ex-West Yorkshire PTE), Busways (ex-Tyne and Wear PTE) and Grampian (ex-Grampian Regional Council).

Minibuses: the user-friendly symbols of deregulation? — In 1958 regulations were introduced permitting conversions of medium-sized mass-produced delivery-van designs into minibuses to be used to carry fare-paying passengers. The physical limitations of these early models restricted capacity to a maximum of 12 passengers. With poor access, severely limited internal headroom, and poorly designed interior layouts, such vehicles were usually limited to small group private-hire work or remote rural services. Thus, until the mid-1980s, the operation of minibuses as public-transport vehicles had been restricted to relatively small numbers. The UK legal definition of a 'minibus' still refers to a vehicle of between nine and 16 seats, but usually public minibuses can carry between 16 and 25 seated passengers. Use of the term 'midibus' is often applied to vehicles of the 25–35 seating capacity range, which confusingly includes shortened versions of standard singledeck vehicles, and stretched versions of minibuses.

From a local experiment in Exeter in 1984, the operation of minibuses by large bus companies grew rapidly, encouraged by the 'Buses' White Paper (Department of Transport, 1984). The changing nature of the labour market also encouraged minibus introduction, as wages in the industry were being depressed through the elimination of national agreements. With three million people nationally unemployed, bus companies outside south-east England were well-placed to offer lower salary levels to minibus drivers. The Transport Act, 1980, reinforced this process by lowering the minimum age for public service vehicle (PSV) drivers from 21 to 18. Indeed, it could be argued that an important underlying aim of large-scale minibus deployment was to depress wage rates. Minibus working agreements in some cases hastened the demise of national pay and conditions of service negotiating frameworks, while in others they followed in the wake of such disintegration. Given the economies of small bus operation, large fleets of minibuses could only have been deployed on an economically viable basis with driver wage levels significantly lower than they had been in the 1970s.

NBC subsidiaries led this rapid development (Turner and White, 1987), with over 500 minibuses purchased by major companies in 1985, compared to 83 in 1984 and just 22 in 1983. Such purchases, coupled with the ending of bus grants in 1984, together with a growing uncertainty of what lay beyond the 1985 Act, saw the beginnings of a slump in new full-size bus purchases: for the 1983–5 period, NBC company PSV deliveries as a whole decreased by 37.8 per cent.

Immediately before deregulation, in October 1986, some 1,800 minibuses were being operated by former or current NBC subsidiaries. By the end of 1986, after the initial impact of deregulation, some 3,000 minibuses were in regular service, and by the end of 1987, the figure for those in operation with major companies was approaching 6,000. By contrast, approximately 55,000 full-size buses remained in operation (White and Turner, 1987).

The reasons for this growth in such a relatively short period reflected the need for companies to establish and maintain a 'competitive edge' in preparation for the new 'market' conditions of deregulation. Minibuses were seen to be able to (Hall, 1988):

(a) increase the frequency of service;
(b) improve service penetration of outlying residential areas along roads which were unsuitable for larger buses. Paradoxically, however, while the use of minibuses helped to improve the geographical accessibility of services, the inherent physical nature of the earlier vehicles employed, with their narrow, difficult entrances, little room for shopping or push-chairs and poor quality of ride – including a high chance of having to stand – did not enhance accessibility for those least able or likely to possess their own transport: the elderly, infirm and young mothers with children;
(c) permit passengers to board and alight anywhere safe along the route through the taxi-like 'hail-and-ride' system. This practice also obviated the need to provide bus-stops, thereby enhancing the potential flexibility of service provision;
(d) provide a faster service through the use of smaller vehicles with good acceleration and fewer passengers to pick up and set down;
(e) provide a friendlier service: smaller vehicles were often marketed in such a way as to encourage personal identification, with regular, familiar and specially trained drivers employed to reinforce a 'customer-friendly' on-board atmosphere.

However, if the government aimed to encourage the small private operator or completely new small-bus businesses, their objective was confounded by the innovatory, entrepreneurial and highly protective actions of the pre-existing major companies in wishing to maintain, and even enhance, their market share. NBC bulk purchases of at least 1,500 minibuses – buying out the complete remaining UK production runs in 1986 – did not represent entrepreneurial enterprise so much as panic buying to keep out the potential deregulated competition. Further, novel, eye-catching liveries and fleetnames were adopted and new forms of 'product marketing' came to the fore.

Consequences

The downward spiral of both the UK bus-operating and manufacturing industries was by no means allayed by the 1985 Transport Act. By April 1990, the total passenger loss for all areas outside Greater London had reached 14 per cent since deregulation, despite an 18 per cent increase in bus mileage in that time. The years 1989–90 saw a decline in passengers of three per cent, with an increase of two per cent in bus mileage. Local-authority spending on bus services also continued to fall, with a 14 per cent real-term reduction over the previous year, leaving overall revenue support expenditure at less than half of that at deregulation. Fares were found to be rising by two per cent per annum in real terms, despite a five per cent reduction in operating costs over the previous 12 months and a one-third overall reduction since deregulation (Department of Transport, 1990b). The number of people employed in the bus industry was reduced and wages fell in comparison to those of comparable workers elsewhere (Association of Metropolitan Authorities, 1990).

Continued uncertainty of operating conditions and the initial chaos of the immediate post-deregulation period in several areas of the country saw bus passengers transfer on to local rail use or take to private cars. Aside from passenger uncertainty, the greatest impact on the British public to emerge from the joint deregulation-privatization process must be the introduction of large numbers of mini and midibuses in urban areas. However, the obvious shortcomings of the 'deregulation' minibuses, with their initial estimated life of five years, have been ameliorated by the production of larger, more sophisticated and more user-friendly vehicles. In many respects this evolutionary phase of the British bus industry has seen history repeat itself, but by doing so has considerably telescoped the timescale involved. In the late 1920s the typical minimum-sized bus was a 14-seater. The 1930s saw 20- and then 26-seaters, while the early post-war period was characterized by the widespread use of 29-seat buses and coaches. The minibus-derived 'midibuses' with seating for up to 33 passengers produced in large numbers from the late 1980s are comparable in size to the standard singledeck bus of the 1950s.

The mass conversion, building and operation of minibuses has distorted vehicle production and operating patterns and 'masked' certain negative aspects of the deregulatory process. The ending of new bus grants in 1984, mass minibus purchase and market uncertainty witnessed a slump in conventional bus building during the early, unstable post-deregulation phases: 1982 UK production of full-size vehicles had been about 4,000, while in 1987 the figure was down to 1,500. The start of an upturn in the market began in 1988, although subsequent high interest rates again depressed both full-sized and smaller vehicle deliveries.

Minibuses helped to counterbalance one of the negative effects of deregulation – fewer buses in urban areas in the evenings and on Sundays, and on a more general level in rural areas – since apologists for deregulation could point to the larger number of vehicles on the roads and the actual increase in road mileage attained through the large-scale use of these small vehicles. Wittingly or otherwise, the use of minibuses provided a masking role for an overall decline in service provision.

Deregulation has generally not seen a flourish of new small operators: predatory action by the pre-existing larger (if smaller than earlier) operators has seen to that. Indeed, several ex-NBC companies have been investigated by the Office of Fair Trading because of the nature of their reaction to competition from smaller operators. Smaller firms have either run commercial services which avoid competition with large operators or have concentrated on tendered services (Association of Metropolitan Authorities, 1990). Neither has the selling-off of NBC subsidiaries witnessed the hoped-for widespread development of share ownership in bus operations: instead it has seen a (limited) number of financial killings, through asset-stripping and other activities not necessarily in the travelling public's interests. The first former-NBC manager to be declared a millionaire at the end of 1987 attained that distinction largely as the result of selling Southampton's city-centre bus-station site for office development. From about that time, the major arena for competition tended to move away from the streets and into the boardroom, with national-holding companies and the public-transport companies buying up other operators. Running the industry on a commercial, profit-orientated basis has made it more difficult for transport authorities to use buses as a means of tackling urban transport and congestion problems.

Conclusion

An impartial observer might argue that by the early 1990s the credit/debit balance sheet for the British bus-deregulation and privatization process was roughly balanced out. Urban trunk routes witnessed more vehicles, greater frequencies and in some cases cheaper fares as the result of competition, but often at the expense of congestion and increased traffic hazards. More especially, just as in the previous US deregulation of air transport or the UK deregulation of coach services, the trade-off saw reduced services in less populated and less- or non-profitable areas. In the case of bus services this has meant poorer rural and off-peak and Sunday urban services. With the threat of diminishing central-government resources for local authorities and PTEs to fund tendered non-commercial services, that situation was likely to deteriorate further. It was exacerbated by the fact that in the early stages of deregulation the larger companies registered more 'commercial' services than those actually making money, simply to keep out the competition. Those non-commercial 'commercial' services were later deregistered, adding to the burden of transport-coordinating authorities.

Supporters of deregulation argued that while the legislation of 1980 and 1985 was helping to make a product-driven industry adjust to the requirements of the market, it would take some time before it became fully market-led. Opponents argued that the 'free market' approach was incapable of dealing adequately with transport planning, integration and regional development, pointing out that growing private monopolies in the bus industry needed to be restrained to protect passengers and staff (Trades Union Congress, 1990). The uncertainties faced by the various components of the UK bus industry would thus continue into the foreseeable future, with little respite.

References

Adeniji, K. (1983), 'Nigerian municipal bus operations', *Transportation Quarterly* 37 (1), 135–43.

Anderson, P. (1990), 'Jamaica's urban bus system: deregulation or public responsibility', *Privatization and deregulation in global perspective*, ed. D. J. Gayle and J. N. Goodrich (London: Pinter), ch. 15.

Ashworth, M. and Forsyth, P. (1984), *Civil aviation policy and the privatization of British Airways* (London: Institute of Fiscal Studies).

Association of Metropolitan Authorities (1990), *Bus deregulation: the metropolitan experience* (London: AMA).

Balcombe, R. J., Hopkin, J. M. and Perrett, K. E. (1988), *Bus deregulation in Great Britain: a review of the first year* (Crowthorne: Transport and Road Research Laboratory, Research Report 161).

Baumol, W. J. (1982), 'Contestable markets: an uprising in the theory of industrial structure', *American Economic Review* 72, 1–15.

Bell, P. and Cloke, P. (eds) (1990), *Deregulation and transport: market forces in the modern world* (London: David Fulton).

Bus Business (Peterborough: EMAP Response Publishing Ltd), fortnightly.

Button, K. J. (ed.) (1991), *Airline deregulation: international experiences* (London: David Fulton).

Button, K. J. and Gillingwater, D. (1986), *Future transport policy* (London: Routledge).

Carter, M. A. (1986), *Bus services in metropolitan areas* (Crowthorne: Transport and Road Research Laboratory, Working Paper WP/TP 18).

Department of the Environment (1990), *This common inheritance* (London: HMSO).

Department of Transport (1984), *Buses* (London: HMSO).

Department of Transport (1987), *Transport statistics, Great Britain, 1976–86* (London: HMSO).

Department of Transport (1990a), *Sale of the National Bus Company: report by the Controller and Auditor General* (London: HMSO).

Department of Transport (1990b), *Bus and coach statistics, Great Britain, 1989–90* (London: HMSO).

Department of Transport (1991), *A bus strategy for London* (Consultation Paper S15/21).

Evans, A. (1988), 'Hereford: a case-study of bus deregulation', *Journal of Transport Economics and Policy* 22 (3), 283–306.

Finch, D. J., Watts, P. F. and Wren, J. (1986), *Innovative public transport services* (Crowthorne: Transport and Road Research Laboratory, Working Paper WP/TP 21).

Fullerton, B. (1990), 'Deregulation in a European context: the case of Sweden', *Deregulation and transport: market forces in the modern world*, ed. P. Bell and P. Cloke (London: David Fulton), ch. 7.

Fullerton, B. and Knowles, R. D. (1991), *Scandinavia* (London: Paul Chapman).

Gayle, D. J. and Goodrich, J. N. (eds) (1990), *Privatization and deregulation in global*

perspective (London: Pinter).
Guardian (1991a), 'When freedom turns out to be licence', *The Guardian*, 11 February (London).
Guardian (1991b), 'Anglo-US deal widens air horizons', *The Guardian*, 12 March (London).
Hall, D. R. (1988), 'The changing nature of minibus technology and deployment in UK public transport strategies with particular reference to the Tyne and Wear area', *Transport technology and spatial change*, ed. R. S. Tolley (Transport Geography Study Group, Institute of British Geographers), 31–78.
Hibbs, J. (1985), 'Bus licensing: arbitration or franchise?', *Implications of the 1985 Transport Bill*, ed. R. D. Knowles (Transport Geography Study Group, Institute of British Geographers), ch. 3.
Higginson, M. (1990), 'Introduction to international transport deregulation', *Deregulation and transport: market forces in the modern world*, ed. P. Bell and P. Cloke (London: David Fulton), ch. 3.
Kahn, A. E. (1988), 'Surprises of airline deregulation', *American Economic Review: Papers and Proceedings* 78, 316–22.
Knowles, R. D. (ed.) (1985), *Implications of the 1985 Transport Bill* (Transport Geography Study Group, Institute of British Geographers).
Knowles, R. D. (1989), 'Urban public transport in Thatcher's Britain', *Transport policy and urban development: methodology and evaluation* (Transport Geography Study Group, Institute of British Geographers) 79–104.
Lewis, P. M. (1990), 'The political economy of privatization in Nigeria', *Privatization and deregulation in global perspective*, ed. D. J. Gayle and J. N. Goodrich (London: Pinter), ch. 18.
Local Transport (London: Kennington Publishing Centre), fortnightly.
Low, L. (1990), 'Privatization options and issues in Singapore', *Privatization and deregulation in global perspective*, ed. D. J. Gayle and J. N. Goodrich (London: Pinter), ch. 19.
Mellor, R. E. H. (1975), 'The Soviet concept of a unified transport system and the contemporary role of the railways', *Russian transport: an historical and geographical survey*, ed. L. Symons and C. White (London: Bell), ch. 3.
Monopolies and Mergers Commission (1987), *BA plc and BCal plc: a report on the proposed merger* (London: Monopolies and Mergers Commission).
Pickrell, D. (1991), 'The regulation and deregulation of US airlines', *Airline deregulation: international experiences*, ed. K. J. Button (London: David Fulton), ch. 2.
Rickard, J. M., Fairhead, R. D. and Watts, P. F. (1988), *Deregulation of local bus services: interim results from non-metropolitan study areas* (Crowthorne: Transport and Road Research Laboratory, Contractor Report 77).
Rimmer, P. J. (1986), *Rikisha to rapid transit: urban public transport systems and policy in South-east Asia* (Sydney: Pergamon).
Sealy, K. (1966), *Geography of air transport* (London: Hutchinson).
Thomson, J. M. (1978), *Great cities and their traffic* (Harmondsworth: Penguin).
Trades Union Congress (1990), *Transport for the 1990s* (London: TUC).
Transport 2000, *Transport Retort*, Transport 2000 Newsletter.
Turner, R. and White, P. (1987), *NBC's urban minibuses: a review and financial appraisal* (Crowthorne: Transport and Road Research Laboratory, Contractor Report 42).
Tyson, W. J. (1988), *The first year of bus deregulation* (London: Association of Metropolitan Authorities).
Tyson, W. J. (1989), *A review of the second year of bus deregulation* (London: Association of Metropolitan Authorities).
White, P. and Turner, R. (1987), 'Minibuses: the way ahead?', paper presented to Planning and Transport Research and Computation (PTRC) conference.

4

Transport, Environment and Energy

John Farrington

The impacts of transport on the environment have a long history. A brief review of this provides the background to a systematic examination of a range of environmental impacts, and ways in which they can be reduced. The energy efficiency of different modes is analysed. Transport policies, which are starting to deal with these issues, are assessed.

Introduction

The impacts of transport on the environment have a long history, a brief review of which gives us a useful background before looking systematically at a range of environmental impacts and at ways in which they can be reduced. Transport is also a significant consumer of energy, and the efficiency of different modes varies. The transport policies of governments are beginning to deal with energy and environmental issues, but many challenges remain.

Historical context

Early objectors

A concern for the impacts of transport development on environments is not new. Canal and railway companies in eighteenth- and nineteenth-century Britain faced opposition from groups and individuals objecting to the routes being followed, on what we would now call 'environmental grounds'. The objectors were not the organized pressure groups, residents' associations or other lobby groups of the present time, but landowners, artists and writers, or sometimes City Corporations. When the objection was successful, the canal or railway company would have to make a concession, such as a route alteration. If a canal or railway caused severance to a landed property by dividing it, an occupation bridge could be built. If the general amenity of a landscape were threatened by the sight and sound of the bargees on a canal or by the passage of steam locomotives, then screening by cutting, tunnelling or even re-routing might be the only way of placating the opposition. A railway company might be obliged to construct an expensive Italianate station building or a mock-Gothic tunnel portal to meet a landowner's demands. More far-reaching consequences resulted from the successful opposition in 1882 by Canon Rawnsley to the building of a railway along the shores of Derwentwater in the Lake District. Rawnsley's experience led him to become one of three founder members of the National Trust (Battrick, 1982).

Photograph 4.1 The Edinburgh and Northern Railway built this station at Markinch (Fife, Scotland) in 1847. The architect, David Bell, produced a restrained Italianate design with awnings on cast-iron columns. The main building material is a cream-coloured siltstone (J. H. Farrington).

Early impacts

The canals and railways of Britain gave rise to considerable environmental impacts, both directly in their construction and operation, and indirectly through the industrial and other land uses they encouraged. Kellett (1979) gives details of the impacts of railways in the five largest cities in Victorian England and Scotland. The main direct impacts on the urban environment were in land-take, severance and housing demolition. The land-take of railways in central areas of cities in 1900 varied between 5.3 per cent in Birmingham and 9.0 per cent in Liverpool. Similar proportions were apparent in US cities, with a high of 12.8 per cent in a major transit point like Kansas City (Kellett, 1979, 290). Large blocks of land were also required for railway use in the areas outside the central districts of cities. Connection between these facilities required many lines, often elevated on embankments, which frequently created a barrier effect between housing areas. Demolition of inner area housing arose from the construction of large urban railway terminals and their approaches. Approximately 20,000 people were displaced by demolition for railway purposes in central Glasgow, and at least 120,000 people in London between 1840 and 1900 (Kellett, 1979, 327). These were mainly in slum areas where tenants had no chance to prevent demolition.

It is clear that transport's environmental impacts, and concern about them, is not new, but our level of awareness is now much greater than it was. More formal systems have been incorporated into decision-making processes. Consultation and public inquiries give the opportunity for impacts to be at least discussed, if not strongly influenced in many cases. These changes have been made possible largely due to the systematic recognition, measurement and assessment of environmental impacts.

Environmental impact

A framework for analysis

The construction and operation of a transport element such as a motorway or an airport is a 'development' which will have direct impacts on the environment, and will also give rise to indirect impacts through its associated land uses and socioeconomic effects. This chapter discusses direct impacts.

In addition to the transport project which is the main, or only, development, transport is usually involved in the construction and operation of other projects such as industrial or housing developments, power generation or mineral extraction. The construction of transport projects themselves will normally also involve related transport impacts of a secondary, but possibly significant, nature. For example, constructing a pipeline involves access by plant and lorry traffic to points along the route, and this traffic will itself give rise to environmental impacts. Finally, it may be necessary to include an abandonment and reinstatement phase in the assessment of impacts. This would apply, for example, to temporary roads, built to serve a construction site and removed when the main site is complete.

This framework for the consideration of transport impacts as phases of projects is summarized in Table 4.1, noting that the 'Project' could equally be the construction or modification of a piece of transport infrastructure, or a development such as a power station or industrial site. Not all phases of 'possible related transport' are involved in every project, but Table 4.1 includes all possibilities, encouraging the comprehensive approach that is an important feature of the systematic assessment of impacts. Environmental impacts have to be assessed for a range of transport modes (Table 4.2). While a more detailed subdivision can be made, the basic distinction in Table 4.2 is appropriate for present purposes, and includes all the modes normally encountered in environmental assessment. Using the above framework, the types of impact arising from particular modes can be discussed.

Table 4.1 Project phases for transport developments

Project phases (including transport access)	*Possible related transport phases (e.g. temporary projects)*
Construction	Construction Operation Abandonment
Operation	(Construction) Operation (Abandonment)
Abandonment*	Construction Operation Abandonment

* In practice this is rarely applied to transport infrastructure projects, but should be considered for projects with a finite lifespan.

Table 4.2 System of transport modes for impact assessment

Passenger	*Freight*
Road: (a) Bus (b) Car	Road
Rail: (a) Diesel (b) Electric	Rail: (a) Diesel (b) Electric
Ferry	Maritime, inland waterway
Air	Air Pipeline

Types of impact

Table 4.3 lists a classification of environmental impacts of transport systems to be used in the subsequent discussion. Before discussing each type of impact in turn, it should be noted that a number of factors affects the specific impacts of any particular transport mode. These include:

1 geographical context, e.g. built-up area, or rural area; terrain and ecosystems of varying types;

Photograph 4.2 A railway bridge dating from 1839, disguised as a fortified gatehouse, carrying the Arbroath and Forfar Railway over the drive to Guthrie Castle in Tayside (Scotland), as requested by the landowner (J. H. Farrington).

Table 4.3 Types of environmental impact

Visual
Noise and vibration
Disturbance of terrestrial and aquatic ecosystems
Atmospheric emissions
Severance, displacement and demolition
Energy consumption

2 traffic levels in the transport system; and
3 characteristics of the vehicles using the transport system.

Also important is the significance attached to a specific impact, and the extent to which mitigation measures can be applied to reduce the level of impact. The assessment of the significance of impacts is a problematic area, which implies the need to make value judgements. This can be at least partly resolved by devices such as the calibration of impact perception to produce a scale, as is done with noise impacts, for example. Based on case studies of people's reaction to the type of noise expected in a particular project, predictions can be made about the probable annoyance that will be experienced. There is an extensive literature on such problems of environmental assessment (see, for example, Pearce et al., 1989).

Visual impacts

Assessment — It is useful to recognize the linear and nodal aspects of transport systems. Most land-based or inland modes are essentially linear; roads, railways and inland waterways all have this characteristic in common. Air and sea transport use airways and shipping lanes for the main movement, but have mainly nodal characteristics as far as visual and other landward impacts are concerned, seen in the terminal installations required to service their traffic. But land-based transport also has strong nodal characteristics in terms of its interchanges and terminals. Such installations often give rise to large-scale visual impact, especially when associated land uses such as

Photograph 4.3 The large quantity of heavy plant requiring temporary access to construction sites is seen in this view of work in progress in the early 1980s on the landfall of a natural gas pipeline between North Sea fields and the St Fergus Terminal in North East Scotland (Geography Department, Aberdeen University).

car parks, warehousing and other commercial land uses are included.

In their linear characteristics transport systems are unique as a land use. While elements of a road or railway, such as a bridge or intersection, can be considered separately in terms of their visual impact, it is often more satisfactory to attempt an overall assessment of impact along the length of the line. However, while measurement of the direct visual effect of a new road or railway is straightforward, the assessment of its impact in a particular context is difficult. The physical dimensions of an embankment, cutting or viaduct can easily be measured and scaled against the surrounding environment. Difficulties arise with the judgement involved in assessing the significance of the visual impact. The perception by individuals and groups of any particular piece of transport infrastructure is likely to vary. These groups include developers, planners and transport operators, local residents, transport users and interest groups. Complete consensus is therefore most unlikely.

The time dimension has a role to play in assessing visual impact. The railways and canals, opposed in the nineteenth and twentieth centuries because of their intrusion in the landscape, are now supported with equal passion by those concerned about the environment, either as valuable landscape components or as modes which are environmentally preferable to others, such as air and road transport.

It is clear that the passage of time can transform the perceived impact of transport systems, and this is most clearly seen in the case of visual impact. A large-scale canal aqueduct or railway viaduct is likely to be valued as a landscape element, while a similar structure for a new road or railway is much less likely to find favour. This is partly because such a new structure would be an additional landscape element, but also because we feel that many of the older engineering and architectural structures form part of our heritage and therefore have more merit than newer structures. An indication of this perspective is given by Pratt (1976) for

Photograph 4.4 This bridge across the M6 near Preston in Lancashire has won design awards. It uses modern materials but attempts to blend with the land contours (J. H. Farrington).

canal architecture, and Biddle and Nock (1983) for railway architecture.

Where the 'time threshold of acceptance' of a structure should be drawn is not clear. Perhaps it has to be threatened with demolition before its visual value increases. Since this is rarely the case as yet with road or air transport, we do not see their merits as visual elements. Roads in particular are also rather ubiquitous and, apart from a few structures such as major bridges and viaducts, have no scarcity value, as well as being relatively recent.

The scale of a structure in relation to its surroundings is another factor to be taken into account. Level terrain, particularly if well vegetated, can absorb a great deal of development without widespread visual impact since intervisibility is restricted. In contrast, relief features tend to impose the need for earthworks and other structures upon the engineers of road and rail routes, and these are potentially widely visible features.

Mitigation — The ability to picture precisely what a new transport system will look like in its environment is a useful tool in assessing impact and evaluating mitigation measures. Artists' or engineers' impressions have long been used for this purpose, supplemented more recently by superimposed photographs. The use of computer simulations has taken the process an important step further. It is now possible to produce a three-dimensional image of a landscape with specified structures located within it, and to examine their appearance from various viewing points. This helps in changing the design of transport structures to mitigate visual impact. Mitigation measures may include resiting the whole project, or perhaps cut-

and-cover construction or tunnelling for new road or rail lines, as planned for the proposed high-speed rail line linking the Channel Tunnel with London, although such measures are costly. Less expensive mitigation measures include alignments following land contours, giving a route more in sympathy with its landscape, or shallow cutting, earth or fence barriers (also useful for noise reduction), retention of trees, reinstatement of habitats, screening by tree-planting, detailed design of structures, and colour selection for clad or painted structures, to harmonize with the landscape as far as possible.

However, visual impacts are bound to remain to some extent. Grade-separated road junctions with their concrete elevated sections and bridges, large road signs and traffic-control systems, overhead wires on electrified railways, extensive car parking at terminals or city centres, are all elements difficult to hide. If we wish to enjoy the benefits of the transport systems of which these elements are a part, their visual impacts must be mitigated and their presence in the landscape accepted.

Noise impacts

Assessment — Noise energy is generated by all forms of mechanized movement. Transport noise forms a large part of the total noise in many environments, both urban and rural. Although overall transport noise levels have tended to stabilize since the late 1970s, it has been estimated that of a total population of nearly 800 million people in OECD countries, about 135 million were in the late 1980s exposed to transport noise regarded by most authorities as unacceptably high (Nelson, 1987, 1.5). While some reduction is expected in the 1990s due to more enlightened treatment of noise generation and mitigation, transport noise remains a significant aspect of transport's environmental impacts.

The main sources of transport noise are road, air and rail systems. In general road-traffic noise affects the most people, particularly in urbanized societies. In the USA about 15 per cent of the total population, and often 50 per cent of city populations, are exposed to traffic noise regarded as unacceptably high. The proportions in the more densely populated UK are probably slightly higher. In most countries only about one to two per cent of the population are exposed to this level of noise from aircraft and railways.

Noise levels can be measured accurately and for most transport noise are usually expressed in terms of the dB(A) scale, in which the basic decibel scale is weighted towards the frequencies most affecting the human ear. This scale (and others with different weightings which may be used at times) are logarithmic, so direct comparison between numbers on these scales is misleading. On the basic decibel scale which measures sound energy, an increase of 3dB represents a doubling in energy output, while for the commonly used dB(A) scale, an increase of 10dB(A) roughly represents a doubling of loudness as perceived by the human ear. It should also be noted that a doubling of traffic on a road, for example, does not result in a doubling of noise levels, since the relationship between traffic levels and noise is not linear. Distance from the noise source is also a vital variable. While much is known about the generation of noise and its behaviour as it travels away from a source, the exact prediction of specific noise levels at a particular location is part of a complex science. A full discussion of these and many other aspects of transport noise will be found in Nelson, 1987.

The measurement of noise is relatively straightforward, but the assessment of what constitutes 'a noise problem' is rather more difficult. Noise has been defined as 'unwanted sound'. There is a threshold at which noise becomes 'unacceptable' or 'a nuisance' and may cause stress and mental tension. At 80dB(A), people standing next to each other would need to shout to be heard, while at 90dB(A) – a typical level for a heavy lorry at seven metres – temporary loss in acuteness of

Photograph 4.5 Although their construction was sometimes opposed by landowners, many canals in the UK are now regarded as landscape assets, providing recreational and habitat resources, as typified by this scene on the Leeds and Liverpool Canal at Parbold near Wigan in Greater Manchester (J. H. Farrington).

hearing would result after a few minutes. At a higher level still there is a threshold (about 130dB(A)) at which noise causes physical pain and ultimately permanent loss of hearing. A jet aircraft taking off would typically give a noise level of 120dB(A) at a distance of 200 metres. A continuous noise level tends to be less disturbing than individual noise events, and a low background (or ambient) noise level can lower the threshold at which an event becomes a nuisance.

The perception of noise nuisance thresholds by individuals can vary widely, but it is necessary to apply specific thresholds for the purposes of evaluation and mitigation such as sound proofing. This has been done in many countries. A level of 65dB(A) is often regarded as an upper acceptable limit, based on experience built up since the 1960s about the noise levels that large samples of people find annoying or unacceptable. The results are not always what might be expected. For example, people in the UK have been shown to be more tolerant of rail noise than the same level of road noise, to the extent of five or ten decibels – a large difference (Fields and Walker, 1982).

Such studies indicate the most useful noise indices for relating noise levels to perception, and these are usually the 18 or 24-hour LEQ (the average of noise levels over 18 or 24 hours) or the L10 (the noise level exceeded for ten per cent of the time). Note that the LEQ tends to understate the noisiest events, which tend to be the most annoying. For aircraft noise, an index reflecting the number of noise events as well as their noise level is often used, such as the Noise and Number Index (NNI) in the UK, or the Day/Night Equivalent Sound Level (DNL) or

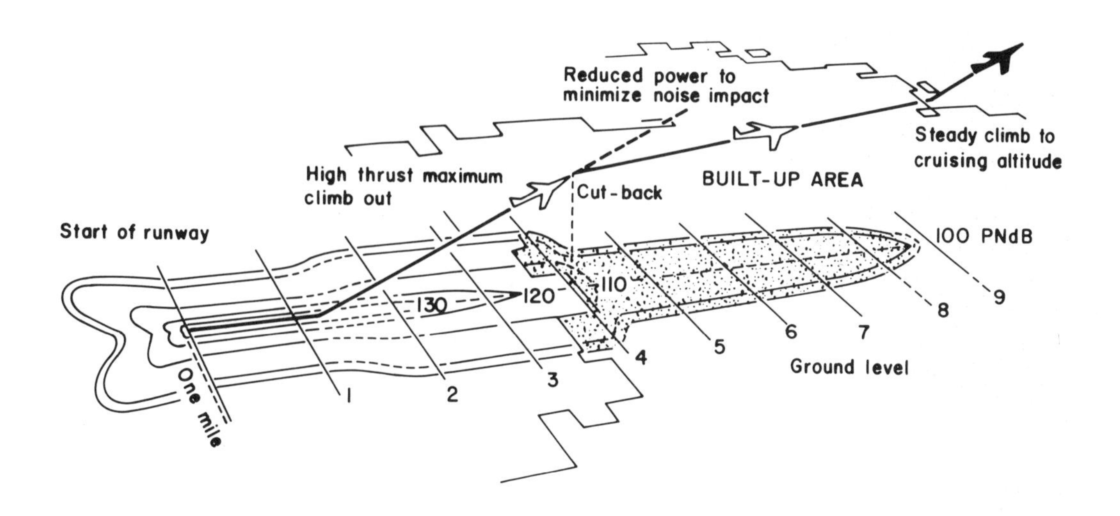

Figure 4.1 Aircraft departure profile used to minimize noise impact. A typical departure profile for aircraft taking off, designed to reduce noise impact in the built-up area beneath the flight path. Such noise-abatement procedures are voluntarily followed by many airlines and are standard procedure at many airports, though they do increase noise levels for people living slightly further away from the airport. Local circumstances have to be considered when devising procedures for each airport.

Source: After Nelson, 1987 (courtesy Butterworth).

Community Noise Equivalent Level (CNEL) in the USA and elsewhere. CNEL includes a greater weighting for noise events during the evening period.

Transport noise affects people to varying degrees, ranging from mild annoyance to mental and physical damage. It can also produce a detrimental effect on neighbourhood environments and may affect property prices. The construction of new roads, high-speed railways and airports introduces new noise sources into areas not previously affected to a significant extent.

Mitigation — As noted above, overall transport noise levels in the developed world have tended to stabilize in the last decade and are expected to fall slightly in the foreseeable future. This is due to mitigation measures, which are now briefly reviewed.

Noise mitigation can be approached under four headings (after Nelson, 1987, 1.4) which are complementary rather than mutually exclusive. Maximum mitigation will be obtained only if all are incorporated in an overall approach.

1 Reduction of noise at the source. This includes vehicle design, traffic management, tunnelling and noise abatement procedures for aircraft (Figure 4.1).
2 Measures to control noise along its transmission path. These consist mainly of barriers such as fences and embankments, and the use of buildings as noise barriers. A two-storey building may reduce noise levels on the 'lee' side by about 13dB(A) (Nelson, 1987, 11.1.5).
3 Measures to protect the observer from noise at the point of hearing. Buildings can be designed to reduce noise impacts on their occupants, for example by locating smaller windows on the noisiest facade. Double or triple glazing and acoustic insulation can have significant benefits, but may require air conditioning to compensate for lack of ventilation.
4 Land Use Planning and Zoning. This

approach is the most effective in reducing noise levels at the larger scale, because it locates housing and workplace further away from noise sources (Stratford, 1974, 51). There is a long time-lag in its application as a corrective measure, but it helps to ensure that new noise nuisances are not created. Planning consent around airports is normally determined by zoning based on current and predicted 'noise footprints', or contour lines of noise levels.

Vibration impacts

Assessment — Vibration from transport sources has two main components:

1 low frequency noise (mainly from exhausts), which can cause parts of buildings, such as windows, to vibrate; and
2 ground vibration transmitted to buildings via their foundations.

The effects of vibration on buildings and on people are not yet fully assessed. It seems likely that rattling window frames and crockery can add to the personal stress induced by noise nuisance. Damage to buildings from traffic vibration is difficult to quantify, but is probably small in relation to other causes of damage such as drought or frost. On the other hand, specific locations such as historic buildings may suffer discernible damage. Sub-surface pipes and cables can also be damaged.

Mitigation — The gradual increase in permitted British lorry weights to the present 38.5 tonnes, and moves to a European Community norm of 40 tonnes, has emphasized the need to reduce vibration by attention to the smoothness of the road surface and to vehicle suspension design. Further research will enable a more precise estimate of vibration impacts in future, and of the compensation or charging levels that might need to be applied to the recipients and sources of vibration if reductions are to be made.

Impacts on terrestrial and aquatic ecosystems

Assessment — Any new construction project will have a direct impact on ecosystems. Land-based transport systems have some unique impacts arising from their linear nature. In particular, they can affect a corridor of habitats, create a barrier effect between habitats and affect drainage systems. However, they can also create a corridor along which species may move, as well as reservoirs of ecological variety in otherwise poorly endowed habitats. Detailed ecological surveys should precede the construction of new infrastructure.

Mitigation — Little can be done to alleviate the direct destruction of habitats, though some relocation may be possible and in the case of pipelines, reinstatement is normal. Careful route alignment can avoid protected sites or areas of special interest. The separation of habitats may be partially overcome by the construction of underpasses for the movement of animal species. Existing drainage patterns may require realignment and this may affect local water tables and habitats.

Quite recently, the positive ecological potential of some transport lines has been recognized. Canals provide aquatic habitats in inland locations. Railway and motorway cuttings and embankments provide thousands of miles of non-agricultural habitats, and British Rail has embarked on a programme of management which will enhance appearance and encourage habitat diversity on its lineside corridors.

Impacts on air quality

Assessment — Transport is a major contributor to air pollution since it depends largely on the combustion of fossil fuels, either in vehicles or at power stations.

The major pollutants emitted are as follows:

Photograph 4.6 A Boeing 727 leaving Hollywood-Burbank Airport in California makes the steep climb from take-off which reduces noise impacts in a distance band away from the airport vicinity (J. H. Farrington).

Carbon Dioxide (CO2)
Carbon Monoxide (CO)
Nitrogen Oxides (NOx)
Hydrocarbons (CxHx)
Sulphur Oxides (SOx)
Lead (Pb)
Suspended particulates

All of these are harmful to flora and fauna including human life, affecting particularly respiratory systems. The hydrocarbon benzene, contained in petrol (leaded and unleaded), is carcinogenic. Traffic fumes contribute about half the average non-smoker's daily intake of benzene. Lead affects human organs including the brain and hence intelligence, especially during childhood. There is also a broader dimension to the impact of transport emissions. Carbon dioxide is a greenhouse gas and nitrogen and sulphur oxides contribute to acid rain.

Cars use nearly two-thirds of the oil consumed in the USA and almost half of that consumed in Western Europe. In Britain, cars produce 20 per cent of total carbon-dioxide emissions and about 40 per cent of acid-rain constituents, as well as most of the airborne lead.

Mitigation — The only effective way of reducing carbon-dioxide emissions is to reduce the amount of fuel consumed. Although internal combustion-engine efficiency continues to increase, the growth in road-vehicle numbers tends to wipe out this gain. A ready technical solution such as battery-powered cars seems unlikely in the immediate future. A shift to more fuel-efficient modes is needed. A movement of traffic from road to electrified railways could achieve significant reductions in carbon dioxide, particularly if the electricity is generated by non-fossil fuels or renewable energy sources.

Catalytic converters fitted to cars reduce emissions of nitrogen oxides, carbon monoxide and hydrocarbons. Lead-free petrol now accounts for about 25 per cent of Britain's consumption and since the platinum in catalytic converters is incompatible with leaded petrol, the scene is set for these two measures to be implemented more widely. From 1993 all new cars sold in the European Community will have to be fitted with catalytic converters. Unfortunately an increase in fuel consumption will result, so that carbon-dioxide emissions will probably increase. Suggestions that compensation can be achieved by applying differential licensing charges to cars, increasing according to engine size, ignore the safety issue since larger cars tend to be safer in accidents.

Severance, displacement and demolition

Assessment — All land-based modes with the exception of pipelines create potential impacts under these headings, and even pipelines carrying volatile materials require corridors of land safeguarded from development for safety reasons. Airports are such large blocks of land that they also create potential severance effects in their particular location.

Some severance effects, notably those of non-motorway type roads, are only partial. Here, the traffic levels determine the degree of severance. Railways, canals and motorway-type roads constitute complete barriers to pedestrian and vehicular cross-movement and can also act as psychological divides, as well as demarcating land-use zones. Severance also occurs in rural areas, where the lines of new roads or railways may cut through farm units and fields, rendering each divided portion uneconomic to work.

Mitigation — Severance effects can be mitigated through the provision of cross-access by means of controlled crossings on non-motorway roads or on railways, or preferably bridges or underpasses (grade separation), or even continuous cross-access by means of cut-and-cover construction or tunnelling. All such measures, especially the latter two, are costly to provide and (except for the same two measures) still give rise to potential inconvenience, perceptual division and safety hazards.

Demolition of property and displacement of people are now generally regarded as undesirable, though some displacement is often the inevitable result of construction. Demolition is avoided as far as possible, especially where the buildings are of significance for architectural or historic reasons. Of course, most homes are of significance to their occupants and compensation at market value, available from 1959 in the UK, cannot fully recompense for the loss of a family home. In the post-war years when housing problems in many British cities were acute and large-scale renewal was taking place in Comprehensive Development Areas, some benefit was seen to be obtained from extensive demolition of residential property. Such approaches were phased out in the 1970s.

Transport and energy

Sources of transport energy

All forms of movement require an input of energy. In most transport systems the source of energy is carried in the vehicle, usually as refined petroleum products such as petrol (gasoline), diesel fuel or kerosene, for burning in internal combustion engines, turbines or jets. Electrically powered transport systems usually involve the collection of power generated elsewhere. Pipelines are an unusual mode in that energy is applied to the system at intervals along its route in the form of pumped pressure.

Other sources of energy for movement consist mainly of wind, human and animal power, coal and coke, and hydrocarbons produced from organic matter. Wind power, mainly used for sea transport, was largely replaced for commercial purposes in most parts of the world by the early twentieth century, but it is still used in some areas. Recent developments suggest that it may have a growing role to play by supplementing mechanical power in ships when winds are favourable.

Human power is still very much with us on foot and on the bicycle. Pedestrian movement dominates in terms of number of trips in urban areas, and the bicycle remains dominant in large areas of the world. Coal and coke have largely been replaced as sources of power on steam-driven vehicles, though coal is widely used indirectly through electricity generation. Ways of using coal more efficiently on steam-powered road and rail vehicles continue to be researched, but as with battery power and hydrocarbons derived from sources other than petroleum (such as sugar cane), technical and economic difficulties remain.

Energy efficiency

The energy efficiency of different transport modes varies considerably. Table 4.4 illustrates this for passenger transport.

The most energy-efficient modes are walking and cycling. The greater speed, convenience and capacity of mechanical modes involve energy penalties which users are clearly prepared to pay for, at least up to a point. The table also shows the generally greater efficiency of buses and trains compared with cars. Double-decker (urban) buses are particularly efficient, and even an express (long-distance) coach is marginally more efficient than the most efficient type of train.

Another point of note is the energy penalty of increased speeds. Air transport is less energy efficient than surface modes. Its advantages of speed and its ability to cross land/sea barriers are dominant factors for many types of journey.

Increasing mobility and energy use — Transport is a large and increasing user of energy in the developed world. In the UK in 1975, 22 per cent of all energy consumed was used in transport. In 1990 transport's share of a much larger energy budget was 30 per cent. The biggest single reason for this increase has been the greater use of cars, travelling a greater distance in total.

Trends in the UK over the decade 1975–85 show that the number of journeys made per head remained about the same (18 one-way trips per week per person, one third on foot), while the amount of time spent in travelling has increased by 10 per cent and, significantly, the distance travelled has increased by one-third (Potter and Hughes, 1990). It appears that we are using technical developments in transport to reduce the relative financial cost of mobility and are therefore travelling more.

Since transport is almost entirely a derived demand (that is, a means to an end), it follows that there should be a greater 'total achievement' of some kind resulting from this increased mobility. Undoubtedly this is true in one sense, in that car users can access a wider range of destinations at greater frequency than those without car use. It is also true, however, that car users tend to live in places which are at a greater distance from facilities, and that many facilities are increasingly centralized. There is, therefore, an element of necessity in this greater mobility.

The car emerges as the greatest single factor in transport's increasing energy use. In the UK it is now responsible for 60 per cent of all transport energy consumption. Improvements in fuel efficiency have been outweighed by increases in total car mileage. Technological progress has been used more to increase performance than to improve fuel efficiency, because in a time of relatively cheap energy, performance is more marketable. Improved roads have, in turn, made the performance more usable.

The evidence of increasing fuel use suggests that we have not yet taken the topic of transport energy sufficiently seriously. Most of our mobility is achieved by the increasing use of cars, which for all their

Table 4.4 Energy efficiency-passenger modes

	No. of persons carried	*Energy used (MJ) per passenger km.*	*Energy used (MJ) per passenger km. (fully laden)*
Petrol car:			
under 1.4 litre	1.5	1.73	0.65
1.4–2.0 litre	1.5	1.99	0.74
over 2.0 litre	1.5	3.08	1.15
Diesel car:			
under 1.4 litre	1.5	1.50	0.56
1.4–2.0 litre	1.5	1.84	0.69
over 2.0 litre	1.5	2.44	0.91
Rail:			
InterCity (100mph electric)	338 (60% full)	0.48	0.29
InterCity 225 (125mph electric)	289 (60% full)	0.65	0.38
InterCity 125 (125mph diesel)	294 (60% full)	0.59	0.35
Spr Sprinter diesel	88 (60% full)	0.55	0.33
Electric suburban	180 (60% full)	0.48	0.26
Double-decker bus	25 (33% full)	0.52	0.17
Single-decker bus	16 (33% full)	0.87	0.29
Minibus	10 (50% full)	0.71	0.35
Express Coach	30 (65% full)	0.38	0.25
Air (Boeing 737)	100 (60% full)	2.42	1.45
Motorcycle	1.2	1.94	1.17
Moped	1	0.81	0.81
Bicycle	1	0.06	0.06
Walk	1 person	0.16	0.16

Sources: Hughes, 1990; Potter and Hughes, 1990.

convenience and flexibility are among the least fuel-efficient forms of transport. The measures that are being, and could be, adopted to achieve more environment- and energy-related transport patterns and practices are discussed briefly in the final section of this chapter.

Measures and policies

Political and economic penalties

Technological measures to reduce impacts and energy consumption are increasingly available. But incorporating them into the policies of governments or international organizations like the European Community is another matter. Many of these measures have a financial cost which may well translate into political and commercial costs.

For example, the use of unleaded petrol and catalytic converters on cars, which can be encouraged by differential pricing or enforced by legislation, has benefits for emission restriction and hence for human well-being. But fuel consumption (and carbon-dioxide production) is increased and the converters add to the cost of the vehicle. Or consider the policy of increasing the price of petrol by taxation to reduce car use. Some personal mobility would be lost, some travel would switch to public transport and less petrol would be used. But rural populations, with longer journeys to work and limited alternative public transport, would be penalized.

Moreover, such policies carry an inflationary element and are therefore politically unattractive. For the same reason, a 'green tax' or 'carbon tax' on the use of fossil fuels has so far been avoided by governments which appear otherwise to be persuaded of the need to reduce such consumption.

On the other hand, an electorate may be ready to accept such policies and their implication of a reduced standard of living (since less disposable income would be available for the purchase of other goods and services). In other words, a higher value may be placed on our environment. (For a discussion of environmental valuation, see Pearce et al., 1989, and Sharp and Jennings, 1976.)

Towards regulation

In fact measures to control impacts have increasingly been adopted by national and international policy-makers since the 1960s, but only recently has effective regulation of transport impacts begun to be implemented. There is now a growing array of legislation in most developed countries controlling impacts to some extent. This is particularly focused on vehicle noise and exhaust emissions, and on planning legislation dealing with the land use, visual and traffic implications of developments. The latter types of control are prone to be patchy, implemented as they are by local authorities which attach different levels of importance to specific types of impact.

Steps towards the reduction of transport's environmental and energy impacts have been taken in the area of environmental impact assessment. For example, the European Community Directive 85/337 of 1985 requires environmental assessments to be made of new projects including motorway, rail and air construction. It seeks to prevent harmful impacts rather than trying to deal with them after they have been created. It encompasses the effects of projects on human health and the quality of life and aims to ensure species diversity and to safeguard the ecosystem 'as a basic resource for life' (Whitelegg, 1988). But procedures for ensuring that this kind of legislation results in a reduction in transport's environmental impacts will have to be rigorous.

A basic problem

A more fundamental problem is the need to see economic and political processes as part of the human system which generates transport use and hence the environmental impacts. There remain deep-seated attitudes which do not help good practice in the sense of mitigating environmental impact and energy consumption. In particular, use of market forces to determine patterns and levels of transport provision may have the opposite effect. For example, Whitelegg (1990) has questioned the pursuit of speed among Europe's railways in their efforts to compete in the market place with road and air transport. He argues that high-speed passenger railways are 'fiscally draining and environmentally ruinous' and that they divert resources away from other, more urgent, transport needs, such as the diversion of freight from road to rail and the provision of better urban transport systems.

In the UK, the procedures for evaluating investment on road and rail transport increase the difficulty of arriving at a more environmentally preferred transport mix, since spending on the two modes is not assessed in the same way and gives little weight to wider environmental or social benefits. It appears that market forces lead to increasing mobility, greater car use, preference for high-speed rail or air travel and dispersed patterns of land use. Together, these have the effect of increasing congestion, air and noise pollution, and land take.

The complete fulfilment of these trends is not compatible with impact minimization. Effective measures to bring about a less motorized, more public-transport orientated and less mobile society are unlikely to be politically acceptable in the immediate future. Gradual moves in this direction are the most that can realistically be expected. Oil price rises seem the most likely way of curbing motorized mobility and encouraging the search for alternative fuels, but the more intransigent problems of dispersed location and sprawling land use which are characteristics of a high-mobility motorized society are less susceptible to short- or even medium-term solution. Despite our best intentions it seems unlikely that the total impact of transport on the environment and on energy consumption can be reduced in absolute terms in the immediate future. In that case the best result we can achieve, short of wholesale economic shrinkage, is to slow the rate of impact growth and to ameliorate impacts as far as is technically feasible and politically acceptable.

References

Battrick, E. (1982), 'The watchdog of the Lake District', *National Trust Magazine* 3, 13–15.

Biddle, G. and Nock, O. S. (eds) (1983), *The railway heritage of Britain* (London: Michael Joseph).

Fields, L. M. and Walker, J. G. (1982), 'The response to railway noise in residential areas in Great Britain', *Journal of Sound and Vibration* 85(2), 177–255.

HMSO, Town and Country Planning Act, 1959.

Hughes, P. (1990), 'Transport emissions and the greenhouse effect', *Universities Transport Studies Group Conference.*

Kellett, J. R. (1979), *Railways and Victorian cities* (London: Routledge).

Nelson, P. M. (ed.) (1987), *Transportation noise reference book* (London: Butterworth).

Pearce, D., Markyanda, A. and Barbier, E. B. (1989), *Blueprint for a green economy* (London: Earthscan).

Potter, S. and Hughes, P. (1990), *Vital travel statistics* (London: Transport 2000).

Pratt, F. (1976), *Canal architecture in Britain* (London: British Waterways Board).

Sharp, C. and Jennings, T. (1976), *Transport and the environment* (Leicester: Leicester University Press).

Stratford, A. (1974), *Airports and the environment* (London: Macmillan).

Whitelegg, J. (1988), *Transport policy in the EEC* (London: Routledge).

Whitelegg, J. (1990), 'The society now departing', *The Guardian*, 15 June 1990, 25.

5
Urban Transport Patterns

Brian Turton

Facilities for the transport of people and freight within urban areas vary widely. The greater share of personal travel is now carried out by private means of transport although public passenger rail and bus undertakings are still of importance for commuter movements. The level of public transport available in Third World cities is much lower than in industrialized countries and private car travel is still largely confined to Western urbanized societies.

Introduction

The focus of attention in this and the succeeding chapter is transport in the world's urban areas, where the concentration of industrial and commercial activity has been accompanied by the growth of intricate patterns of movement. The demands currently made by both freight and passenger traffic upon urban transport infrastructures are invariably greater than the available capacity, producing a situation in which most of the world's towns and cities require programmes of transport improvement. Consequently much of the content of contemporary urban transport geography has a problem-orientated approach with a particular concern to identify the spatial aspects of transport-planning programmes. These approaches are discussed in Chapter 6, but as a prelude this chapter reviews the relationships between traffic and land-use patterns, the principal types of movement and the attributes of urban transport modes. Variations in the pattern of urban land uses will induce alterations in the volume, type and directions of personal and freight movements; the distribution of traffic between road and rail transport and between the public and private sectors is also influenced by changes in social and economic conditions both at the national and local scales (Webster, 1988).

Each mode is recognized by transport planners as having particular advantages with respect to specific types of intra-urban movement and to its contribution to the overall efficiency of the urban transport system. However in many instances personal preferences and perceptions can combine to produce an allocation of travel between modes whereby the least suitable modes are often the most heavily patronized but others, principally in the public sector, experience ever-declining levels of patronage (Jones et al., 1983; Sheppard, 1986).

Land use and traffic generation

The mapping of reliable patterns of traffic movements within towns first became possible with the advent in the 1950s of comprehensive land-use transportation plans designed to promote the development of major cities and conurbations in North America and western Europe. These ambitious programmes involved the collection of data on contemporary traffic generation and transport facilities as an essential prelude to the prediction of future flows and the formulation of new or amended transport systems to cater for these changing movement patterns. Although the overall pattern of intra-urban flows comprises both freight and personal movements, emphasis has been firmly upon the latter in most of the major transportation surveys, and data are collected relating to both individual and household tripmaking behaviour and the broader travel structures associated with specific zones within a city.

At the individual or disaggregate level information on travel is collected for a specific time period, usually up to one week, with regard to trip purpose, trip time, destination, mode of travel and route selected. Particular note is taken of multipurpose and multimodal journeys as these are becoming of increasing significance within the overall pattern of movements (Barber, 1986).

Freight-traffic generation by industrial and commercial premises has received much less attention than personal travel but sufficient studies have been made to relate different scales and categories of industry to certain levels of traffic (Starkie, 1967; Bartlett and Newton, 1982). Data on tripmaking obtained from surveys of individual households and industrial premises can be augmented with information collected from vehicle drivers passing through traffic cordons drawn at various radii from city centres. These data are then mapped on a zonal basis; the urban area is sub-divided into traffic districts and zones and the size and shape of each division is determined by physical urban structure and land-use patterning. An aggregate picture of trips with origins and destinations in each traffic zone can then be produced and most of the major transportation surveys incorporate these interzonal flow maps. The basic map depicting all categories of trip may be supplemented by maps devoted to modal choice, journey purpose or time; these trip characteristics may then be analysed in the context of the dominant land use within individual traffic zones.

Taking the UK as a representative example of a highly urbanized industrial nation, one-fifth of all road traffic is associated with built-up areas with a similar proportion of private cars and light vans. A large share of the national bus and coach fleet is allocated to urban services and 29 per cent of the total traffic is recorded in towns and cities. In contrast only 14 per cent of the heavy goods traffic is found in urban areas and this proportion has been decreasing since the 1970s (Table 5.1) (Department of Transport, 1990a).

Personal travel

Travel surveys

One drawback of many transportation surveys is their disregard of pedestrian journeys and the inclusion only of trips of over one-quarter of a mile made by a motorized mode, a selective exercise that omits the walking which can account for up to 30 or 40 per cent of all personal travel. Information on personal tripmaking is obtainable from national surveys and, at the local level, from regional- and urban-transportation studies. In the UK the National Travel Survey was launched in 1964 and the most recent was completed in 1985. Data were obtained primarily on a household basis. Nine categories of journey purpose are defined associated with work, education, escorting (work and education), shopping, personal business requirements, social and entertainment trips and holiday and related activities. In practice many urban trips fulfil several purposes, with one car, for example, being used by a number of occupants, each with different travel

Table 5.1 Great Britain: traffic on major urban and non-urban roads

(a) *Traffic distribution by vehicle types (percentage of total vehicle – kms.)*

	Cars/taxis	Buses and coaches, excepting minibuses	Heavy goods vehicles	All vehicles
Motorways	22.4	17.8	38.1	23.7
Major roads in built-up areas	32.7	47.6	18.4	31.7
Other major roads	44.9	34.6	43.5	44.6

(b) *Traffic by average daily flows (000s vehicles)*

	1979	1984	1989
Motorways	31.19	34.04	54.44
Trunk roads in built-up areas	14.56	13.25	18.10
Other trunk roads	8.14	10.48	14.39

Source: Transport Statistics Great Britain 1979–89, 1990.

objectives (Hanson and Schwab, 1986). The timing of a particular journey will be strongly influenced by its purpose while the choice of mode used will be determined by socioeconomic factors such as income levels, access to a private vehicle, the quality of public transport and the size of the urban area. In the British National Travel Survey respondents complete daily travel diaries over a week together with a record of the use made of cars available to household members. Many journeys, and particularly those making use of public transport for work trips, involve several modes and these can also be associated with multipurpose trips (Table 5.2).

The journey to work

Commuting dominates the overall pattern of regular personal movements and a more detailed review of this traffic illustrates the complexity of travel within major cities. The scale and extent of travel-to-work is dependent upon the relative locations of workplace and residence, household income, the level of public transport, car availability and a variety of personal behavioural trends and has been studied in depth in many industrialized cities (Daniels and Warnes, 1983; Dasgupta et al., 1990).

Journey-to-work patterns are becoming increasingly complicated with the expansion of major employment centres on the urban periphery and by the fact that commuting now extends far beyond the limits of the built-up area so that the movements of these rurally-based workers are superimposed upon those of employees resident within the urban area (Plane, 1981). However a study of commuting in principal EEC cities showed that 71 per cent of workers lived within a 10-km. radius of their work and that only 7 per cent travelled further than 26 km. Journey-to-work trips were accomplished in less than 30 minutes by three-quarters of all employees and only one in five took more than two hours for each trip. Although there is evidence that the average length of each journey is increasing, travel time is being reduced with the use of faster modes of transport. In West Germany, for example, the percentage of workers whose journey occupied only up to 15 minutes rose from 21 to 41 between 1960 and 1978. The European survey identified a wide range in the proportions of those using the private car for journeys to work, with a variation from 36 per cent in Italy to 49 per cent in Belgium. Public transport accounted on average for just over one-fifth of all journeys but was of

Table 5.2 Great Britain: journeys by method of travel and purpose

			(percentage of all journeys)			
			Purpose			
Main mode of travel	*Work*	*Education*	*Shopping*	*Social/ entertainment*	*Leisure*	*All other purposes*
Car (driver and passenger)	23	2	18	29	4	24
Local bus services	24	13	29	20	1	13
Rail	51	6	12	15	3	13
Cycle/motorcyle	46	9	11	20	4	10
Walk	13	12	21	19	19	16

NB: Only journeys of over one mile are included.

Source: National Travel Survey 1985/6.

more importance in the UK. Overall 18 per cent of all trips were carried out on foot and in The Netherlands cycling and motorcycling was used by one-third of all commuters (Pickup and Town, 1983) (Figure 5.1).

The relative importance of different modes for work trips, however, can vary in response to increasing road traffic congestion or to improvements in the quality of public transport, and especially rail services. Between 1978 and 1988, for example, the total volume of workers entering Central London during the morning peak rose by 7.3 per cent to 1.157 million, with public transport accounting for a large share of this increased traffic. Travel-to-work journeys on the London Underground increased by 38 per cent and on suburban railways by 21 per cent between 1982 and 1989, whereas road-based commuting by bus fell by 26 per cent and in cars by 18 per cent (Department of Transport, 1990b).

The commuting pattern also includes many examples of multipurpose journeys involving shopping and education, and the routes followed between home and workplace are often indirect in order to complete these various tasks. A survey of commuting in Reading indicated that eight per cent of all work trips involved additional objectives of this nature (Rigby and Hyde, 1977).

Within major cities in the USA, journey-to-work trips account for one-quarter of all household travel and the private car is firmly established as the dominant mode despite the comprehensive public-transport systems available in many of the larger urban areas (Fielding, 1986) (Table 5.3). However, between 1977 and 1983 the largest percentage increases in work trips were recorded in smaller American cities of less than 250,000 population and in the largest (three million plus) centres the increase has been only just over seven per cent.

Recent USA personal-travel surveys also indicate that the number of non-work trips is now increasing more rapidly than the commuting journeys, even during peak travel periods, and work trips are becoming shorter as the suburbs attract more employment centres (Gordon et al., 1988). Commuting patterns in many major industrial cities are also beginning to show the effects of the redistribution of population associated with urban renewal, with an increase in the proportion of low-income and less-skilled workers who are resident in inner-urban areas. In Manchester an increased dependence upon public transport in the inner city has been reported as low levels of car ownership among the less skilled has limited their opportunities for employment in peripheral urban areas (Dasgupta, 1983).

By 2000 AD 20 out of 24 cities with populations in excess of 11 million will be in the Third World and the commuting patterns in these and in lesser cities differ markedly from those in industrialized states. Deficiencies in the public-transport system, the

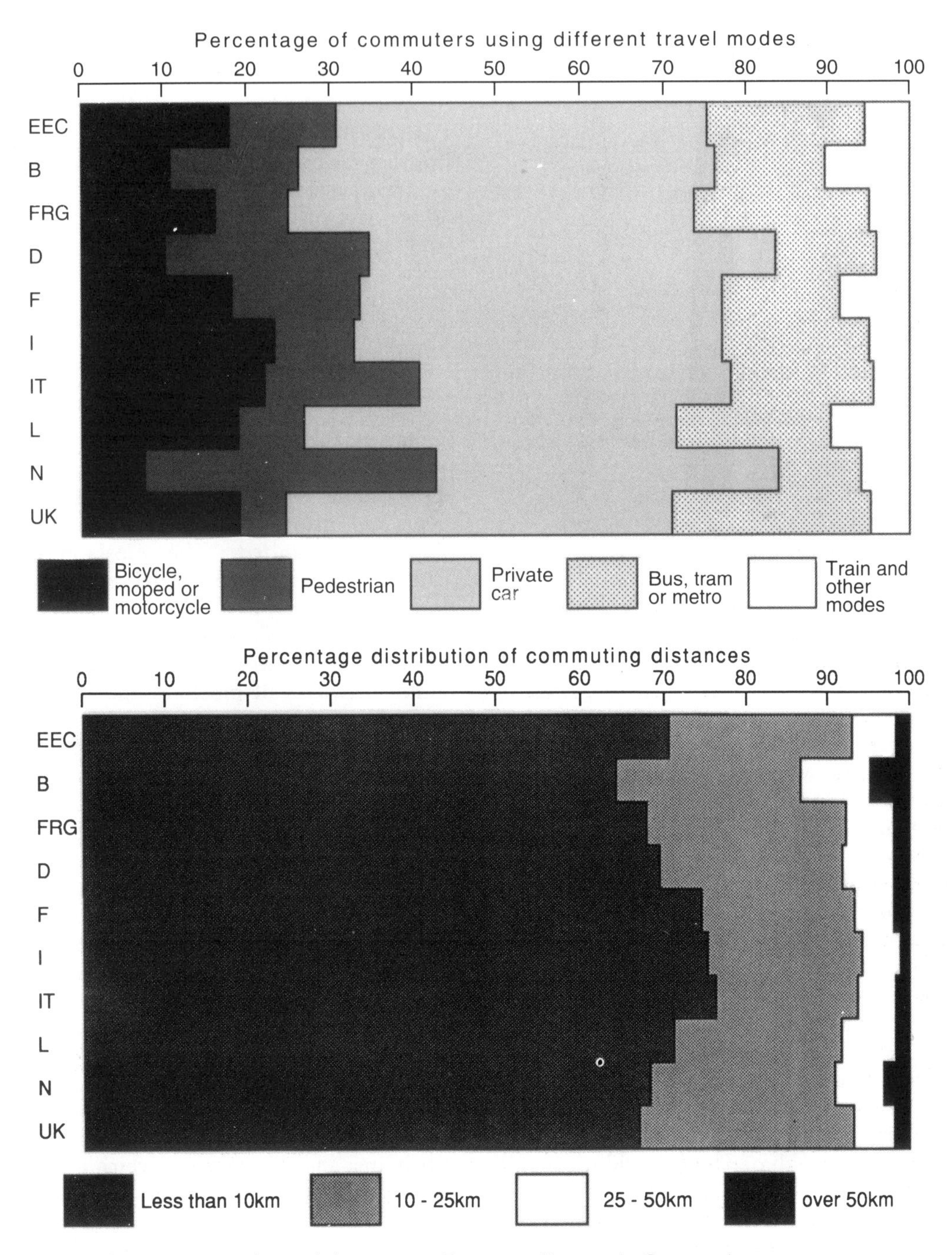

Figure 5.1 Commuting modes and distances in European Economic Community states.

EEC	European Economic Community	I	Ireland
B	Belgium	IT	Italy
FRG	Federal Republic of Germany	L	Luxembourg
D	Denmark	N	Netherlands
F	France	UK	United Kingdom

Source: Pickup and Town, 1983.

Table 5.3 Commuting in selected United States cities

(a) *Travel in selected cities 1976*

	Mode of transport			
	Private car	Public transport	Walk	Other modes
New York	47	44	8	1
Baltimore	83	12	5	1
Houston	93	4	3	1
Raleigh	95	2	3	1
Oklahoma City	95	1	3	1
Seattle	87	8	3	1
Denver	87	5	5	2

(b) *Changes in proportion of commuting trips into Central Business Districts made by public transport (percentage)*

	1970	1980
Boston	60.6	58.5
New York	60.6	60.2
Chicago	74.9	74.1
St Louis	29.7	26.7
San Francisco	47.9	52.4
Los Angeles	21.4	23.8

Source: Hanson, S. (ed.), *The geography of urban transportation*, 1986.

presence of workforces dominated by low-income groups and the distribution of low- and high-income residential areas combine to produce journey-to-work patterns where non-motorized modes and informal public-transport services assume a far greater significance than in Western urban society (Rimmer, 1986). Whereas in the latter the choice in general lies between motorized private and public transport, in the developing countries the urban commuter has the option of using grossly inadequate scheduled public services, a range of informal transport facilities or of walking or cycling to work. The rapid growth of Third World cities, with the expansion of peripheral low-grade housing areas, has resulted in an increasing separation of home and workplace for many of the lower-paid workers, who in Rio de Janeiro, for example, can spend up to 25 per cent of net income on fares for bus journeys. Third World urban workers also suffer a time penalty when using public transport. A recent survey of journeys-to-work showed single trip times of 89 minutes in Lima, Peru, and 77 minutes in Ibadan, Nigeria, compared with an average of 49 minutes for a sample of 44 United States cities and 48 in British provincial cities. In Delhi the regular use of a bus for commuting can account for 36 per cent of the disposable income of low-paid workers and in many Indian cities journey-to-work travel costs, expressed in terms of the percentage of GDP per head, can be almost fourfold those in Britain (White, 1990).

Freight traffic

Several different categories of goods movement and distribution within urban areas may be recognized. The most complex activity involves the regular delivery of essential commodities to individual households but in most advanced nations much of this is now achieved by means of permanent distribution networks of pipelines and cables for water, sewage, gas and electricity. Road transport is still necessary however for solid refuse, mail, milk and some other food supplies and these regular but often slow-moving services account for a large part of urban freight traffic. Goods pickup and delivery trips in a sample 11 United States towns represent 41 per cent of all urban commercial vehicle movements, with peak traffic occurring between the daily commuting periods (Barber, 1986) (Table 5.4). A second category of freight traffic is the transfer of finished and semifinished manufactured goods between plants, where the level of movement depends upon the degree of dispersion of industrial activity within the urban area and the extent to which manufacturing is confined to specific zones. The transport of freight between urban industrial premises and destinations in other parts of the country also contributes to the overall pattern of goods movements, as does commercial traffic in transit through a town. The types of vehicle used for these various movements is determined by the size and nature of the consignments carried. The

Table 5.4 Urban truck traffic in the United States (based on data from 11 US urban areas)

Trip purpose at final destination	Percentage of total daily trips
Return to home base	19.3
Personal use	9.1
Pick-up and delivery operations	41.1
Mail and express services	6.1
Construction	4.9
Maintenance and repair	8.0
Business use	7.2
All other	4.3

Source: Hanson, S. (ed.), *The geography of urban transportation*, 1986.

frequency of use is similarly related to the needs of the market, with daily household deliveries being at one end of the scale and occasional visits of large commercial vehicles to major industrial plant at the other. A study of goods movements in Hull showed that road transport accounted for 81 per cent of inbound and 88 per cent of outbound traffic, with rail and waterways carrying the remainder (Wilbur Smith, 1977).

The principal modes of urban travel

Walking

Until the successive introduction of horse buses, electric trams and motor buses, walking was the dominant mode in towns and cities of the Western world. Today it can still account for up to 35 per cent of all journeys in these cities and in addition almost all trips which are classified as primarily car- or public transport-based do of necessity incorporate some walking at the beginning and end of the journey. Regular walkers are obviously concentrated within those groups who lack access to a car or are unable to drive through reasons of age or infirmity; trips to school in particular are largely carried out on foot. Walking is still the dominant mode for urban shopping trips but for most other activities such as work, entertainment and leisure it ranks below the private car but above public transport. Most trips are made in city centres, involving journeys between car parks, bus and rail stations and workplace and between office premises, shops and restaurants.

However, it is in Third World cities that walking assumes the greatest importance as a means of personal mobility and a large proportion of all journeys are made by pedestrians. In India the lower income groups depend upon walking for almost 60 per cent of all urban journeys and in Delhi two-thirds of urban squatters walk to work, often over distances in excess of 10 km. (Rao and Sharma, 1990). Residents in African cities are less dependent upon walking than their Indian counterparts and make more use of bus services, although survey evidence indicates that at times when money is in short supply other financial commitments take precedence and walking replaces the bus trip.

Table 5.5 Third World city travel by car, bus and informal urban transport

(a) *Motorized transport modes*

	Percentage of urban trips by:		
	Car	*Bus*	*Informal transport*
Bangkok	29	59	12
Hong Kong	30	39	31
Jakarta	29	49	22
Karachi	16	63	21
Manila	29	22	49

(b) *Travel by bus and informal sector*

	Percentage of urban trips by:	
	Bus	*Informal transport*
Delhi	78	22
Hyderabad	49	51
Bangalore	48	52
Kampur	6	94
Jaipur	18	82
Agra	13	87
Baroda	45	55
Chieng Mai	7	93

(Data from various surveys made during period 1970–80)

Source: Dimitriou, H. (ed.), *Transport planning for Third World cities*, 1990.

Non-motorized vehicles

In Western cities the pedal cycle is the principal vehicle in this category but its use is only of limited significance in the overall travel pattern. In the less-developed countries, however, the cycle and related forms of transport assume considerable importance as a means of mobility and as with walking it is the lower income groups who depend upon them for many of their essential journeys (Table 5.5). The principal vehicles in use are cycle-rickshaws, horse-drawn carts such as the tonga and the hand-drawn rickshaw, although the latter is now confined to Calcutta. In Indian cities of between 500,000 and one million population 60 per cent of households make use of one or more pedal cycles and one in five urban inhabitants own a machine. The ownership level in Chinese cities is about 460 cycles per 1000 (or about one in two) persons and in both China and India the level of access to cycles is increasing (Cai Jun-shi, 1988). In contrast the cycle ownership rate in many African cities is only about 20 per 1000 (or one in 50) persons, a level which is possibly explicable in terms of costs in relation to disposable income or by the fact that the cycle does not enjoy as high a status as a means of transport as it does in Indian cities (Maunder and Fouracre, 1988). In the latter, cycling can account for up to 30 per cent of all urban trips whereas in Africa the proportion is usually below two per cent. In major Chinese cities the cycle is used for up to 80 per cent of all trips and surveys of traffic in Indian cities show that cycles can represent up to one-half of total flow, with volumes of 7,500 machines on some principal roads in Delhi. Low-income commuters in the Indian capital make use of cycles for 40 per cent of all work trips involving distances of up to 8 km. but for longer journeys the bus assumes greater importance.

In South-East Asia the cycle rickshaw, with a capacity for two persons, is used extensively as a means of public transport and in some Indian cities can account for up to one-fifth of all trips and form a similar proportion of traffic on main roads. Rickshaw costs inhibit their use by the lower paid, however, and the average distance travelled by passengers is usually less than those journeys made by cycle.

Private cars

In the industrial world the rapid expansion in the ownership and use of the private car has been responsible for more changes in the structure of personal travel in urban areas than any other means of transport (Figure 5.2). In Europe and North America the car is now the leading mode for all categories of journey and with the growing trend towards decentralization of many urban activities which have traditionally been located at the city core private transport is becoming of even greater significance. The popularity of the car within the urban environment is explicable in terms of its flexibility and personal convenience which enables the user to organize the daily travel programme with a minimum of constraints. Trips that start from home can usually terminate at a point conveniently close to the required destination and in many urban areas workplaces are now equipped with car-parking facilities which minimize the walking element in the overall journey. Shopping in city centres will usually involve a choice of car park and some walking; but suburban and out-of-town retail, employment and leisure centres are able to offer car parks whose capacity is far in excess of those in central areas.

Although a large proportion of car trips are made by the driver only, shared vehicles are of particular significance during the peak periods for work journeys. What is described in the USA as carpooling is the second most important means of commuting and one-fifth of all workers who commute now do so in shared cars. The overall trip time will usually be longer than that of the single driver but it is considerably less than the time required for the equivalent journey by public transport and travel costs per person contributing to a carpool are reduced by at least 50 per cent. Carpooling can offer particular benefits

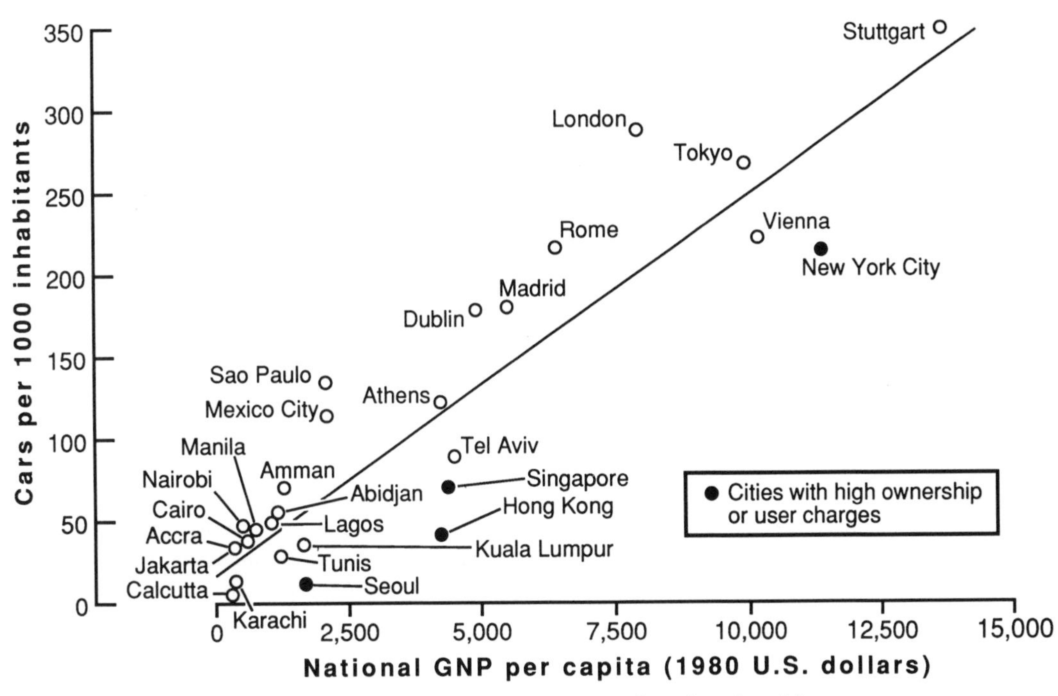

Figure 5.2a Private car ownership and GNP per capita in selected major cities.

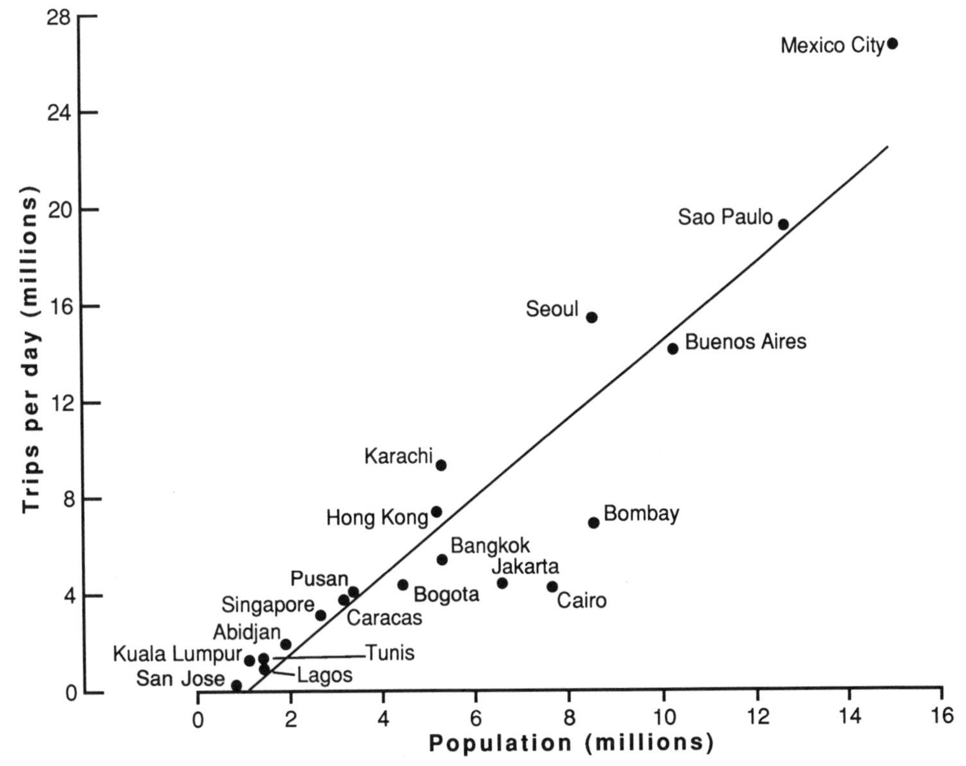

Figure 5.2b Private car trips per day in selected major cities.
Source: Dimitriou, 1990.

when all occupants are from the same household, as is the case with 35 per cent of all carpoolers in the USA (Teal, 1987). It is workplace location rather than residence location that usually influences the pattern of car sharing, as the vehicle can be parked during the day at a point convenient to most of the occupants. Carpooling is not used so extensively in the UK although a recent survey of office workers showed that the percentage of employees from the same office who shared cars for the journey to work varied from 32 to 65 per cent, with the average vehicle occupancy ranging between 2.38 and 4.44 persons (Daniels, 1980).

The perceived advantages of car travel in Western urban society have resulted in continued increases in the number of households with one or with two or more vehicles, although considerable variations between towns do exist in terms of household-car ownership. The low rates of car ownership in Third World cities are a direct reflection of income levels and a comparison of the world's major cities on the basis of cars and GNP per capita reveals some marked contrasts (Dimitriou, 1990). Whereas London, New York, Tokyo and Rome have at least 200 cars per 1,000 inhabitants and a national GNP per capita in excess of $6,000 (1980 level) centres such as Calcutta, Manila, Nairobi and Seoul have fewer than 50 vehicles per 1,000 and a GNP of below $2,500 (Figure 5.2).

Public transport

Urban public road transport in the industrialized world dates from the nineteenth century introduction of horse-drawn buses and trams, superseded in the 1890s and 1900s by electric trams. These were complemented and eventually largely supplanted by the motor bus in the twentieth century although tram systems still play an important role in many European cities. Railborne services for the urban traveller were first introduced as adjuncts to inter-urban networks but by the mid-nineteenth century urban railways specifically designed to link city centres and outlying residential areas were being constructed. New inner-city lines also appeared following the successful opening in 1863 of the sub-surface railway linking the London mainline terminus at Paddington with the city. The perfection of electric traction in the 1880s enabled the building of deep 'tube' railways. Complex systems were established in London and many other European cities, although in the USA there was often a preference for elevated systems along existing streets. These railways and road-passenger services together offered the urban population a previously unknown facility for rapid and generally cheap transport, especially for work journeys, and provided the only mechanized forms of travel until the advent of the private car.

Many urban railway systems have a distinctive radial form as they were initially designed for travel into city centres which contained the principal sources of employment. Facilities for travel between peripheral areas are normally more limited, although the London and Paris networks possess inner loops which allow for some direct movements not involving the central area (Figure 5.3).

Bus networks are more flexible in their operation although here again the emphasis is still firmly upon providing services into and out of central business districts. Another basic distinction between public rail and road transport is that access to the former is confined to stations that are generally spaced at least one kilometre apart whereas bus boarding and alighting points are usually at intervals of about 400 metres. In larger urban areas most of the population is within 500 to 750 metres of a bus route.

There is a general correspondence between city size and the level of provision of public transport in industrialized nations, but in the less-developed world the dramatic expansion of towns and cities has rarely been matched by the provision of what can be seen by Western standards as even a minimal level of such facilities. Urban rail networks in

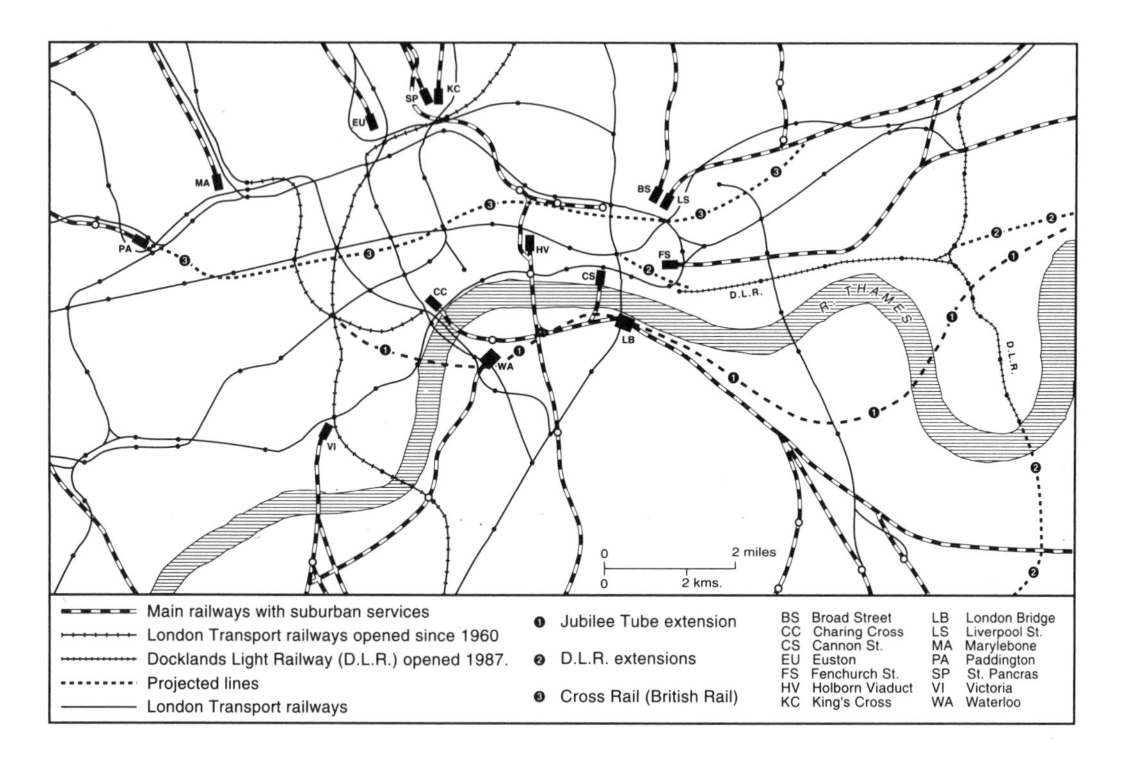

Figure 5.3 Existing and projected surface and sub-surface passenger railways in central London, 1990.

Sources: British Rail and London Underground maps; *Modern Railways*.

particular are much more restricted in their distribution than in Europe or the United States and buses carry the bulk of public travel in most cities. The eight cities in India with populations of between one and four million have no suburban rail services and the bus, which is the only available form of public transport, carries up to 40 per cent of all intra-urban trips. Within India there is also a marked imbalance between the distribution of urban centres and that of bus undertakings, with these eight large metropolitan areas containing nine per cent of the urban population but 21 per cent of the national bus fleet, whereas the 204 cities between 100,000 and one million population are provided with only 16 per cent of all buses to serve a demand created by 27 per cent of the total urban population (Nagakaja, 1988).

Singapore's population of 2.6 million is served by over 250 bus routes operated by two major private undertakings and several smaller companies. The bus network within this densely built-up area is within a five-minute walk of most inhabitants and the fleet of 3,200 vehicles provides a frequency of ten minutes or less on 80 per cent of the services (Rimmer, 1986). Such high frequencies are uncommon, however, and in many Third World cities bus users must adapt to much longer service intervals. Buses are generally used more intensively in cities of the less-developed world than in Europe and high fares, as seen in the context of disposable incomes, act in conjunction with the inadequate capacity of the fleets as a whole to limit the overall market (White, 1990).

Many cities in the less-developed world are largely dependent upon informal bus services which, together with various forms of shared taxis and non-motorized modes, provide

Table 5.6 Basic characteristics of urban transport modes

Facility	*Maximum capacity in persons/hour*	*Average speed (kms p. hour)*	*Interval between access points (kms)*
Bus on conventional road network	9–10,000	16–24	0.2–0.5
Bus using reserved lane on express highways	20,000	56	0.8–1.6
Urban railway	50–60,000	32–48	1.6
Light rapid transport system	40,000	26–38	0.5–1.3
Private car on conventional road network*	1,000	19–40	
Private car on urban motorway network*	3,000	72–80	

* Assuming 1.5 occupants per vehicle.

Source: World Bank Reports.

what is known as paratransit. The small-capacity vehicles which usually characterize this sector, however, are less efficient as carriers than conventional buses. The Turkish dolmus, for example, with its five seats is able to convey only 1,800 passengers per hour compared with the 5,000 capacity of a standard bus. However, the services offered by paratransit vehicles do satisfy a market whose demands cannot be fully satisfied by the formal public-transport sector in many Third World cities (White, 1990).

The urban taxi plays only a minor role in the public urban-transport system, constrained in its use by cost and limited capacity. However, it is of significance within inner cities for short-distance trips made for shopping, business or in connection with other public-transport modes and the demand for taxi services is rising. In London the vehicle fleet increased by 18 per cent in the 1977–87 period to a total of 14,800 taxis, but this excludes the large numbers of minicabs which are not subject to the strict controls imposed upon taxis. Apart from trips within cities the taxi is frequently used on a shared basis for central city to airport journeys or between main-line rail termini (White, 1989). In the Third World, however, taxis make up a significant part of the informal public-transport sector, with unofficial capacities of up to 12 or 14 per vehicle.

Modal choice

The selection by the consumer of what is perceived as the most desirable and convenient mode of travel within urban areas is influenced by personal income, degree of access to a car, journey purpose and time, and availability of public transport. The choice of means of transport is particularly critical during the commuting periods. The capacity of individual modes varies widely and some consideration of this attribute must be made at this stage (Hanson and Schwab, 1986). Table 5.6 summarizes the speeds and carrying capacities of the principal modes and it can be seen that maximum effective capacity ranges from 1,000 to 60,000 persons per hour with speeds varying from 10 to 50 miles per hour. Assuming an average loading of 1.5 persons the private car is the least efficient mode with a capability of transporting only 1,000 persons per hour on each lane of a conventional road network, increasing to 3,000 persons on an urban motorway. The bus, with a capacity of up to 80 or 100, can carry up to 10,000 people per hour at maximum frequency within cities but if reserved lanes on an urban expressway are available this total can rise to 20,000. Modern tramway networks can cater for up to 20,000 passengers per hour but a light rail-transport system is capable of carrying up to 40,000 persons under the most favourable operating circumstances. Underground, surface or elevated rail services can cope with up to 65,000 passengers per hour although the level of access to such systems is much lower than that for bus or tram networks.

The choice between public and private transport is especially influenced by the variations in frequency of buses and rail services which are usually at a maximum during the journey-to-work periods. In smaller urban areas the choice will normally be between bus and car and final selection will also be strongly influenced by journey purpose (Mitchell, 1980). For shopping trips, walking remains the dominant mode and still ranks second after the car for work, social and leisure trips. Public transport by either bus or train only appears as the leading choice in larger cities where journeys-to-work are involved, although difficulties are associated with the peak hours and the selection of the car as the favoured mode for most non-work trips may be explained by its attractions in terms of flexibility, adaptability and comfort.

An analysis of office commuters in the UK indicated substantial differences between male and female employees when means of travel were examined. Between 84 and 94 per cent of all male workers were able to drive whereas only between 54 and 62 per cent of women had licences. Women driving to work represented only one-quarter of all car drivers and almost all car passengers were female. As a result women employees made up the large majority of bus passengers and although public transport was used by only 32 per cent of all workers in the survey this proportion increased to between 65 and 80 per cent for women commuters (Daniels, 1980).

In the Third World, however, the choice of personal transport is often inevitable since public facilities are so frequently incapable of meeting demand or have fares which low-income groups cannot afford and an alternative mode must of necessity be selected. In major Chinese cities the pedal cycle is used for over one-half of all journeys whereas public transport accounts for only 30 per cent in Guang Zhou and 43 per cent in Beijing, both major cities with populations of 3.29m and 5.86m respectively (Cai Jun-Shi, 1988).

Conclusions

Journeys within urban areas are made to satisfy a set of well-defined and regularly patterned needs linked with social and economic activities so that the general nature of travel does not vary greatly from city to city. However, there are significant differences in the overall structure of journey patterns and in the means of transport which are used for personal travel and for freight carriage. In particular the levels of personal mobility associated with access to the private car differ greatly between industrialized and developing countries, and the quality and quantity of public transport available also displays substantial variations.

There are trends in almost all areas towards increases in the volume and complexity of intra-urban movements which impose considerable strains upon the road and rail infrastructure. In particular the decentralization of many urban functions, such as retailing and leisure, which have traditionally been sited in central areas, has led to a reorientation of many trip destinations. The following chapter examines the nature of the urban congestion and interrelated problems which these changes have caused, together with the remedies that have been applied or that are currently under active consideration.

References

Barber, G. (1986), 'Aggregate characteristics of urban travel', *The geography of urban transportation*, ed. S. Hanson (New York: Guilford), ch. 4.

Bartlett, R. S. and Newton, W. H. (1982), *Goods vehicle trip generation and attraction by industrial and commercial premises* (Crowthorne: Transport and Road Research Laboratory, LR 1059).

Cai Jun-shi (1988), 'Public transport in China'

(Singapore: International Union of Public Transport, conference report).
Daniels, P. W. (1980), *Office location and the journey to work* (Aldershot: Gower).
Daniels, P. W. and Warnes, A. M. (1983), *Movement in cities* (London: Methuen).
Dasgupta, M. (1983), *Employment and work travel in an inner city context* (Crowthorne: Transport and Road Research Laboratory, SR 780).
Dasgupta, M., Frost, M. and Spence, N. (1990), *Mode choice in travel to work in British cities, 1971–81* (London School of Economics and Political Science, Department of Geography, Research Paper).
Department of Transport (1990a), *Transport statistics, Great Britain, 1979–89* (London: HMSO).
Department of Transport (1990b), *Transport statistics for London* (London: HMSO).
Dimitriou, H. T. (ed.) (1990), *Transport planning for Third World cities* (London: Routledge).
Fielding, G. J. (1986), 'Transit in American cities', *The geography of urban transportation*, ed. S. Hanson (New York: Guilford), ch. 10.
Gordon, P., Kumar, A. and Richardson, H. W. (1988), 'Beyond the journey-to-work', *Transportation Research* 22 A/6.
Hanson, S. (ed.) (1986), *The geography of urban transportation* (New York: Guilford).
Hanson, S. and Schwab, J. (1986), 'Describing disaggregate flows', *The geography of urban transportation*, ed. S. Hanson (New York: Guilford), ch. 6.
Jones, P. M., Dix, M. C., Clarke, M. T. and Heggie, I. G. (1983), *Understanding travel behaviour* (Aldershot: Gower; Oxford Studies in Transport).
Lomax, D. E. and Downes, J. D. (1977), *Patterns of travel to work and school in Reading* (Crowthorne: Transport and Road Research Laboratory, LR 808).
Maunder, D. A. and Fouracre, P. R. (1988), 'Non-motorized transport in Third World cities' (Singapore: International Union of Public Transport, conference report).
Mitchell, C. G. (1980), *The use of local bus services* (Crowthorne: Transport and Road Research Laboratory, LR 923).
Nagakaja, V. (1988), 'Public transport in India', (Singapore: International Union of Public Transport, conference report).
Pickup, L. and Town, S. W. (1983), *Commuting patterns in Europe* (Crowthorne: Transport and Road Research Laboratory, SR 796).
Plane, D. A. (1981), 'The geography of commuting fields', *Professional Geographer* 32, 182–8.
Rao, M. S. and Sharma, A. K. (1990), 'The role of non-motorized urban travel', *Transport planning for Third World cities*, ed. H. T. Dimitriou, ch. 4.
Rigby, J. P. and Hyde, P. J. (1977) *Journeys to school* (Crowthorne: Transport and Road Research Laboratory, LR 776).
Rimmer, P. J. (1986), *Rikisha to rapid transit: urban public transport systems and policy in Southeast Asia* (Oxford: Pergamon).
Sheppard, E. (1986), 'Modelling and predicting aggregate flows', *The geography of urban transportation*, ed. S. Hanson (New York: Guilford), ch. 5.
Starkie, D. N. (1967), *Traffic and industry* (London School of Economics and Political Science: Geographical Paper 3).
Teal, R. F. (1987), 'Carpooling: who, how and why?', *Transportation Research* 21 A/3.
Webster, F. V. (ed.) (1988), *Urban land use and transport interaction: policies and models* (Aldershot: Gower).
White, P. (1989), 'Fighting off a mini-challenge', *Transport* 10(1), 13–15.
White, P. (1990), 'Inadequacies of urban public transport systems', *Transport planning for Third World cities*, ed. H. T. Dimitriou, ch. 3.
Wilbur Smith and Associates (1977), *Hull freight study* (Crowthorne: Transport and Road Research Laboratory, SR 315).

6

Urban Transport Problems and Solutions

Brian Turton and Richard Knowles

In towns and cities of the developed world the principal transport difficulties are caused by the dominance of the private car, leading to severe congestion on urban roads, together with a steady decline in the patronage of public transport. In Third World cities public transport facilities are grossly inadequate and inner areas are congested with a mixture of motorized, animal-drawn and pedestrian traffic. Solutions to transport problems include new road construction, traffic-management schemes, rail-based rapid-transit systems and transport coordination programmes.

Introduction

Almost all the world's urban areas face difficulties in accommodating the complex variety of movements described in the previous chapter. Major cities in particular provide an arena in which the principal transport modes exist in a competitive environment and create what can often be seen as intractable problems. Devising and implementing solutions to ensure a more efficient and acceptable use of the urban-transport infrastructure has been the principal concern of planners and policy-makers. Many of the issues involved, such as environmental conflict, transport policy and administrative control, are discussed at greater length elsewhere in this book. In this chapter the leading difficulties and remedies are examined. These two are interrelated in what is termed the urban transport-planning process, involving the identification and categorization of travel patterns, the forecasting of future movements in the context of urban developments and the modes of transport available, and the preparation and implementation of plans to cater for this traffic in the most acceptable manner (Dimitriou, 1990a). It is not the intention here to describe all aspects of these processes and attention will be concentrated upon the spatial aspects of the principal problems and of the policies and plans advanced to alleviate them. The nature and scale of transport problems vary with the size of urban area, the balance between the use of private and public transport and the level of highway and public transport infrastructure available. One major contributory factor is the tendency for increasing distances to be established between homes and the principal destinations of daily trips, resulting in a higher level of personal tripmaking and lengthier journeys. These in turn impose greater demands upon road and public-transport systems already nearing or surpassing their capacity (Hanson, 1986).

Traffic congestion is thus one of the most basic difficulties experienced to some extent by almost all towns and cities. But more

Table 6.1 Basic transport data for selected Third World cities

City	*Population (1985) in 000s*	*Cars per 1000 popn. in 1980*	*Cars annual growth rate 1970–80 (per cent)*	*Bus per 1000 popn. in 1980*
Bangkok	6,100	71	7.9	1.22
Bogata	4,500	42	7.8	2.13
Bombay	10,100	21	6.1	0.36
Buenos Aires	10,900	53	10.0	1.20
Cairo	7,700	32	17.0	1.10
Calcutta	11,000	10	5.6	0.33
Hong Kong	5,100	39	7.4	1.83
Jakarta	7,900	33	9.8	0.72
Karachi	6,700	35	8.4	2.32
Manila	7,000	45	8.0	5.30
Mexico City	17,300	105	–	1.23
Rio de Janeiro	10,400	104	12.1	1.20
Sao Paulo	15,900	151	7.8	1.28
Seoul	10,300	15	11.7	1.55

Source: Dimitriou, H. (ed.), *Transport planning for Third World cities*, 1990.

recent additions include atmospheric pollution and other environmental issues and the complex problem of equity in terms of personal accessibility for different groups within urban society.

The problems

Congestion

Traffic congestion may be simply defined as the situation that arises when road and rail networks are no longer capable of accommodating the volume of movements that occur on them. The location of congested areas within a town is determined by the physical transport framework and by the patterns of land use and their associated trip-generating activities. The level of traffic overloading will also vary in time, with a particularly well-marked peak during the daily journey-to-work periods. Although congestion on urban roads is largely attributable to overloading there are other aspects of this basic problem that also require solutions. In industrialized nations increasing volumes of private car, public transport and commercial vehicle traffic have exposed the inadequacies of urban road systems, particularly in city centres where street patterns have often survived largely unaltered from the nineteenth century and earlier. The intricate nature of many city centres makes motorized movement difficult and long-term car parking almost impossible. In the Third World the problem is particularly acute in Indian and South-East Asian cities whose cores are composed of a mesh of narrow streets often accessible only to non-motorized traffic.

The rapid growth in private car ownership and use in cities of the Western world in the second half of the twentieth century has rarely been matched by a corresponding upgrading of urban road systems; these increases will probably continue into the next century. In urban areas of the less-developed countries car ownership is at a much lower level but there is evidence of an increased rate in recent years, especially in South America and South-East Asia (Table 6.1) (Rimmer, 1986). Satisfactory definitions of the saturation level of car ownership vary but if a figure of 50 cars to 100 persons is taken, then in several United States cities the level is now 80 per 100 whereas in cities of South-East Asia the figure rarely exceeds 10 per 100. One factor contributing to congestion in Third World cities is the uncontrolled intermixing of motorized and non-motorized vehicles. The proliferation of pedal and motor cycles causes particular difficulties.

Public transport decline

Congestion on public-transport systems is a major problem but is only one of a group of interrelated factors initiating a cycle of deterioration that results in reduced services and declining revenue. Buses and tramcars contribute to the overall problem of excessive vehicle flows and there can also be overloading and unacceptable conditions for passengers within individual vehicles. Urban railway networks also experience congested conditions when, even at maximum frequency and capacity, trains are incapable of meeting demands and overcrowding becomes a permanent feature of rail travel during peak morning and evening periods.

One common problem directly attributable to the growth of car ownership and usage is the decline in public-transport patronage. Reduced revenues are countered by lower frequencies and higher fares which in turn discourage the use of public services and produce a cyclic effect which, unless it is arrested, can result in the virtual elimination of public transport in smaller urban areas and a substantial rationalization in larger cities. There are marked differences in the levels of patronage of public transport throughout the working day, and the concentration of traffic into the peak periods creates serious problems for operators when attempting to provide the most efficient services in terms of staff and vehicle deployment. Difficulties are also encountered at other periods when the levels of patronage are often insufficient to justify the provision of services.

Car parking

The provision of carparking space within or on the margins of central business districts for city workers and shoppers is a problem that has serious implications for land use planning. A proliferation of expensive and visually-intrusive multistorey storage facilities can only provide a partial solution and supplementary on-street parking can only compound traffic congestion. The extension of pedestrian precincts and similar features in city centres is intended to provide more acceptable environments for shoppers and other users in central areas. But such traffic-free zones in turn produce problems as they create new patterns of access to commercial centres for car-borne passengers and users of public transport, while the latter often lose the advantage of being carried directly to the shopping zone (Roberts, 1981).

Changing land-use patterns

Contemporary moves towards locating retailing complexes, leisure centres and business parks on the periphery of major urban areas have also resulted in transport problems. Bus routes which traditionally focus upon centrally located shopping centres rarely offer convenient access to the new out-of-town complexes. The larger stores often find it necessary to introduce their own bus routes to attract custom from suburban residential areas. Car-borne shoppers who make use of these new centres on the urban fringe often create traffic problems as they drive through suburban areas and similar cross-city journeys are made by patrons of leisure centres and workers in business and science parks (Hall, 1989). Although the development of these various peripheral retail and employment centres is beneficial in that they reduce the volume of traffic focusing upon urban cores, the traffic patterns they have created need to be considered in future transport-planning programmes.

Third World city problems

Deficiencies in public-transport services may be identified in almost all the world's towns and cities but the problems are particularly severe in urban areas of the less-developed countries. Rapid rates of urbanization in these nations frequently involve the extension of low-income unplanned settlement on city peripheries and workers from these areas are exerting increasing pressure on already inadequate public-transport facilities. Although rising car ownership is a feature of Third World urban expansion, the majority of journeys are still made either on foot or by bus, and to a more limited extent by railway. This problem of expanding demand would be greater if it were not for the fact that low disposable incomes prevent many workers from making regular use of public transport (Roth, 1984). For example, in India the costs of public transport, expressed in terms of GDP per capita, can be four times those in the UK and in Delhi up to 36 per cent of disposable income can be expended on bus travel (White, 1990). Those that can afford to make regular use of buses face long, slow, uncomfortable journeys on poorly maintained vehicles and heavily congested roads, and wherever practicable urban migrants try to locate their households as near as possible to their workplaces. The purchase of new buses and the maintenance to a satisfactory standard of existing vehicles is frequently hampered by a lack of foreign currency. It is not unusual for between 60 and 80 per cent of a bus fleet to be out of service because of a lack of spare parts and skilled labour, yet a 70 per cent level of availability is generally regarded as the minimum acceptable to ensure satisfactory services (Dimitriou, 1990b). As a result of the shortage of both new vehicles and of reliable existing buses, few undertakings in the less-developed countries are able to meet the growth in demand for services in existing urban areas and the creation of new demands in recently built-up zones. It is forecast that users of public transport will experience increasing difficulties, particularly in the cities of Brazil, China, Indonesia, India and Nigeria in the next century as a result of rapid population rises.

A comparison of the availability of buses illustrates the difficulties faced by many Third World cities. Whereas in the UK there are 0.9 buses per 1,000 population the ratio in the less-developed nations is about 0.6 per 1,000, falling to only 0.3 per 1,000 in some Indian cities. However, the relative smallness of the bus fleets is countered by a much more intensive usage of vehicles, with a total of 70,000 kms per annum in some Third World cities compared with only 45,000 kms per annum in British urban areas (White, 1990). The inadequacies or poor representation of the formal public-transport sector have encouraged the expansion of paratransit modes, which flourish on routes alongside conventional vehicles and in those city areas where larger buses are physically prevented from operating because of narrow streets. Minibuses now carry a substantial share of the market and although these vehicles and hand-drawn and motorized rickshaws and allied vehicles meet a need for transport they do in turn create serious difficulties as a result of slow speeds and frequent stops to pick up and set down passengers. Because of the small scale of the businesses that own paratransit vehicles they often concentrate their carrying activities in areas where revenue is likely to be maximized; other areas where a demand for public transport exists tend to be ignored or poorly served.

The transport-deprived groups

Investigations into the travel habits and potential requirements of individual household members indicate that there are several clearly defined groups within the urban community who experience considerable difficulty in securing an acceptable level of mobility and accessibility to essential daily or periodic facilities. The elderly, the sick and disabled, those on very low incomes and those below the legal driving age are those most commonly disadvantaged in this respect. Recent research indicates that the numbers of persons who suffer from transport-deprivation and have been described as the 'transport poor' are increasing. In the UK six million people, or about

12 per cent of the adult population, suffer from some form of disability and by 2025 AD 20 per cent will be aged over 65, with two million persons over 80 years (Oxley and Benwell, 1985). In the USA an official survey of Worcester, Massachusetts, indicated that the proportion of persons aged over 65 rose from 11.9 to 13.4 per cent between 1960 and 1980 and that the percentage of families defined as below the poverty line increased from 5.4 to 7.5 in the decade 1970–80 (Hanson, 1986). Although many people in these categories are able to drive, to be driven or to make use of specially adapted cars, eventually their level of mobility is bound to decline and they become dependent upon relatives and friends or upon public-transport facilities. Bus and rail undertakings recognize the needs of the disabled traveller and modifications such as wheelchair access ramps at terminals and buses with low-level access have been introduced but these aids are not widespread at the present time.

Surveys of the elderly in urban areas indicate that this group generally makes much less use of cars than the employed and that they encounter more difficulties in travelling by public transport or on foot. Certain facilities used on a regular basis, such as post offices and food shops, are often accessible by walking but longer trips which necessitate bus travel present problems in terms of physical access to the bus and services are not always convenient. Unavoidable trips, such as those to hospitals, clinics and doctors, can cause particular problems for those elderly persons who cannot rely upon assistance from car-driving friends and relatives and who cannot afford the use of taxis. Distances travelled overall by the elderly are shorter than the average, indicating that their radius of action and hence the range of facilities available to them is also limited. The problems of this group are progressive with age as mobility is gradually diminished by events such as the death of the car-driving partner or increasing physical infirmity (Hopkin et al., 1978). Those under the legal age for car-driving also form a transport-deprived group although many social and leisure trips are made in a family unit. In many advanced nations journeys to school and college are confined to acceptable distances by a planned distribution of institutions and catchments throughout the built-up area and by the use of free or subsidized school-bus services but in Third World cities access to education can present serious problems. In the UK most children who live within one km. of their school either walk or cycle but traffic accidents and other hazards are associated with these journeys.

Environmental problems

The detrimental impact that many aspects of transport can exert upon the environment are fully discussed elsewhere. It is sufficient at this point to state that within urban areas both the problems associated with transport systems and the solutions advanced to combat them can have severe environmental implications. Excessive flows where heavy commercial vehicles form a large proportion of the total traffic cause atmospheric pollution, high noise levels, vibration which can progressively undermine older structures and visual intrusion. Construction of new urban highways and some types of rail networks can in turn lead to community disruption and create for adjacent built-up areas the same conditions of excessive noise produced by traffic on the routes they are intended to supplement.

Road safety

Within the UK and many other industrial nations the greater proportion of serious accidents occur in urban areas. Roads in built-up zones display an accident rate up to three times greater than in other road categories. Pedestrians and cyclists are especially vulnerable and 95 per cent of pedestrian accidents in Britain are recorded in urban areas with one half of these occurring in town centres. For the young the problem of accidents is highly localized: 90 per cent of incidents involving the under-five group is recorded within a quarter mile of home, often on minor roads in residential areas. The vulnerability of the young to traffic accidents in the UK is closely related to the problem of the 'transport deprived' in that children account for almost one-third of pedestrian trips and form 39 per cent of all pedestrian casualties (Whitelegg, 1987).

The solutions

The urban transport planning process can produce a wide variety of proposals designed to alleviate problems associated with freight and passenger movements. Comprehensive plans for transport-improvement programmes over periods of up to 20 years for major towns and cities are now an accepted procedure but there is frequently an urgent need for more immediate short-term proposals which can provide temporary relief for difficulties such as traffic congestion in localized parts of an urban road system. Although it is possible to categorize solutions according to their principal objectives and methods of operation, it must be recognized that most urban-transport problems are interrelated and plans to solve them are similarly interlinked. The implementation of a scheme to alleviate congestion on one part of an urban-road network may well create difficulties elsewhere: the more ambitious plans involving the construction of motorways or new rapid-transit systems can have great deleterious effects upon the physical and social structure of urban communities bisected by these new lines of communication (Buchanan, 1963; Schaeffer, 1975). The costly schemes introduced in Western industrial cities also contrast markedly with the limited extent to which Third World cities have been able to combat their transport problems.

Investment in additional road capacity

One of the most commonly adopted methods of combating congestion in small towns or in districts of larger urban centres is the construction of bypasses to divert through traffic. In Britain these date from the 1930s and the extension of the principle to the orbital or sub-orbital road has produced routes such as the M25 motorway around Greater London or the M42 ring-road which interconnects the M5, M6 and M40 motorways converging on Birmingham. Although an effective bypass will remove a large part of the traffic from a town centre there will usually be some routes carrying through traffic which continue to cause congestion within the urban area. Orbital routes themselves can also soon become overloaded if the initial forecasts of traffic volumes have been inaccurate. The inadequacies of the M25 motorway opened in 1986 are now being corrected with the construction of additional carriageways in both directions.

The solution to congestion on intra-urban road networks was also seen by mid-twentieth-century planners as the provision of additional capacity in the form of new or improved highways. Since the pioneer transportation studies of the 1950s and 1960s were carried out in US metropolitan areas where the needs of an auto-dominated society were seen to be paramount, this construction of additional road capacity was generally accepted as the most effective solution to movement problems and urban freeways were built in many large cities such as Chicago, San Francisco and Los Angeles (Dunn, 1981). Transport planners in Western Europe incorporated many of their American counterparts' proposals into their own programmes and urban motorways became a leading component in many road-improvement schemes (Muller, 1986). However, it soon became evident that the extra capacity gained from road construction was rapidly filled with additional traffic attracted to the new facility. Construction of urban motorways and their complex junctions with the conventional road system requires large areas of land and the demolition of tracts of housing and commercial properties. Planners and policy-makers came to accept that the investment of massive amounts of capital in new highways dedicated to the rapid movement of motor traffic was not necessarily the most effective manner of solving transport problems (Starkie, 1982). Although US cities contain many examples of complex freeway networks, urban authorities in Western Europe have tended to be more selective in their adoption of motorways and in the UK in particular road-building programmes have been substantially modified from the original design. What has been described as 'heroic structural change' has been replaced by a concern to make the most efficient use of existing facilities. In London, for example, the ambitious plans for inner-orbital motorways which aroused such fierce opposition in the early 1970s were abandoned and replaced by a programme based upon more selective road improvements (Bayliss, 1977).

Traffic management measures

Temporary and partial relief from road-traffic congestion may be gained from the introduction of traffic-management schemes, which involve the reorganization of traffic flows and directions without any major structural alterations to the existing street pattern. Among the most widely used devices are the extension of one-way flow systems, the phasing of traffic-light controls to take account of traffic variations and restrictions on parking and vehicle-loading on major roads. In Glasgow, for example, an experimental computer-controlled scheme to coordinate traffic at inner city intersections increased peak hour speeds by 16 per cent. On multilane highways which carry heavy volumes of commuter traffic, certain lanes can be allocated to incoming vehicles in the morning and to outgoing traffic in the afternoon, producing what has been described as a tidal-flow effect.

Traffic management has received particular attention within urban residential areas, where excessive numbers of vehicles create problems of noise, vibration, atmospheric pollution and, above all, accident risks, especially to the young. The concept of 'traffic calming' has been introduced to many cities in Europe and involves the creation of an environment in which cars may travel but where priority is accorded to the pedestrian. Carefully planned street-width channelling, parking restrictions and speed-control devices such as ramps are combined to secure a safe and acceptable balance between the vehicle and the pedestrian (Tolley, 1990).

Bus priority and allied proposals

Many transportation-planning proposals have been directed specifically towards increasing the speed and schedule reliability of bus services and most large European cities have adopted bus-priority plans in an attempt to boost the attractions of public transport. Bus-only lanes, with or against the prevailing direction of flow, may be designated in heavily congested roads in order to secure time savings although such savings can be dissipated when buses enter inner-city areas where priority lanes are absent. Buses may be accorded priority turns at intersections and certain streets may be reserved for buses only, particularly in pedestrianized shopping areas.

An effective use of buses may be made by incorporating bus-only lanes within new highways, allowing both private and public traffic to benefit from the new route. In the USA this method is the main means of providing bus-priority measures. It dates from the early 1970s, when bus lanes were provided on freeways approaching Washington, New York and San Francisco. In its first year the exclusive bus lane provided on the link between the New Jersey Turnpike and the Lincoln Tunnel carried a daily average of 34,000 persons in 800 buses but these totals are nowhere near the maximum capacity of a bus lane, which could be 60,000 persons per hour in 1,200 vehicles. The latter total is far in excess of current urban demand; bus lanes are therefore not fully utilized but they do provide a highly efficient means of conveying bus passengers into city centres (Westwell, 1983).

Where entirely new towns are planned there is an opportunity to incorporate separate bus systems within the urban road network and thus enable buses to operate to schedules unaffected by conventional road-traffic conditions. In the UK, Runcorn New Town, built as an overspill centre for the Merseyside conurbation, was provided with a figure-of-eight busway linking the shopping centre, major industrial estates and residential areas. About 90 per cent of the town's population were within five minutes walk of the busway and operating costs were 33 per cent less than those of vehicles on the conventional road network (Vincent et al., 1976). Although the system is not used to the extent originally envisaged, it successfully illustrates the integration of public-transport planning with urban development. The introduction of bus-only roads also permits the use of vehicle-guidance systems, whereby the bus is not steered but controlled by lateral wheels; conventional control is used when the public-road system is re-entered. Such systems are in use in Adelaide and experiments have been carried out in many other cities (Adelaide, 1988). The bus can also be given additional advantages in the redevelopment of major retailing and transport complexes in city centres. Rebuilding or remodelling of commercial centres to accommodate covered malls or precincts can provide the opportunity to site bus termini in more convenient locations for shoppers. Major reconstruction of rail stations and termini can also allow bus stations to be integrated more closely with rail facilities. The 'park and ride' scheme, now adopted by many European cities, is intended to reduce the number of cars entering central areas, particularly at weekend peak-shopping periods. Large open spaces on the urban fringe act as temporary carparks and drivers are carried by bus into city centres at an overall charge that matches or betters central area parking costs. The advantages of the bus as an efficient carrier can be secured and the costs of providing parking accommodation are considerably lower on the outskirts than in city centres. Commuters can also be catered for in a similar manner with the provision of large capacity carparks adjacent to suburban rail stations.

Unconventional bus services

Many towns and cities have attempted to promote bus transport by increasing its flexibility and level of response to market demand. In suburban areas the dial-a-ride system has met with partial success, with prospective passengers booking seats by telephone within a defined area of operation. Such vehicles typically serve the residential areas around a district shopping centre and capacity is limited, so they are best suited to operations in areas of low demand or in off-peak periods. Fares are higher than on conventional buses since about one third of revenue is required to finance the control centre which receives booking calls and despatches buses to travellers' homes or close by (Martin, 1978). Before the widespread introduction of minibus services following deregulation, experimental services were introduced with small-capacity vehicles which could be hailed in the same way as a taxi and which could negotiate the complex street patterns of housing estates more easily than larger buses. As with the dial-a-ride system, however, these hail-stop minibuses could only cope with a limited demand.

Vehicle restraint schemes

These priority measures designed to enhance the efficiency of bus services can also be combined with plans to restrain the use of inner urban streets by private motorists. A filter system can be applied whereby cars are only allowed into congested inner-city zones if vehicles are fully loaded, thus promoting a more efficient use of cars than is usually the case. In the UK the Nottingham experimental zone and collar scheme introduced in the 1970s was an attempt to restrain morning peak-hour traffic originating in two suburbs from penetrating the city centre by the use of specific zone exits and park-and-ride services into the core (Collins, 1975). Other methods are based upon fiscal restraint and involve the levying of premium tolls or taxes on drivers wishing to enter inner zones with payment made either on entry or exit. Another variation is based on an electronic-metering programme whereby roadside computers can determine the journey lengths made in restricted zones and despatch accounts to the drivers. These schemes can make substantial contributions to the reduction of road congestion but their implementation by local urban authorities can meet with strong resistance from the motoring lobby, with implications for the political survival of the local authority responsible.

Any plans that involve the introduction of priority measures for fully laden vehicles, area licensing or vehicle filtering should favour the expansion of carpooling which has already been noted as a significant element in American commuting patterns (Teal, 1987). Although the method is not popular with many commuters, an expansion of car-restraint schemes could stimulate its adoption in the future.

Rail rapid transit

Investment in rail-based rapid-transit schemes has been used to encourage suburban development, to provide an alternative to congested urban roads, and more recently to help regenerate the declining economies of city centres, inner cities and derelict docklands (Church, 1990; Roberts, 1985; Williams, 1985).

Trains and trams (streetcars) were the earliest nineteenth century mass-transit modes and enabled the first complete separation of place of residence from workplace (Kellett, 1969; Ward, 1964). Underground railways are mainly twentieth-century developments, although they started with London's Metropolitan Railway in 1863 and London's first deep-level electrified tube line in 1890. From the 1930s, mass car ownership encouraged suburbanization and urban dispersal to occur on a much larger scale and in a less concentrated form. Motor vehicles now competed for congested urban road space with trams which ran on the road, except for a few segregated suburban lines. Trams were regarded as an outdated mode of transport and the British response from the late 1930s onwards, encouraged by government advice from 1946, was to replace trams with buses in all towns and cities. This was typically completed by Manchester in 1949 and London in 1952, although trams survived in Sheffield until 1960 and Glasgow until 1962 and were retained along the Blackpool seafront. In contrast, the German response was to modernize trams into Light Rail (Stadtbahn) systems and put them underground in congested city centres and inner cities to avoid conflict with road traffic. Cologne, Essen and Hannover provide good examples (Hall and Hass-Klau, 1985).

An alternative response was to develop and/or electrify suburban railway lines as in Copenhagen, Glasgow and London, or underground Metros as in London, Munich and Stockholm. This was particularly successful where land-use zoning powers were used to concentrate high-density suburban development around railway stations. The three Scandinavian capitals provide notably successful postwar examples with Copenhagen's five suburban rail corridors developed from its famous 1947 Finger Plan, and Stockholm's numerous metro lines and Oslo's four metro lines developing similar suburban corridors (Fullerton and Knowles, 1991).

In the UK the most widespread response to urban congestion in the early 1960s was to try and provide more roadspace for the sharply increasing volume of cars by building or widening roads, and by traffic management.

Table 6.2 Typical characteristics of urban rail systems

	Streetcars	*Light rail*	*Suburban rail*	*Metro*
A *Urban size*				
Population	200K–500K	100K–1 million	Over 500K	Over 1 million
CBD employment	Over 20K	Over 20K	Over 40K	Over 80K
B *Route characteristics*				
Route length from CBD	under 10Km	under 20Km	under 40Km	under 24Km
Track	On street	over 40% segregated	segregated	segregated
CBD access	surface	surface	surface to CBD edge	underground
Station spacing in suburbs	350m	1 Km	1Km–3Km	2Km
Station spacing in CBD	250m	300m	–	500m–1Km
Maximum gradients	10%	8%	3%	3%–4%
Minimum radius	15m–25m	25m	200m	300m
Engineering	minimal	light	medium	heavy
C *Rolling stock*				
Carriage weight	16 tonnes	under 20 tonnes	46 tonnes	33 tonnes
Number of carriages	1 or 2	2 or 4	up to 12	up to 8
Carriage capacity	50 seats 75 standing	40 seats 60 standing	60 seats 120 standing	50 seats 150 standing
Carriage access	step	step or platform	platform	platform
D *Performance*				
Power current	DC 500–750V	DC 600–750V	DC 600–1.5KV or AC 25KV	DC 750V
Power supply	overhead	overhead	overhead or 3rd rail	3rd rail
Average speed	10–20 Kph	30–40 Kph	45–60 Kph	30–40 Kph
Maximum speed	50–70 Kph	80 Kph	120 Kph	80 Kph
Typical peak headway	2 minutes	4 minutes	3 minutes	2–5 minutes
Maximum hourly passengers	15,000	20,000	60,000	30,000

Source: Knowles and Fairweather, 1991.

American transport consultants ignored rail-transit investment and advised city after city to build extensive and expensive urban-motorway networks (Starkie, 1982). Urban-motorway plans were soon abandoned or curtailed because of cost and an environmental backlash and transport consultants started advocating Mass Rail Transit schemes. Recommendations for rail electrification and reopening the Argyle Line in Glasgow in 1968, a Link and Loop rail connection underneath central Liverpool in 1969 and the Tyneside (Light Rail) Metro in 1971 were all accepted and built within ten years, partly financed with government grants (Fullerton and Openshaw, 1985; Halsall, 1985; Robinson, 1985; Westwell, 1983). In London the Victoria and Jubilee underground lines were opened in 1968 and 1979 respectively, the first such lines since the 1920s. However, Manchester's Picc-Vic underground link was rejected on cost grounds in 1975 (Knowles, 1985).

Metros need a large volume of potential users to justify the expense of tunnelling and the long period of disruption to city streets during construction. Metros are, therefore, rarely built in cities of under 500,000 people, and are more typical of million plus cities. The Soviet Union actually has a policy of building a metro when a city has grown to a million people (Jackson, 1989). Metros are usually fully segregated from other rail traffic with stations 0.5 km. to 1 km. apart in city

Table 6.3 Metro systems in operation 1991

Country	*Number of systems*	*City*
Europe 28		
Austria	1	Vienna
Belgium	1	Brussels
Czechoslovakia	1	Prague
Finland	1	Helsinki
France	4	Lille; Lyon; Marseille; Paris
Germany	6	Berlin; Cologne; Frankfurt am Main; Hamburg; Munich; Nuremburg
Greece	1	Athens
Hungary	1	Budapest
Italy	2	Milan; Rome
Netherlands	2	Amsterdam; Rotterdam
Norway	1	Oslo
Portugal	1	Lisbon
Romania	1	Bucharest
Spain	2	Barcelona; Madrid
Sweden	1	Stockholm
United Kingdom	2	Glasgow; London
North America 12		
Canada	2	Montreal; Toronto
United States	10	Atlanta; Baltimore; Boston; Chicago; Cleveland; Miami; New York/Newark; Philadelphia; San Francisco/Oakland; Washington DC
Soviet Union 13		Baku; Dnepropetrovsk; Gorky; Kharkov; Kiev; Kuibyshev; Leningrad; Minsk; Moscow; Novosibirsk; Tashkent; Tbilisi; Yerevan
Asia 17		
China	2	Beijing (Peking); Tianjin
Hong Kong	1	Hong Kong
India	1	Calcutta
Japan	9	Fukuoka; Kobe; Kyoto; Nagoya; Osaka; Sapporo; Sendai; Tokyo; Yokohama
North Korea	1	Pyongyang
Singapore	1	Singapore
South Korea	2	Pusan; Seoul
Latin America 7		
Argentina	1	Buenos Aires
Brazil	2	Rio de Janeiro; Sao Paulo
Chile	1	Santiago
Colombia	1	Medellin
Mexico	1	Mexico City
Venezuela	1	Caracas
Africa 1		
Egypt	1	Cairo

Source: Knowles and Fairweather, 1991.

centres, and about 2 km. apart in the suburbs. They can carry up to 30,000 people per hour in one direction and extend up to 24 km. from the city centre (Table 6.2). Metro systems are widespread with 77 operational worldwide, 27 of them in Europe and a further 36 under construction or in design (Knowles and Fairweather, 1991) (Table 6.3). The World Bank advises Third World countries not to invest in metros in their burgeoning capitals unless cheaper and more flexible road-based public-transport systems cannot cope.

Suburban rail provides frequent local passenger services on main-line surface railways up to about 40 km. from the city centre as in Dublin, Manchester, London and hundreds of other cities in dozens of countries in all five continents (Knowles and Fairweather, 1991). Suburban rail is sometimes separated from long-distance rail routes and can be extended under city centres through short sections of tunnel, as in Copenhagen, Glasgow, Liverpool and Munich. Suburban trains are longer, with stations more widely spaced, gradients shallower and speeds higher than for metro systems (Table 6.2).

The revival of metro and suburban rail investment in Europe and North America gave way, in the 1980s, to light rail investment. Electrified light rail systems offer many of the advantages of metros and suburban rail, but at much lower cost as the routes are less heavily engineered and the lighter carriages can travel up steeper gradients and around tighter curves. Light rail requires at least 40 per cent of its track to be segregated from road traffic to avoid road-traffic congestion, and this differentiates it from trams or streetcars (Table 6.2). A further distinction needs to be made between light rail systems such as London Docklands Light Railway, which are fully segregated and can be automatically driven, and those such as the Manchester Metrolink which are partly segregated. Light rail can operate up to 20 km. from city centres and carry up to 20,000 passengers per hour. They are usually found in urban centres with between 100,000 and 1 million people, such as Charleroi in Belgium (200,000) and Hiroshima in Japan (900,000). When they occur in larger urban areas such as Vienna (1.5 million) they normally complement a metro system (Knowles and Fairweather, 1991).

Light rail systems are operational in 100 cities, mainly in Europe and North America, but also in the Third World (Table 6.4). Light rail systems are being built, planned or considered in hundreds of cities worldwide to relieve urban road congestion or help regenerate run-down areas, including more than 40 in Britain (Figure 6.1). Regeneration routes include London Docklands Light Railway and the Don Valley Route Two of the Sheffield Supertram. Light rail systems, especially in Germany, are sometimes upgraded from streetcars, as in Stuttgart, while others such as London Docklands run on mainly new alignments. A final group, including Manchester Metrolink, Los Angeles, Tyne and Wear, and Vancouver Sky Train, utilize old railway routes. An increasing number of newer light rail systems, such as Manchester Metrolink, Sheffield Supertram and Zürich, cross the city centre on surface routes on wholly or partially segregated streets. Here they are driven 'on sight' with priority at traffic lights. This is much cheaper than tunnelling and is also more accessible and safer for passengers.

Streetcars are still found in about 250 towns and cities worldwide but they are mainly in the Soviet Union and other former communist countries, where road congestion is less significant as private car ownership has been severely restricted (Table 6.2). It is expected that most of the streetcar systems will either be upgraded into light rail or abandoned as urban road congestion increases.

Transport coordination

Successful transport planning depends to a large extent upon the integration of the various modes of urban transport and the coordination of their operations. The establishment of transport authorities by major metropolitan councils represents a significant stage in the programme to secure an efficient system. In the USA legislation to establish city mass-transit undertakings dates from the 1940s, with the creation in 1945 of the Chicago Transit Authority to purchase the existing subway system and the surface rail suburban lines. In 1952 the Authority also acquired the Chicago motorcoach system to become one of the first city-transport organizations in the world to exercise control over the majority of its public-transport network (Smerk, 1968).

In the UK the 1968 Transport Act enabled the formation of Passenger Transport Authorities and Passenger Transport

Table 6.4 Light rail systems in operation 1991

Country	*Number of systems*	*City*
Europe 62		
Austria	1	Vienna
Belgium	3	Antwerp; Brussels; Charleroi
Czechoslovakia	2	Bratislava; Brno
France	5	Grenoble; Lille; Marseille; Nantes; St Etienne
Finland	1	Helsinki
Germany	20	Bielefeld; *Bochum-Herne; Bonn; Braunschweig; Bremen; Chemnitz; Cologne; *Dortmund; *Duisburg; *Düsseldorf; *Essen-Mulheim; Frankfurt am Main; Freiburg; *Gelsenkirchen; Hannover; Karlsruhe; *Krefeld; Mannheim-Ludwigshafen; Stuttgart; Wurzburg
Hungary	1	Budapest
Italy	3	Genoa; Rome; Turin
Netherlands	4	Amsterdam; Rotterdam; The Hague; Utrecht
Norway	1	Oslo
Poland	1	Czestochowa;
Romania	6	Brasov; Cluj; Constanta; Craiova; Poleisti; Resita
Spain	1	Valencia
Sweden	2	Gothenburg; Stockholm
Switzerland	6	Basle; Berne; Geneva; Lausanne; Zürich; Neuchatel
United Kingdom	4	Blackpool; London Docklands; Manchester**; Tyne and Wear
Yugoslavia	1	Sarajevo
North America 20		
Canada	4	Calgary; Edmonton; Toronto; Vancouver
United States	16	Baltimore; Boston; Buffalo; Cleveland; Detroit; Fort Worth; Los Angeles; Newark; New Orleans; Philadelphia; Pittsburgh; Portland; Sacramento; San Diego; San Francisco; San José
Australia 1		Melbourne
Soviet Union 3		Krivoy Rog; Naberezhyne-Chelny; Volgograd
Asia 6		
Hong Kong	1	Hong Kong
Japan	2	Hiroshima; Kyoto
Philippines	1	Manila
Turkey	2	Istanbul; Konya
Latin America 4		
Brazil	1	Rio de Janeiro
Mexico	3	Guadalajara; Mexico City; Monterrey
Africa 3		
Egypt	2	Alexandria; Helwan
Tunisia	1	Tunis

* = Stadtbahn Rhein-Ruhr ** = opening 1992

Source: Knowles and Fairweather, 1991.

Executives in major conurbations, charged with providing integrated and efficient public passenger-transport systems within their areas of operation. Basic policy-making and funding is the responsibility of the PTA while the PTE is concerned directly with the planning and operation of transport facilities provided by a variety of different operators such as bus companies and British Rail. In many cases the PTEs inherited extensive bus networks from the local authorities and were responsible for their operation alongside those of other undertakings. Following the reorganization of local government areas in 1974 the PTAs and PTEs became responsible for public-transport policy and planning in major conurbations which, for the first time, were administered by one overall authority,

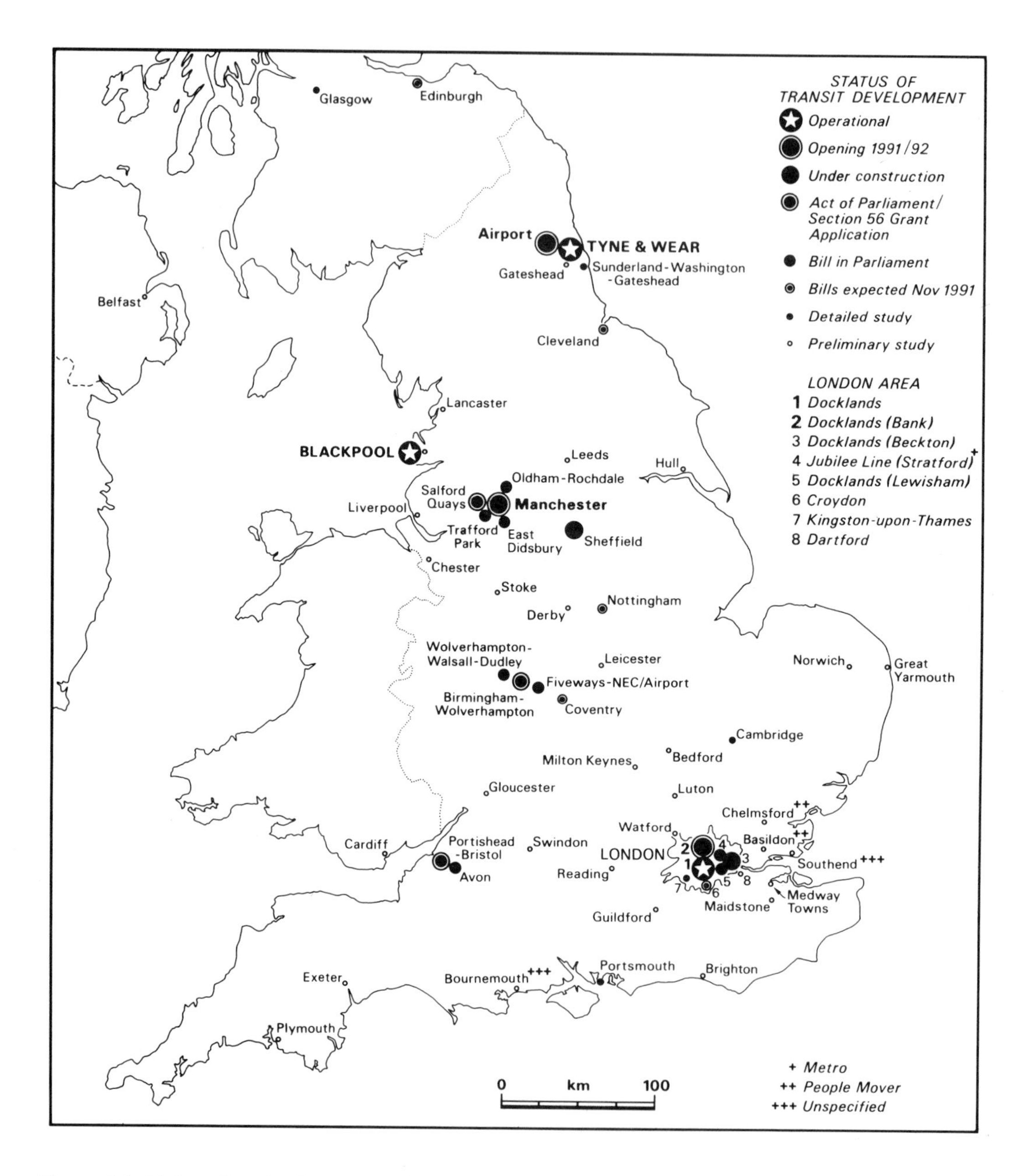

Figure 6.1 Light rail and other transit developments in the UK, 1991.

the new Metropolitan County Councils, which existed until their disbandment in 1986. The extent to which these conurbation-transport bodies achieved their objectives is closely influenced by the socioeconomic nature of their respective areas and the character of the road and rail networks subject to their control. In the USA the Mass Transit Authorities which corresponded roughly to the British PTAs and PTEs had legal powers to acquire rail and bus undertakings within their operating areas if such action was thought to benefit the travelling public. The authorities in New

York, Philadelphia and several other large cities exercised this right. In contrast the British PTAs were unable to purchase local rail lines but were obliged under the 1968 Act to enter into contractual agreements with British Rail in respect of rail services within each PTA area. Whereas the PTAs are responsible for providing funds for new or modernized stations and rolling-stock, British Rail retain the rights to operate the services.

One example of how this division of responsibility hindered the task of securing effective integration of all conurbation transport was that British Rail revised rail timetables only once a year whereas PTAs often saw the need for more frequent alterations in order to coordinate bus and rail facilities. A similar situation existed with respect to bus services, in that the PTAs subsidized undertakings which operated buses within the conurbations but without possessing any influence or control over the operating policies of the parent companies. However, the PTAs can be credited with several achievements such as the introduction of through-ticketing facilities, joint road-rail fare structures and the publication of passenger information incorporating all modes of travel available within the conurbation. They were also responsible for introducing new facilities such as the Tyne and Wear Metro, the central Liverpool underground rail loop and the extension of the Clydeside electric suburban rail network, although in many cases plans for these schemes had been initiated before the establishment of PTAs and PTEs. Although the PTAs still exist as joint boards, together with PTEs, their functions today are more limited. They have lost control over commercial bus services and the powers to promote transport integration which they possessed between 1969 and 1986.

'Non-transport' solutions

Throughout the world's cities a wide range of policies, strategies and plans have been devised and implemented in order to solve the problems of urban transport in the private and public sectors. The acceptance by transport planners that conventional patterns of socioeconomic activity, with their emphasis upon standardized working hours, dictate corresponding demand patterns for transport services creates a situation in which the development of an efficient transport system capable of meeting all requirements proves almost impossible. Many physical and transport planners now actively support proposals for changes to these established patterns of activity as a means of securing greater measures of success in their planning programmes. One of the most obvious targets is to reduce the volume of travel during the journey-to-work periods by encouraging the spread of job start and finish times over much longer timescales (Plane, 1986). The establishment of what may be described as a shift system for most of an urban workforce would be unpopular and difficult to initiate but the extension of each journey-to-work period over four or five hours rather than two could lead to a substantial and effective reduction of the congestion attributable to commuting. Already the widespread adoption of 'flexitime' in the service sector has helped to flatten the commuter-travel peaks in many major cities during the 1980s.

Another category of 'non-transport' solution to the congestion problem lies in the growth of home-based economic activity made possible by the expansion of telecommunications, personal-computing facilities and information technology. Many jobs in the financial and commercial sector traditionally carried out in city-centre offices can be accomplished as effectively at home, with computer linkages with headquarter offices. Significant increases in the numbers of employees who could work at home in this way would again contribute to a reduction in the demand for public transport or road space within urban areas. An extension of the opening hours of retailers in city centres during the week could also encourage more workers to shop later and delay their journey home.

Solutions for the 'transport deprived'

Various remedies have been suggested and implemented to meet the special needs of the 'transport deprived' in urban areas. For the disabled who require some form of public transport to make journeys beyond the range of small electric wheelchairs the taxi is the most acceptable mode but costs are high in comparison with other alternatives. Buses may be frequent and provide transport to most of the required destinations but access problems may be insuperable and few undertakings have vehicles adapted for use by the disabled. Where disabled persons are members of special clubs, minibuses are often provided for trips into town centres and other locations and can carry wheelchairs. These facilities are generally superior to any offered by public transport. Rail journeys can present even greater difficulties but many more urban stations are now provided with ramp or lift access to platforms and much of the rolling-stock incorporates space for wheelchairs. In Manchester the Metrolink light rail transport system is fully accessible to the disabled but at an extra cost of £10 million for the first phase of the project.

Case-studies

This chapter concludes with a series of case-studies of individual cities where transport problems have been analysed and plans introduced in an attempt to solve the more outstanding difficulties. Examples have been drawn from North America, Europe, the Far East and the Third World to illustrate the complexity of the interaction between the demand for transport facilities and the response from private- and public-sector enterprises.

Glasgow – urban deprivation and transport innovation

The Clydeside conurbation, with Glasgow as its core, contains almost half of the population of Scotland in a region characterized by some of the highest unemployment rates in the UK, low levels of car ownership and extensive areas of inner-city dereliction which are currently undergoing urban renewal. When the Passenger Transport Authority and Executive were created in 1972 they became responsible for a complex system of public transport comprising a city-owned bus network, bus services operated by over 40 private or state-owned companies, an underground railway and an extensive network of local-surface railways which is the largest outside Greater London. The responsibilities and functions of the Glasgow PTA and PTE were then assumed by the Strathclyde Regional Council in 1975 following the reorganization of local government in Scotland.

In preparing a comprehensive transport policy for Glasgow the PTE had to take account of several trends in social and economic conditions. The population of the inner city fell by almost one-third between 1961 and 1981 and large numbers were accommodated in new housing estates on the periphery (Law et al., 1984). Demolition of many former residential and industrial areas bordering the Clyde had released land for new road building and the proportion of car-borne commuters was rising steadily. Travel by public transport declined by one-third in the 1970s and although the level of government support increased, operating costs also rose, creating a situation that could only be remedied by withdrawal of services, higher fares or revenue support from local rate-payers.

Electrification of the suburban railways north and south of the Clyde began in the 1960s. They were linked in 1979 with the reopening of the Argyle line passing under Glasgow Central Station (Figure 6.2). The circular underground line, opened in 1897 to serve the city centre and lower Clydeside as far west as Govan, was refurbished and reopened in 1980, with new bus or surface-rail interchange facilities at each of its 15 stations and park-and-ride facilities at several. The coordination of bus and rail,

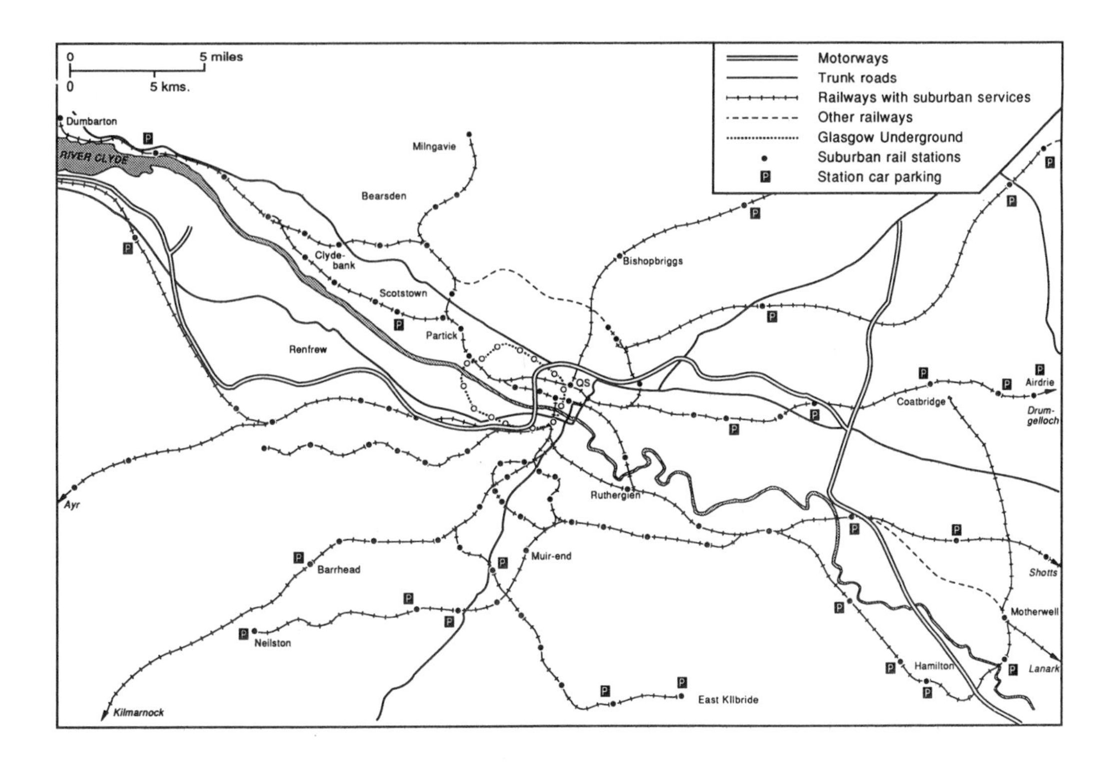

Figure 6.2 Principal roads and railways in the Strathclyde conurbation.

and of bus services provided by different undertakings, was a keystone of Glasgow's PTE policy; what was seen as undesirable competition was fought vigorously. For example, the PTE gave financial support to the newly electrified Glasgow-Ayr line, which carried a substantial commuter traffic, and during the 1970s it successfully opposed the introduction of parallel bus services. However, the 1980 Transport Act sanctioned such competition and Ayr-Glasgow bus services have eroded some of the rail traffic (Westwell, 1983). Although the injection of government capital enabled the completion of the two major rail projects in the late 1970s, patronage continues to decline and 90 per cent of all passengers travel on the bus network. Since deregulation in 1986 the appeal of the bus system, especially in the outer housing estates, has been increased with use of small-capacity vehicles which are more flexible in their operation than the double-deckers. But competition between different companies for passengers in the inner city has resulted in congestion and the levels of service coordination achieved by the PTE during the first decade of its life have been severely reduced. The most recent proposals for the conurbation include additional bus-priority measures, new stations on the rail network and options for an LRT system (Strathclyde PTE, 1990).

The South-East Asian city – private car restraint and rapid mass transit

Kuala Lumpur and Singapore have both grown rapidly since the 1960s and their expansion as commercial and administrative centres has generated formidable problems of urban transport. Programmes of public-transport improvement are seen as the most acceptable solutions to these problems but to date Singapore has achieved more progress and is noteworthy in its application of car restraint policies (Table 6.5). Proposals for transport development in Kuala Lumpur

Table 6.5 Public and private transport facilities in Kuala Lumpur and Singapore 1980

	Singapore	*Kuala Lumpur*
Population (millions)	2.3	1.0
No. of stage buses (over 50 seats)	2,900	660
No. of minibuses (14–16 seats)	800	400
No. of taxis	8,100	3,000
Private cars (thousands)	140	50
Buses per 1000 persons	105	70
Cars per 1000 persons	61	50

Source: Rimmer, *Rikisha to rapid transit*, 1986, 320.

made in 1972 and 1976 and incorporated into the 1984 Master Plan for the city are based upon the completion of inner urban ring-roads, an expansion of bus capacity, the construction of an LRT system and the adoption of area traffic-control systems (Rahim, 1988). The latter would accord priority on major highways to buses, minibuses and fully-loaded taxis and private cars and a limited form of area licensing was proposed. It was estimated that the reduction of road congestion would require a 40 per cent shift from private to public transport but the securing of this objective is highly unlikely and the emphasis upon car usage remains (Figure 6.3).

Singapore, with a population of 2.6 million, has experienced a planned expansion in the form of satellite towns along radial corridors from the city centre (Figure 6.4). Given the high density of population and the demand for efficient public and private transport, a heavy investment in a new mass-transit rail system has been coupled with an area-licensing scheme for private cars requiring access to the central commercial core. Although the decision to construct a 67 km. rail mass-transit network was made in 1982, a corporation to operate the system was not established until 1985, followed in 1987 by the creation of the Public Transport Council to approve and regulate fares and services on bus routes, taxis and the mass rapid-transit system. The latter has been opened in stages between 1987 and 1992 and is eventually expected to account for one-third of all public-transport travel. The two major bus operators, subject to the Public Transport Council policy, will have routes rationalized and new feeder services introduced as the railway network is completed; currently these two companies have 2,900 vehicles on 250 routes. It is estimated that most of Singapore's population is within five minutes walk of a bus route and 80 per cent of services have frequencies of ten minutes or less. Peak-hour car and taxi traffic since 1975 has been controlled by the Area Licensing Scheme, which requires supplementary licences for access to the city centre during the morning peak period except for fully-laden vehicles. The success of the scheme may be judged by the fact that incoming car traffic has decreased by 20 per cent since 1975, despite an estimated 30 per cent rise in the number of city employees. The principal issue in the future development of an integrated transport system is the requirement that all public-transport undertakings must be financially self-supporting; the new mass-transit network in particular must be able to maintain high levels of patronage and to minimize operating costs (Rimmer, 1986; Gray, 1988).

The African city – remedies for the current inadequacies of public transport

Although car ownership is steadily rising in many African cities, the majority of urban travellers will continue to depend upon public transport for motorized trips and the major problem facing transport planners is to increase the efficiency of existing bus, and in a few cases, rail services. Many cities in Nigeria and other West African states have either no or only poorly developed conventional bus services and most passenger transport is provided by minibuses or shared taxis (Adenji, 1983). The half-million population of Benin City in Nigeria is currently served by 600 minibuses on 20

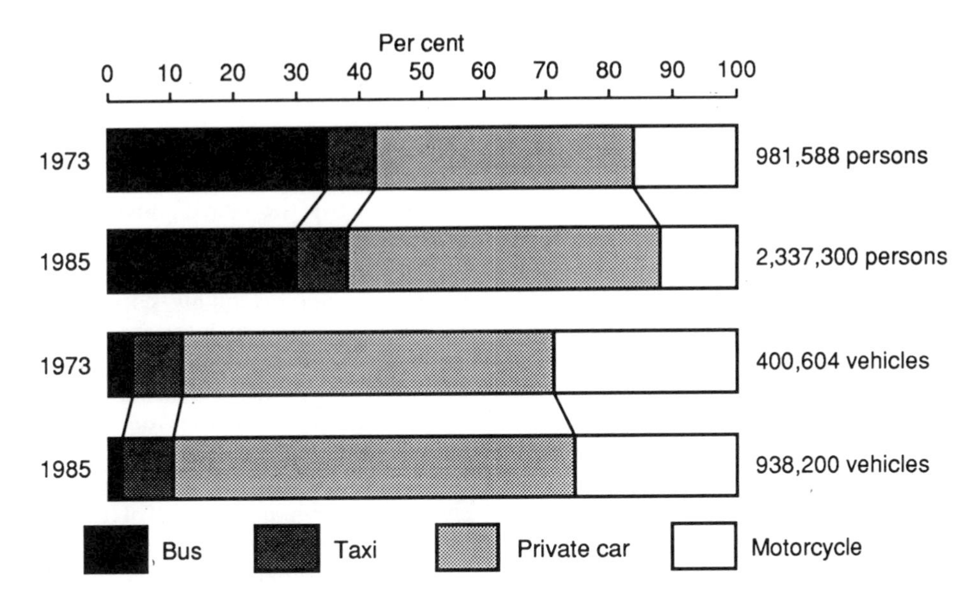

Figure 6.3 Kuala Lumpur: changes in the use of transport modes and in numbers of vehicles, 1973 and 1985. *Source:* Rahim, 1988.

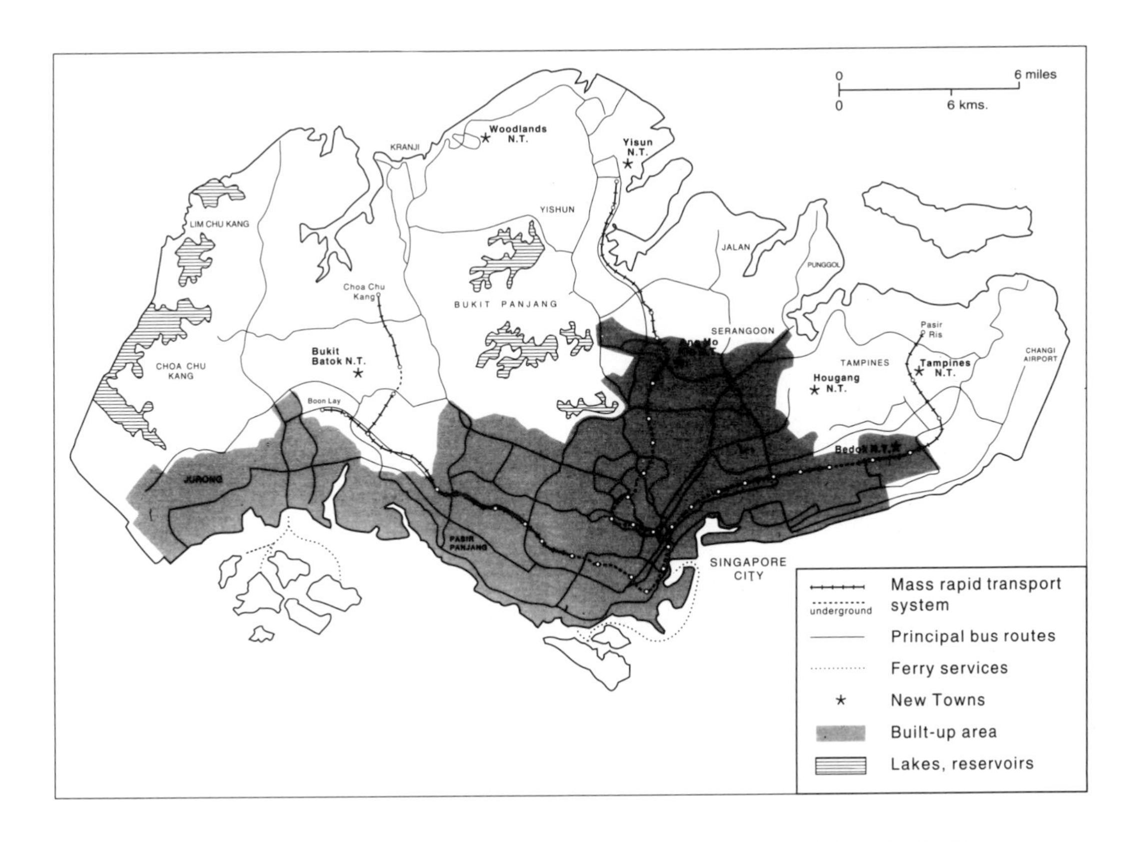

Figure 6.4 Singapore: the mass rapid-transit system lines open in 1991 and the principal bus routes.

routes operated by many small undertakings with no route-licensing system. Most of the services radiate out from the city centre and many connect with intercity taxi routes on the urban fringe. This minibus fleet accounts for about 38 per cent of all urban trips; shared taxis take a further 40 per cent of the market, catering especially for journey-to-work and social travel (Wiredu, 1989). In larger metropolitan centres such as Lagos, however, private-car movements have increased to a critical level which has prompted the application of experimental vehicle-restraint schemes similar to those in Singapore (Ogunsanya, 1984).

Harare in Zimbabwe, with a 1982 population of 900,000, has the bulk of its African workforce resident in high-density suburbs on the periphery of the city and almost entirely dependent upon public transport for access to the commercial centre and the industrial zones. A widespread network of conventional bus services is supplemented by a fleet of shared 'emergency taxis' which operate on fixed routes from the city centre to suburban termini. Severe shortages of spares creates operating problems which result in regular overcrowding and lengthy waits for vehicles. The rapid growth of the satellite town of Chitungwiza (1982 population: 170,000) has presented particularly severe problems, as workers rely upon bus services for the 20 km. journey into Harare (Atkinson, 1984). Proposals to improve public transport include a scheme for mass rail transit, combining the use of existing railway routes into the city with construction of two new lines linking the centre with the airport, the large industrial estate at Southerton and Chitungwiza. However, it is unlikely that this scheme will be implemented and investment in conventional bus services is seen as the more realistic short-term solution.

The Californian city – a reappraisal of public transport in auto-dominated societies

The conurbations of San Francisco and Los Angeles typify the development of an intra-urban pattern of travel which was dominated by the car during the mid-twentieth century. Both cities possess highly complex freeway networks designed to meet the ever-growing demand for road communications (Figure 6.5). In San Francisco the location of the central business district on a peninsula between the Pacific coast and San Francisco Bay and the growth of suburbs along the bay shores to north and south of the core has resulted in the construction of bayside freeways interlinked by three toll bridges. The Los Angeles metropolitan region is polycentric in form and the central business district contains only 8 per cent of all jobs in the county. The need for interconnection between the numerous settlements within the region was met by a network of freeways which carry some of the highest traffic volumes in the world. By the mid 1960s road congestion within and on the approaches to central San Francisco had become so acute that the existing suburban rail network was reconstructed as the Bay Area Rapid Transit (BART) system, with suburban car parks and, wherever possible, coordinated bus services (Higgins, 1981). Traffic on the system has yet to reach the levels predicted when it opened in 1972 but BART has improved urban mobility in the region and carries more than one half of all CBD-bound journeys to work. The contribution of the Los Angeles mass-transit system to commuting trips is much lower, at about 24 per cent, but both networks are seen as making an important contribution to the reduction of urban freeway congestion. The investment in public transport in these two cities exemplifies the first of two major changes that have been made in transport policy within America since the 1960s: namely, a reappraisal of plans continually to increase urban road capacity and the adoption instead of mass-transit systems (Orski,

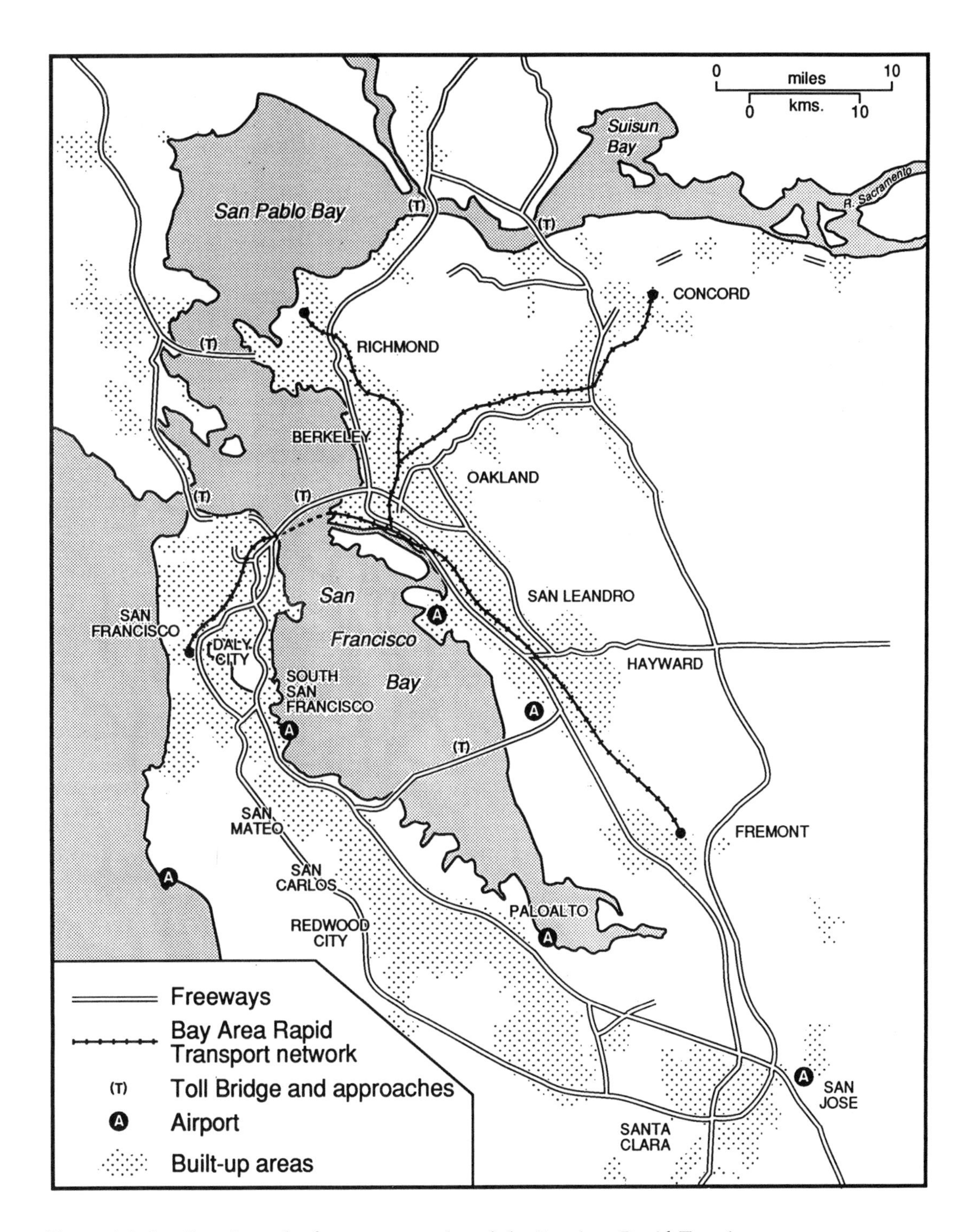

Figure 6.5 San Francisco: the freeway network and the Bay Area Rapid Transit system.

1982; Fielding, 1983). Although the latter have succeeded in capturing a proportion of the car-borne market, the enormous investments involved have been questioned. During the 1980s more emphasis was placed upon securing a more efficient use of existing facilities at lower costs.

Conclusions

Transport planners face an ever-increasing array of problems. Although a wide range of solutions have been devised and applied with various levels of success, an acceptable resolution of many of the basic difficulties has yet to be achieved. The reconstruction of existing roads and the building of additional expressways has alleviated congestion in many urban areas; but it has brought in its wake additional difficulties created by the inevitable generation of further volumes of traffic. The eventual realization that the demands for road space created by steadily growing private-car traffic could never be adequately met except at enormous cost to the urban environment stimulated a reappraisal by transport planners of the potential of the public sector where, it was hoped, new investment could revitalize existing rail and bus systems, increase their attractiveness to the motorist and thus raise their share of the total urban travel market.

This process of public sector modernization is still a feature of many contemporary urban-transport plans in cities of leading industrial nations although the proportion of the total market which new mass-transit and similar schemes acquire is rarely at the forecast level. Compromise proposals involving physical and fiscal restraint of private motoring coupled with improvements in the efficiency of existing public transport services, and, where appropriate, the provision of additional capacity in the form of light-rail transport are now increasingly being incorporated within transport-planning programmes.

Cities in the Third World have yet to face the problems posed by excessive volumes of cars and here the principal concern is to ensure that public transport, represented almost entirely by buses, can be improved in order to offer a sufficient level of mobility to populations which still depend to a large extent upon this mode of travel for all basic-journey purposes. Whereas bus services in the industrialized world, until recent moves towards deregulation, have been subject to rigid licensing procedures, the position in the developing nations is of a much more informal nature, with conventional buses sharing the traffic with paratransit services. An inability to secure financial resources generally prevents the larger Third World cities from installing the elaborate mass-transit systems common in Europe or North America. Many analysts believe that a lower level of technology based upon established bus transport would be the more effective way of solving the present imbalance between public transport demand and supply.

References

Adelaide Transit Authority (1988), *Report on guided busways* (Adelaide).

Adenji, K. (1983), 'Nigerian municipal bus operations', *Transportation Quarterly* 37/1.

Atkinson Williamson Partnership (1984), *Study of the Zimbabwe road passenger transport industry: final report vol. 1* (Harare: Ministry of Transport).

Bayliss, D. (1977), 'Urban transportation research priorities', *Transportation* 6/3, 4–17.

Buchanan, C. D. (1963), *Traffic in towns* (Harmondsworth: Penguin).

Church, A. (1990), 'Waterfront regeneration and transport problems in London Docklands', *Port cities in context: the impact of waterfront regeneration*, ed. B. S. Hoyle (Transport Geography Study Group, Institute of British Geographers), 5–37.

Collins, B. J. (1975), 'Transportation in Nottingham', *Journal of the Chartered Institute of Transport* 36/10.

Dimitriou, H. T. (ed.) (1990a), 'Transport problems of Third World cities', *Transport planning for Third World cities*, (London: Routledge), ch. 5.

Dimitriou, H. T. (ed.) (1990b), 'The urban

transport planning process', *Transport planning for Third World cities*, (London: Routledge), ch. 2.

Dunn, J. A. (1981), *Miles to go: European and American transportation policies* (Cambridge: Lexington).

Fielding, G. J. (1983), 'Changing objectives for American transit', Parts I and II, *Transport Reviews* 3, 287–99.

Fullerton, B. and Knowles, R. D. (1991), *Scandinavia* (London: Paul Chapman).

Fullerton, B. and Openshaw, S. (1985), 'The Tyneside Metro in full operation', *Rapid transit systems in the UK: problems and prospects*, ed. A. F. Williams (Transport Geography Study Group, Institute of British Geographers), 27–45.

Gray, M. G. (1988), 'Planning and public transport issues in Singapore', International Union of Public Transport, *conference report* (Singapore).

Hall, D. R. (1989), 'The Metrocentre and transport policy', *Transport policy and urban development*, ed. R. D. Knowles (Transport Geography Study Group, Institute of British Geographers), ch. 4, 105–29.

Hall, P. and Hass-Klau, C. (1985), *Can rail save the city?* (Aldershot: Gower).

Halsall, D. A. (1985), 'Rapid transit in Merseyside: problems and policies', *Rapid transit systems in the UK: problems and prospects*, ed. A. F. Williams (Transport Geography Study Group, Institute of British Geographers), 76–91.

Hanson, S. (1986), 'Dimensions of the urban transportation problem', *The geography of urban transportation*, ed. S. Hanson (New York: Guilford), ch. 1.

Higgins, T. J. (1981), 'Coordinating bus and rapid rail in the San Francisco Bay area', *Transportation*, 10/4.

Hopkin, J. M., Robson, P. and Town, S. W. (1978), 'Transport for the elderly', (Crowthorne: Transport and Road Research Laboratory), SR 419.

Jackson, C. (1989), 'Metro plans span the world', *Developing metros 1989, Railway Gazette International Supplement.*

Kellett, J. R. (1969), *The impact of railways on Victorian cities* (London: Routledge & Kegan Paul).

Knowles, R. D. (1985), 'Rapid transit in Greater Manchester', *Rapid transit systems in the UK: problems and prospects*, ed. A. F. Williams (Transport Geography Study Group, Institute of British Geographers), 46–75.

Knowles, R. D. and Fairweather, L. (1991), *The impact of rapid transit*, Metrolink Impact Study Working Paper 2 (Department of Geography, University of Salford).

Law, C. M., Grundy, T. and Senior, M. L. (1984), *The Greater Glasgow area* (University of Salford).

Martin, P. H. (1978), 'Comparative assessment of unconventional bus systems', (Crowthorne: Transport and Road Research Laboratory), SR 387.

Muller, P. O. (1986), 'Transportation and urban form', *The geography of urban transportation*, ed. S. Hanson (New York: Guilford), ch. 2.

Ogunsanya, A. A. (1984), 'Improving urban traffic flow by traffic restraint: the case of Lagos', *Transportation*, 12/2.

Orski, C. K. (1982), 'The changing environment of urban transportation', *Journal of the American Planning Association* 48, 309–14.

Oxley, P. R. and Benwell, M. (1985), 'An experimental study of the use of buses by elderly and disabled people', (Crowthorne: Transport and Road Research Laboratory), RR 33.

Plane, D. A. (1986), 'Urban transportation: policy alternatives', ch. 16 in Hanson, S. (ed.), *The geography of urban transportation* (New York: Guilford).

Rahim, M. N. (1988), *Public transport planning in Malaysia* (University of Keele, Department of Geography, Occasional Paper 14).

Rimmer, P. (1986), *Rikisha to rapid transit: urban public transport and policy in Southeast Asia* (Oxford: Pergamon).

Roberts, J. (1981), *Pedestrian precincts in Britain* (London: TEST).

Roberts, J. (1985), 'Light rail: what relevance for London's Docklands?', *Rapid transit systems in the UK: problems and prospects*, ed. A. F. Williams (Transport Geography Study Group, Institute of British Geographers), 98–137.

Robinson, S. E. (1985), 'Tyne and Wear Metro: development of an integrated public transport system', *Rapid transit systems in the UK: problems and prospects*, ed. A. F. Williams (Transport Geography Study Group, Institute of British Geographers), 4–26.

Roth, G. (1984), 'Improving the mobility of the urban poor', *Basic needs of the urban poor*, ed. P. J. Richards and A. M. Thomson (London: Croom Helm), ch. 9.

Schaeffer, K. and Sclar, E. (1975), *Access for all: transportation and urban growth* (Harmondsworth: Penguin).

Smerk, G. M. (1968), *Readings in urban transportation* (Bloomington: Indiana University Press).

Starkie, D. (1982), *The motorway age: road and road traffic policies in post-war Britain* (Oxford: Pergamon).

Strathclyde PTE (1990), *Public transport development study* (Glasgow).

Teal, R. F. (1987), 'Carpooling: who, how and why?', *Transportation Research* 21 A/3, 203–16.

Tolley, R. S. (1990), *Calming traffic in residential areas* (Tregaron, Dyfed: Coachex).

Vincent, R. A., Layfield, R. E. and Bardsley, M. D. (1976), *Runcorn busway study* (Crowthorne: Transport and Road Research Laboratory), LR 697.

Ward, D. (1964), 'A comparative historical geography of streetcar suburbs in Boston, Massachusetts and Leeds, England, 1850–1920', *Annals of the Association of American Geographers* 54, 477–89.

Westwell, A. R. (1983), 'Public issues of transport in the west of Scotland', *Public issues in transport*, ed. B. J. Turton (Transport Geography Study Group, Institute of British Geographers), 5–30.

White, P. R. (1990), 'Inadequacies of urban public transport systems', *Transport planning for Third World cities*, ed. H. T. Dimitriou (London: Routledge), ch. 3.

Whitelegg, J. (1987), 'The geography of road accidents', *Transactions of the Institute of British Geographers* 12 (2), 161–76.

Williams, A. F. (ed.) (1985), *Rapid transit systems in the UK: problems and prospects* (Transport Geography Study Group, Institute of British Geographers).

Wiredu, Y. K. (1989), 'Minibuses in an African city', *Paratransit* 4, 22–3.

7

Inter-urban Transport

Brian Turton

Inter-urban transport accounts for a large proportion of national freight and passenger movements and involves road, rail, waterway and air-transport facilities. The distribution of traffic by mode and between the private and public sectors varies greatly but in most industrialized nations road transport carries the majority share of freight and passengers. The private car in particular accounts for a substantial proportion of inter-urban passenger travel.

Introduction

Studies of inter-urban transport are a well-established component of general transport literature although the term 'inter-urban' may be interpreted in various ways. A literal acceptance of the term would involve all roads, railways, waterways and airways connecting areas classified as urban, but most studies concentrate upon principal cities and their links. A national transport system may be sub-divided into its inter-urban, urban and rural components but whereas the latter two have recognizable areal forms the links between urban centres are essentially linear in character and are more conveniently examined on a regional or national scale.

However, there is often overlap between these three categories as certain sections of inter-urban passenger networks can also carry commuter services which are integrated into metropolitan journey-to-work patterns, and a long-distance rail route may have local passenger services linking rural settlements. There is also overlap between studies of inter-urban transport networks and flows and research into the broader field of inter-regional linkages, and these in turn can often be related to questions of regional planning, particularly in the context of transport improvement as an integral part of regional economic development projects (Gwilliam and Mackie, 1975). Much of the research into inter-urban transport is promoted by government agencies and international funding bodies, such as the World Bank, in order to investigate the case for investment in existing routes or the construction of additional capacity. The ambitious proposals in the national plans of less-developed countries for investment in the transport sector have often been criticized for their shortcomings in the area of inter-urban transport: the relative merits of long-distance road and rail transport as effective contributors to economic growth and social integration are particularly open to debate.

Inter-urban freight and passenger traffic involves all the principal means of inland transport and the importance of individual modes is influenced by many factors related to freight rates, passenger fare-charging policies, levels of economic activity and government policy on transport. Certain modes are accepted as being especially suited to particular types of traffic so that the allocations of inter-urban flows between modes may be discussed in the context of the costs of movement, the implications for energy consumption and environmental considerations.

The evolution of inter-urban transport networks has been extensively documented and modelling techniques have been enlisted to aid the explanation of growth rates and directions. The Taaffe, Morrill and Gould model has been applied to transport networks in East and West Africa and rail simulation models have been used in the explanation of railway development in New England and Italy (Haggett and Chorley, 1969). Most inter-urban rail networks were largely completed by the early twentieth century and the construction of national motorway systems is now well advanced in many industrialized states. Political changes, and in particular the creation of new states, can have a profound effect upon transport systems, however. The division of Germany in 1949 was followed by many adjustments in the inherited road and rail networks but after four decades of separation the unification of the East and West states in 1990 has made a programme of transport integration essential, with a particular emphasis upon infrastructural improvements in the former Democratic Republic.

Characteristics of inter-urban modes

Railways

Railways played a dominant role in linking the growing manufacturing centres of Western industrialized nations during the nineteenth and the first half of the twentieth centuries and also in developing European colonial possessions. The advantages possessed by the railway for the long-distance carriage of freight and passengers gave it a near monopoly of inter-urban traffic in many industrial nations until it was first seriously challenged by road motor transport in the 1920s. In many countries, however, the railway is now the minority partner in inter-urban transport. It survives by concentrating on the carriage of the types of freight to which it is best suited and through the receipt of state subsidies to meet the shortfall in freight and passenger revenues.

Railways provide one of the few examples of transport undertakings which own their routeways and have exclusive control over the operation of their traffic. Although this enables services to be organized without encountering the problems of congestion that road transport frequently experiences, the costs of maintenance and repair of track, signalling equipment, freight depots and passenger stations fall exclusively upon the railway. These fixed costs form a high proportion of total expenses. The rates charged to freight and passenger traffic must reflect this, with the costs of transport over any distance being based both upon terminal loading and unloading expenses and the variable costs of train operation. With increasing competition from road transport in both industrialized and developing nations, railways now concentrate upon the carriage of low-value bulk commodities such as coal, metal ores and heavy chemicals, with increasing lengths of haul providing the lowest rates to the consumer. Freight rates can also be reduced by the introduction of high-capacity wagons, usually dedicated to a specific commodity, which enable trainloads of over 10,000 tonnes gross to be operated.

Inter-urban rail passenger traffic levels have generally fallen. Increasing shares of the long-distance market have been gained by the private car and, to a lesser extent, by domestic air and coach services. Over distances of 300–400 km., however, rail can still offer the fastest journey timings between city centres, and the progressive electrification of many inter-urban networks has resulted in the recapture of much traffic lost

Table 7.1 National railway networks and electrification (countries with over 10,000 km. track in 1988)

Country	*Rail km. in operation (000s)*	*Percentage of network electrified*
USA	205.0	<1
USSR	146.7	35.6
Canada	62.8	<1
India	61.8	12.2
China	54.0	15.2
Australia	39.5	4.9
France	34.6	34.0
German Federal Republic	30.4	39.8
Brazil	30.4	7.6
Japan	27.2	52.0
Poland	26.6	37.6
South Africa	23.6	35.7
Italy	18.8	54.1
United Kingdom	16.9	26.0
Mexico	15.1	<1
Spain	14.3	47.5
German Democratic Republic	14.0	25.0
Czechoslovakia	13.1	29.0
Sweden	11.6	64.6
Romania	11.0	21.3

Source: Transport Statistics Great Britain 1979–89, 1990; *Jane's World Railways 1989–90*, 1990.

to competing modes (Table 7.1). In the UK, for example, the Intercity sector of British Rail operates 750 trains daily over 17 routes on 4,800 km. of track and accounts for 80 million of the 700 million long-distance passenger journeys made each year (Prideaux, 1988). The speed advantage offered by rail over road transport is now being further exploited by the use of more powerful electric locomotives and the construction of entirely new railways permitting average speeds of over 190 kph (Figure 7.1). In 1983 the Train à Grande Vitesse was introduced by the French railways and the new Paris–Lyon service, with speeds of up to 270 kph on an entirely new track, has captured one million passengers from the roads and two million from domestic airlines.

Despite these advances, however, the railway will probably continue to occupy a subordinate role in Western industrial nations. But in China and the USSR, where private-car ownership levels are low, the railway still carries the bulk of inter-urban traffic although the quality of service is often very poor by Western European standards. The railway has the particular advantage over other modes of inter-urban transport of offering the greatest capacity for traffic in terms of the amount of freight or passengers that can be conveyed through a given distance in a given time period. Although this capacity is no longer fully utilized on many inter-urban routes the railway does retain this ability to adjust to fluctuating traffic levels. The most heavily-used trunk railways in Western Europe and North America originally responded to demands for additional capacity by installing double and then quadruple track, and the introduction of electric traction then enabled a more intensive use of the existing lines. In terms of energy consumption the railway is also more efficient than other modes in the movement of both passengers and freight and the extension of electrification can bring environmental benefits which road haulage cannot match (Table 7.2).

Roads

Inter-urban road systems, unlike rail, are provided and operated by a variety of interests. Routes are built and maintained by the state, or less often by private companies, and freight haulage and passenger services are run by undertakings in both the public and private sectors. The European turnpike roads of the eighteenth and nineteenth centuries were the first attempts to improve inter-urban transport facilities but were soon superseded by canals and later by railways. It was only with the introduction of reliable internal combustion-engined vehicles in the early twentieth century that road transport again contributed effectively to long-distance communications and subsequently developed to achieve the commanding position it now occupies within the transport sector of most modern economies. Whereas railway-pricing

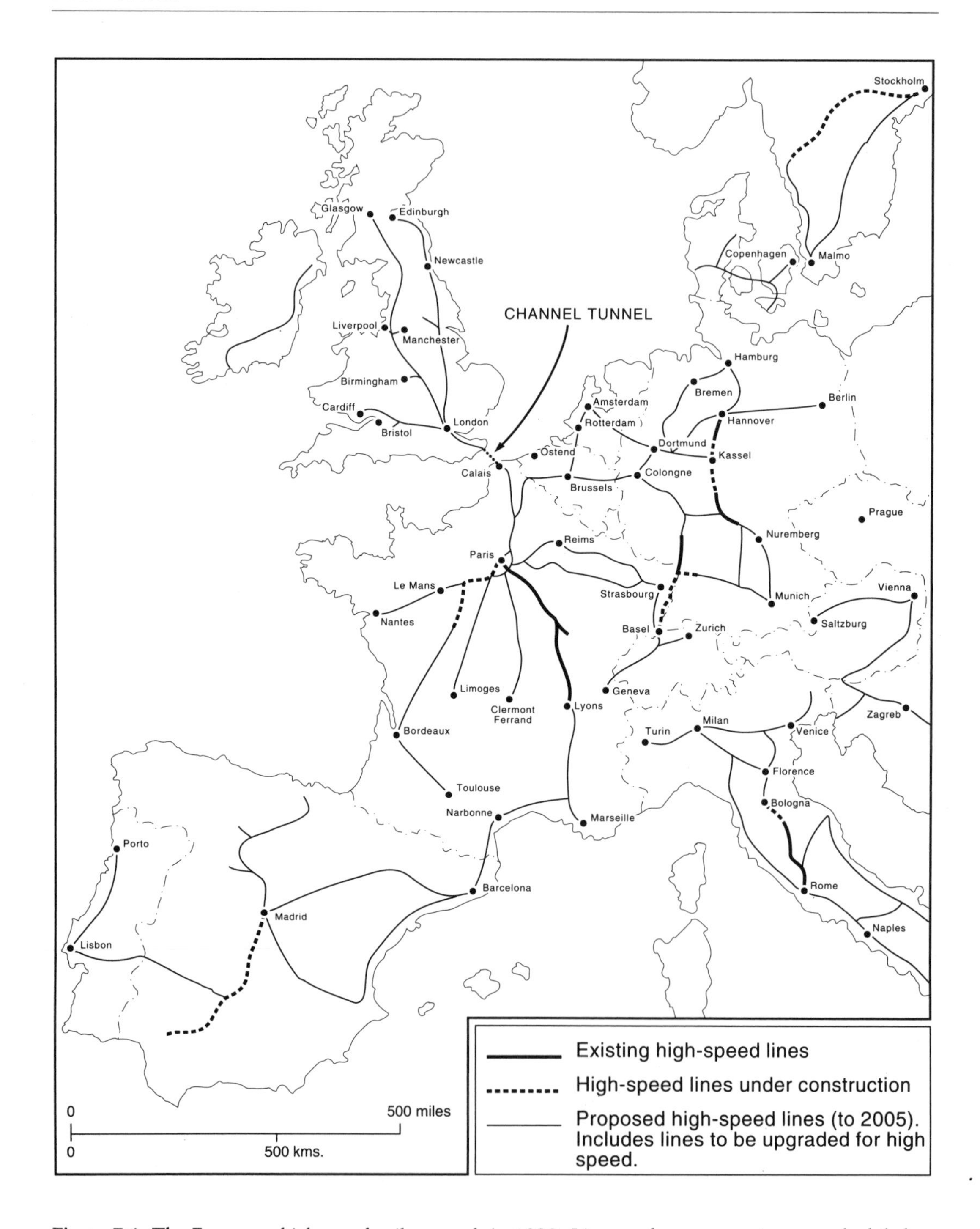

Figure 7.1 The European high-speed rail network in 1990. Lines under construction are scheduled for completion at various dates during the 1990s and proposed new routes are planned for completion by the early 21st century. Through-rail services using the Channel Tunnel will probably begin in mid-1993.

Sources: Modern Railways; International Railway Gazette.

Table 7.2 Energy efficiency of principal transport modes

(a) Passenger transport

		Passenger-miles per gallon of fuel
Small private car	(1)	42
Standard private car	(1)	15
Bus	(2)	110
Commuter rail	(2)	100
Small jet airliner	(3)	24
'Jumbo' jet airliner	(3)	28

(1) Assuming 1.4 passengers per vehicle
(2) Assuming 50% occupancy rate
(3) Assuming 70% occupancy rate

(b) Freight transport

	British Thermal Units per ton-miles
Air	14,700–42,000
Truck	1,110–2,800
Rail	330–670
Waterways	680
Pipelines	450

Source: Bevilacqua, O. M. (1978), *Transportation Planning and Technology*, 4.

structures have to take into account track and ancillary property maintenance costs as well as line-haul expenses, road rates are made up primarily of vehicle movement costs, with terminal and depot expenses representing only a small proportion of the total. As a result, road services can offer strongly competitive rates in the market for freight and passenger traffic over distances of up to 300–400 km., journey lengths which account for a high percentage of the total number of inter-urban links in many states.

Commercial goods vehicles now form a high proportion of total trunk-road traffic and many users benefit from a flexibility of operation which the railways cannot match. The majority of inter-urban passenger traffic consists of private cars; long-distance transport by coach and bus now accounts for only a small minority of all personal journeys.

Express highways date from the pioneer Italian autostrada of the early 1920s. The term expressway denotes a divided highway where some junctions with feeder roads are at the same level whereas freeways or motorways are designed with all intersections at two or more levels, allowing an uninterrupted route for through-traffic. Although the effectiveness of such roads in meeting the increasing demands for medium- and long-distance travel by freight and private passenger vehicles is frequently questioned, most advanced nations still continue to construct motorways as the most acceptable alternatives to heavily congested conventional road networks (Table 7.3).

Domestic airways

During the seven decades in which they have been operating, inter-urban air-passenger services have expanded to offer a substantial challenge to the railway for longer hauls. In very large territorial units such as the USA, Canada and Australia they are now the dominant means of travel, despite the fact that rates can be much higher than those of rail. Where the pattern of urban centres is not only dispersed, but also separated by tracts of undeveloped or sparsely inhabited land with difficult terrain, roads and railways may not always exist to provide direct surface links between towns. Air services are thus the only alternative. The growth of internal air transport has also been fostered by the expansion of international services, as the same terminal will often be used for both types of flight. It offers passengers with a foreign origin or destination a more convenient means of travelling than that of road or rail, where an additional journey between city terminal and airport is necessary.

Although domestic air transport has the great advantage of much higher journey speeds than land-based modes, this benefit can only be realized on routes where the overall journey time from city centre to city centre, including the trips to and from airports of departure and arrival, is less than that by rail. For distances of around 400 km. the latter mode can frequently offer the shorter time, but as links between city centres and airports are constantly being improved internal flights will become more

Table 7.3 Motorways and trunk-road networks (countries with over 1000 km. of motorway in 1988)

	Length of motorway network (km.)	*Principal road network including motorways (km.)*	*Motorways as percentage of principal road network*
USA	83,214	650,387	12.8
German Federal Republic	8,715	39,915	21.8
Canada	7,445	69,209	10.7
Italy	6,083	51,880	11.7
France	5,400	35,250	15.3
Japan	4,280	50,941	8.4
United Kingdom	2,981	15,406	19.3
Spain	2,142	20,563	10.4
Netherlands	2,060	4,118	50.0
South Africa	1,781	52,504	3.4
Belgium	1,567	14,468	10.8
South Korea	1,550	13,805	11.2
Switzerland	1,486	19,893	7.5
Austria	1,405	11,675	12.0
Venezuela	1,200	36,307	3.3

Source: International Road Federation: *World Road Statistics 1984–88*, 1989.

competitive over these shorter distances. Many United States airports, such as O'Hare (Chicago), Hartsfield (Atlanta) or Fort Worth (Dallas) handle over 40 million domestic passengers annually and rank among the world's busiest terminals on this basis, whereas international services are far less important. Airlines in the USA and the former USSR currently carry the largest shares of internal long-distance passenger traffic. In the former USSR domestic air services account for 93 per cent of all air passenger travel, compared with only 67 per cent in the USA where international travel is much more strongly established.

Pipelines

This highly exclusive means of freight transport has evolved to satisfy the demand for the convenient, safe and economic movement of petroleum and its refined products in large volumes. Whereas rail, trunk road and domestic air networks normally reflect the distribution of major population centres, pipelines are generally confined to routes linking major coastal oil terminals and refineries with inland industrial centres and thus play only a limited part in the overall freight-distribution system. Pipelines are only constructed where the flow of crude or unrefined petroleum is sufficiently high to justify the high capital costs involved. In consequence the more complex networks are confined to Anglo-America, Europe and the former USSR. Where regular transfers of over 10 million tonnes of oil annually are made, pipeline diameters of up to one metre are possible; but with quantities of less than one million tonnes small bore pipes are adequate and on some routes inland waterways or coastal tankers can offer strong competition.

Inland waterways

The constraints imposed by terrain upon canal building or the improvement of natural waterways result in this mode of transport being far less extensive than road or rail systems in respect of its inter-urban connections. However, the deadweight of individual barges can range up to 1500 tonnes on major navigable rivers and this capacity, coupled with the relative cheapness of transport by water as compared with road or rail, makes inland waterways particularly suited to the carriage of bulky, low-value commodities such as coal, iron ores and minerals where speed of transit is not of critical importance. High transhipment costs at interchange depots with road and rail transport, however, do restrict the use of water transport to longer hauls, as these fixed expenses account for a high proportion of the total costs of operation. The Rhine and its associated rivers and canals still act as an indispensable means of inter-urban freight transport in Western Europe. These waterways carry over 500 million tonnes annually in Belgium, Holland, France and the former German Federal Republic, where the annual total was 233 million tonnes in 1988.

Although waterway transport is much more restricted in its extent than road and railway systems, the Rhine does serve many of the principal heavy industrial centres of Western Europe and carries substantial shares of the total freight in individual EEC nations. France has the most extensive network and also possesses the highest length of waterway capable of accepting barges of over 3,000 tonnes, whereas just under half of Germany's waterways are restricted to craft of up to 1,500 tonnes. The Rhine, which is navigable upstream to Basle, carries over 250 million tonnes of freight annually, with about 40 per cent of this classed as international traffic moving between ports in the Low Countries, Germany and France. Within France freight can be distributed from the Rhine axis to ports on the Moselle, Marne and Rhone rivers and on the Rhine-Marne and Rhine-Rhone canals. River ports on the Weser, Elbe and Main in Germany are similarly linked to the Rhine by major canals although the bulk of German waterborne traffic is concentrated upon the Rhine in the Ruhr region. On these European rivers and canals the average length of haul for goods carried exclusively by water is 217 km. in the former German Federal Republic and 330 km. in Czechoslovakia but hauls of 500 km. are common on the Rhine (Bierman, 1988).

European inter-urban transport

The investigation into the future of European passenger transport carried out by the OECD in the mid-1970s is one of the few detailed contemporary accounts of inter-urban facilities and illustrates the contrasts between the various modes and the markets which they serve (OECD, 1977). Although it involves international rather than national linkages the level of interaction between cities in member states of the EEC is sufficiently high to justify viewing movement within the EEC as a whole as an example of inter-urban travel patterns. The network was defined as linkages between all national capitals and cities with populations of over 750,000, together with what are identified as important transport centres, regional development centres and regional capitals. This definition encompassed all cities of over 500,000 population and some smaller but significant centres, giving a total of 77 nodes, which were linked by 49,447 km. of road, 47,000 km. of rail and 259 direct air services. Motorways and other high-speed trunk routes accounted for 39 per cent of the road network and 64 per cent of the rail network had two or more tracks. Individual states displayed wide variations in the proportion of their intercity road systems which had been built or upgraded to motorway standard: the highest percentages were recorded in Belgium, Holland, Italy and the former German Federal Republic. An air-transport network was also constructed based on 77 airports and 259 direct intercity links which in most cases had at least ten scheduled weekly flights. London, Paris, Frankfurt, Rome, Milan, Amsterdam, Zürich and Madrid were identified as the dominant focal points and these eight together accounted for almost half of all passenger movements over the network.

Traffic distribution between modes

There are substantial differences in the distribution of inter-urban traffic between

Table 7.4 Distribution of traffic by mode for selected countries 1986

(a) Passenger traffic (percentage of total passenger – km.)

	Private road transport	*Public road transport*	*Rail*
France	82.7	6.4	10.9
German Federal Republic	83.1	10.0	6.9
United Kingdom	85.0	7.7	7.3
Italy	84.8	7.2	8.0
Netherlands	87.2	7.1	5.7
Spain	72.2	19.2	8.6
Yugoslavia	45.8	38.8	15.4
USA	98.4	1.0	0.6
Japan	51.0	11.5	37.5

(b) Freight traffic (percentage of total tonne – km.)

	Road	*Rail*	*Waterway*
France	70.3	25.9	3.8
German Federal Republic	57.6	22.5	19.9
United Kingdom	61.3	9.4	29.3*
Italy	89.1	10.8	0.1
Netherlands	70.7	3.6	25.7
Spain	89.6	10.4	–
Yugoslavia	41.0	50.6	8.4
USSR	10.8	83.7	5.5
USA	31.9	47.0	21.1
Japan	50.5	4.6	44.9*
China	15.0	65.0	20.0*

*Inclusive of coastal shipping.

Source: International Road Federation: *World Road Statistics 1984–88*, 1989.

transport modes and the density of freight and passenger traffic also varies according to levels of economic development. It is difficult to present reliable accounts of specific inter-urban patterns: national statistics rarely include separate data on long-distance transport, so comparisons between different countries can only be based upon aggregate data which usually include movements in rural and urban areas as well as that between cities and towns. Recent international data, however, do show the wide variations that exist in the proportional use made of the different modes for freight and passenger movements. In most countries road and rail are the main competitors for traffic but waterways still play an important part in the transport of freight in China, the USSR, the USA and several Eastern European countries (Table 7.4).

Personal travel

Analyses of the choice of mode for personal travel involve a complex range of factors and employ highly sophisticated methodologies. It is only possible in this chapter to consider the basic issues. Mode selection for inter-urban travel is influenced by journey purpose, overall trip length, travel costs and the range of transport options available. The business and commercial trips which account for a substantial proportion of all inter-urban travel involve complex movement patterns and the private car is invariably selected as the most convenient mode. Journeys to specific destinations from a base city can make use of high-speed rail or air services where distances of over 300 km. are covered, although even here the car can still be a viable option despite the large amount of the working day that is inevitably devoted to travelling. Social and recreational trips at the inter-urban scale rely heavily upon the car, as the costs for a family of four will be much lower than those for travel by rail or coach if car depreciation, maintenance, insurance and tax are excluded from the selection exercise. However, several railway undertakings are successfully expanding their share of the social and recreational travel market with the issue of concessionary tickets for families, students and the retired. Long-distance express coach services usually cater for those without access to cars and without the financial resources or desire to make use of the faster rail or air facilities. When the distribution of passenger travel is examined it must be remembered that urban traffic is usually included in the published data so that an accurate picture of inter-urban patterns can rarely be constructed. Moreover, information relating to private passenger traffic in many developing countries and those of the communist bloc is unavailable so comparisons cannot be easily made. Data for

Table 7.5 Great Britain: movement of principal types of freight by mode in 1989 (percentages of total tonne – km.)

Commodity	*Road*	*Rail*	*Pipeline*	*Water**
Petroleum and by-products	7	4	15	74
Coal and other solid fuels	30	50	-	20
Ores	64	36	-	-
Chemicals	94	4	-	2
Minerals and building materials	75	12	-	13
Machinery and other manufactured goods	95	1	-	4

*Including coastal shipping.

Source: Transport Statistics Great Britain 1979–89, 1990.

Table 7.6 United States intercity traffic changes 1970–1986

(a) Freight traffic by mode (percentage of total ton – miles)

	1970	1980	1986
Rail	39.8	37.5	35.8
Road	21.3	22.3	25.1
Inland waterway	16.5	16.4	15.7
Oil pipeline	22.3	23.6	23.1
Domestic airways	0.1	0.2	0.3

(b) Passenger traffic by mode (percentage of total passenger – miles)

	1970	1980	1986
Private cars	86.9	83.5	80.4
Domestic airways	10.1	14.1	17.6
Bus (excluding urban services)	2.1	1.7	1.3
Rail	0.9	0.7	0.7

Source: Statistical abstract of the United States 1988, Table 979, 1988.

intercity travel in the USA indicates the capture by private cars and domestic airways of the once dominant railway traffic and the relegation of the latter to bottom place, below bus and coach transport in the ranking (Table 7.6).

Freight haulage

The allocation of inter-urban freight flows between modes is determined by an interlinked set of factors related to costs, length of haul, value and bulk of the commodity and the quality of service required by the customer in respect of speed, safety and particular needs such as refrigeration (Hay, 1973; Fowkes et al., 1987). The introduction of container facilities, which have had a substantial impact upon long-distance freight haulage, are discussed more fully elsewhere (Chapter 10) and the rationalization of much of the distribution network associated with wholesaling and retailing has also affected inter-urban goods movements (McKinnon, 1989; Quarmby, 1989).

Since data on inter-urban freight movements are rarely recorded separately, the following discussion is based on statistics which usually include freight flows in urban areas as well as on trunk routes. Within the UK, for example, freight haulage is now dominated by road transport, which accounted for 82 per cent of all goods by tonnage in 1988. The principal bulk commodities, however, still make substantial use of rail, water and pipeline services (see Table 7.5) and in the case of petroleum products road accounts for less than one-tenth of all movements. Rail continues to carry the greatest proportion of coal and coke but for other bulk commodities such as metal ores, chemicals, fertilizers, minerals and building materials road transport takes the greatest shares (Department of Transport, 1990). The USA is one of the few countries to publish separate data on inter-city traffic and the present situation indicates the importance of pipelines and waterways in freight transport, these two jointly carrying more traffic than the railway network. The latter, however, is still the leading mode but pipelines and roads are now of almost equal importance (Table 7.6).

Information on inter-urban freight flows is rarely available for developing countries but road transport is generally continuing to expand its share of the market. In 1960, when Nigeria became independent, road transport carried 77 per cent of all freight, but by the mid-1980s, following an ambitious road-improvement programme, this share had increased to 94 per cent. Railways carry most of the remaining traffic and the Niger-Benue river system, which in 1950 carried 8 per cent of all freight, is now largely defunct (Ezeife, 1984). In Zimbabwe, however, the railway still plays a significant part in the haulage of coal, iron ore and agricultural products, and it is estimated that road services carry only about ten per cent of all medium- and long-distance freight. Of the principal industrialized nations the USSR, Poland and Czechoslovakia rely most heavily upon rail transport for freight haulage and in China it is estimated that about two-thirds of all freight traffic is dealt with by rail (Table 7.4). Inland waterways carry over 20 per cent of all freight traffic in China and Japan (principally coastal services) and play a significant role in the USA, Brazil, Holland, Belgium, Germany and Hungary. Elsewhere the majority of freight is attracted to road networks which carry at least two-thirds of such traffic in the UK, Italy, Spain, France, Holland, Finland and Sweden.

The inland-transport system of Japan has experienced some of the most dramatic changes in the allocation of freight traffic between road and rail in recent years. In 1955 railways still accounted for 82 per cent of all inland freight movements but within the following 20 years road transport expanded to capture 73 per cent of the market and by 1987 its share was 92 per cent. Since 1955 railborne traffic has fallen from 43.3 to 20.7 million tonne-kms whereas that carried by road has risen from 9.5 to 226.4 million. Coastal shipping also makes a major contribution to the overall freight-transport system, its share of the total rivalling that carried by road (Saito, 1989).

With the expansion of cargo unitization in the late 1960s, and especially the development of container services since 1970, a greater measure of coordination and integration between the principal freight-carrying modes has been achieved. The establishment of container terminals at strategic coastal and inland locations has allowed unitized freight to be transferred from road to rail and back to road for final delivery but such operations still account for only a small proportion of all inter-urban freight movements.

Contemporary inter-urban transport

Contemporary inter-urban transport systems illustrate the influences of national policy as well as those associated with the social and economic nature of market demand. In particular the rapidity with which road traffic and the resultant congestion have increased during the second half of the twentieth century has stimulated many governments to produce and implement policies designed to secure a more equitable and more efficient distribution of long-distance freight and passenger traffic between transport modes (Starkie, 1976). In the developing world, the scene for many ambitious transport-investment plans, governments have been especially concerned with allocating funds within the transport sector in such a way as to achieve the most effective improvements on what are often poorly endowed communication systems (Hoyle, 1973). Legislation and policies of subsidization have been introduced in attempts to encourage the use by medium and long-distance traffic of what is seen as the most suitable mode. During the 1980s these moves have been given added impetus with the recognition of the vital role the transport sector can play in energy and environmental conservation. In Anglo-America and Western Europe transport policy has focused especially upon the consequences of the fundamental shift of passenger and freight traffic from rail to road in terms of increasing highway congestion and the redundancy of many sections of the rail network (Dunn, 1981). With the loss by railways of their former majority share of traffic in many industrial nations they have undergone a series of technical, spatial and organizational changes in order to retain and consolidate those sections of the market in which they still effectively compete.

Very few systems, however, can survive

commercially without state subsidies and massive investment programmes but current transport policy demonstrates an increasing acceptance of the need for energy conservation and of the fact that improved rail-passenger services can contribute towards the relief of congestion on inter-urban roads (Department of Transport, 1977). Transport policy in most Western industrial economies can usually only be implemented through selective subsidization or public ownership of certain transport undertakings, but in nations such as China, the USSR or until recently the Eastern European countries, coordination and integration of the principal inter-urban facilities have been carried out as part of much wider programmes of economic planning. Since in many of these states the expansion of mining and heavy industry has been a priority objective, railways have received special attention and the low levels of private car ownership have made road improvements on the scale of those in Western Europe or the USA unnecessary (Howard, 1990). Within the Third World, investment in inter-urban routes has to compete for limited resources which are also urgently needed for improvements to rural and urban transport infrastructures (Addus, 1989). Several countries have inherited good quality long-distance road and railway routes from the former colonial governments; current efforts often focus upon the demands of rural agricultural regions and of the rapidly expanding towns and cities. The more detailed examination of individual states which follows illustrates the diversity of policy responses in the area of inter-urban transport.

The United Kingdom

Although substantial parts of the inter-urban transport system in the UK have been owned and operated by state organizations since the late 1940s, it was not until 1977 that the general aims of a national transport policy were published (Department of Transport, 1977). This report came towards the end of a period in which coordination and the subsidization of much of the rail and bus network had been major objectives of government-transport policy, but during the 1980s public investment in the transport sector has been severely reduced (White and Doganis, 1990). Competition between and within inter-urban modes has been encouraged and the private sector is becoming more closely involved in new road- and rail-construction programmes. Although the government was strongly urged by motoring and industrial interests in the 1930s to emulate Italy and Germany by building a national motorway framework, firm proposals were not published until 1946; the first route, the Preston bypass section of the M6, was not completed until 1958. Subsequent plans envisaged that a primary road network of 5,600 km., including 3,220 km. constructed to motorway standard, would be completed by the late 1980s but this was not achieved (Starkie, 1982). By 1990, 2,993 km. of motorway were open, accounting for 19 per cent of the inter-urban principal road network but carrying 16.5 per cent of all traffic and over twice the volume of vehicles that were handled in 1975. In marked contrast to the French and German systems, however, the UK motorway network connects only a limited number of major centres, and the Scottish cities of Glasgow and Edinburgh will not be linked to the national framework until the mid-1990s.

The road-investment programme of 1989 recognized the deficiencies of the contemporary motorways. It includes the widening of many of the existing roads and the building of duplicate routes, where possible with private funding, in the west midlands and northwest England (Department of Transport, 1989). Inter-urban rail-passenger services are provided by an InterCity Sector, based on high-speed routes between principal population centres, a Regional Sector, linking less important towns and cities and including many cross-country connections, and Network Southeast, which combines inter-urban routes with an intensive commuting system focusing upon London.

The Intercity network is now operated on a commercial basis but the other two sectors continue to receive state subsidies first sanctioned by the 1968 Transport Act, passed to protect the railways from further drastic reductions in both routes and services as first proposed in the 1963 Beeching report.

Passenger-traffic growth on the inter-urban network has been stimulated by the extension of electrification and a general reduction in long-distance journey times. Although the network is 30 per cent smaller than it was in 1962, traffic volumes have recovered to the levels carried at this earlier date. More through services between major northern and midland cities and those in southern England have been introduced, eliminating the requirement for train changes in London, and traffic on these routes is likely to increase with the projected opening of the Channel Tunnel railway in 1993. Although the number of passengers carried on domestic air services increased by 60 per cent between 1979 and 1989 to 11.8 million this total is still only one per cent of all passenger traffic, a proportion that has not changed in the last decade. London and most large provincial cities are linked by frequent flights. The London-Glasgow route in particular is the scene of some of the most vigorous competition on Europe's domestic air system: fares (1990 levels) vary between £184 and £89, the latter being currently less than the first-class rail fare (Figure 7.2).

Germany

As a pioneer of the motorway concept, by 1939 Germany had built a nationwide network of autobahnen focusing upon Berlin, the Ruhr and the Baltic port of Hamburg. With partition in 1949 the two new states were obliged to reorientate their inter-urban networks and the Federal Republic embarked upon a massive programme of road reconstruction and extension to meet the demands of its new territory. Whereas the emphasis in the prewar system was upon east-west links, the new West German network required the strengthening of north-south connections between cities in Bavaria, the industrial Rhine valley and on the Baltic coast. The core of 2,100 km. of autobahnen inherited from the 1930s had been extended to a total of 8,800 km. by 1989 and all major urban centres are now located on the motorway system or the supplementary framework of 3,000 km. of federal trunk highways. Recent Federal Infrastructure Plans have in fact placed more emphasis upon developing the latter roads rather than autobahnen and funds originally intended to provide 7,000 km. of additional motorway have been diverted to improve other trunk routes (Scott, 1985) (Figure 7.3).

Over 6,000 km. of the autobahnen are incorporated in the European 'E' route network so that with the expansion of EEC trade many of the domestic inter-urban facilities are related to the connections of the German motorways with neighbouring systems in France, Switzerland and the Low Countries. Following political unification of the two Germanies in 1990 and the reinstatement of Berlin as the national capital, substantial revisions of the road network will be required and the former East German motorways will need to be upgraded to the standards of their Western counterparts. During the 1980s a greater emphasis was laid upon railway development in Federal Germany: the Deutchesbundesbahn received 29 per cent of funds from the transport sector's allocation in the 1980s as compared with 16 per cent in the previous decade. Although nine per cent of the rail network was closed between 1980 and 1985, several new high-speed routes have been opened or planned between major industrial centres, reflecting similar programmes in France and Italy (Figure 7.1). These new services are intended to run over 2,000 km. of improved or newly constructed track by 2000 AD. Speeds of up to 250 km. per hour will be possible on the Hannover-Wurzburg and Mannheim-Stuttgart lines, although access to these routes from intermediate towns will be much more limited than on the existing railways. Thus the new

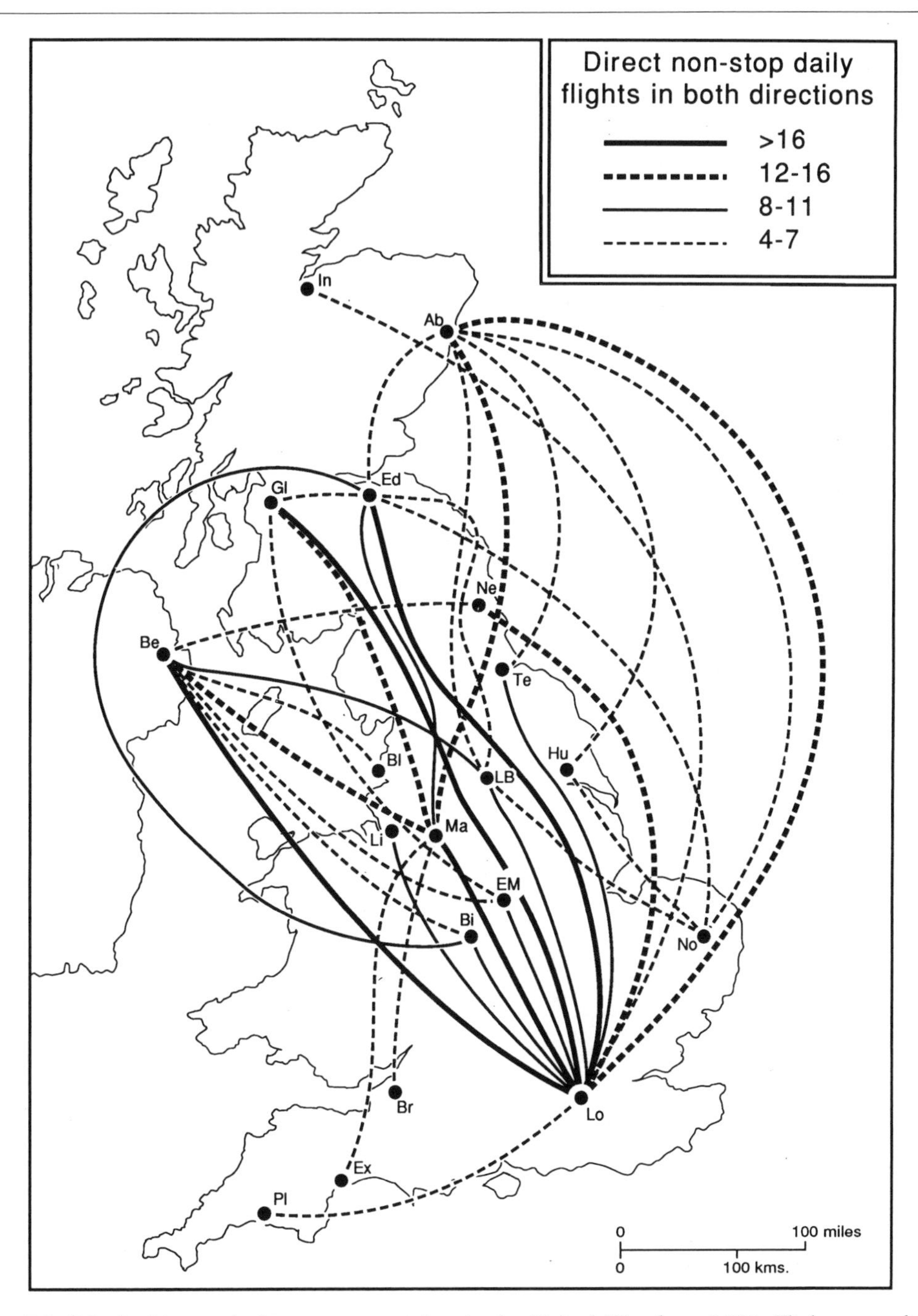

Figure 7.2 Principal internal air-passenger services in the United Kingdom, 1990. Flights to and from London make use of Heathrow, Gatwick and Stansted Airports.

Ab	Aberdeen	EM	East Midlands	Li	Liverpool
Be	Belfast	Ex	Exeter	Lo	London
Bi	Birmingham	Gl	Glasgow	Ma	Manchester
Bl	Blackpool	Hu	Hull	Ne	Newcastle
Br	Bristol	In	Inverness	No	Norwich
Ed	Edinburgh	LB	Leeds–Bradford	Te	Tees-side

Sources: Current airline timetables.

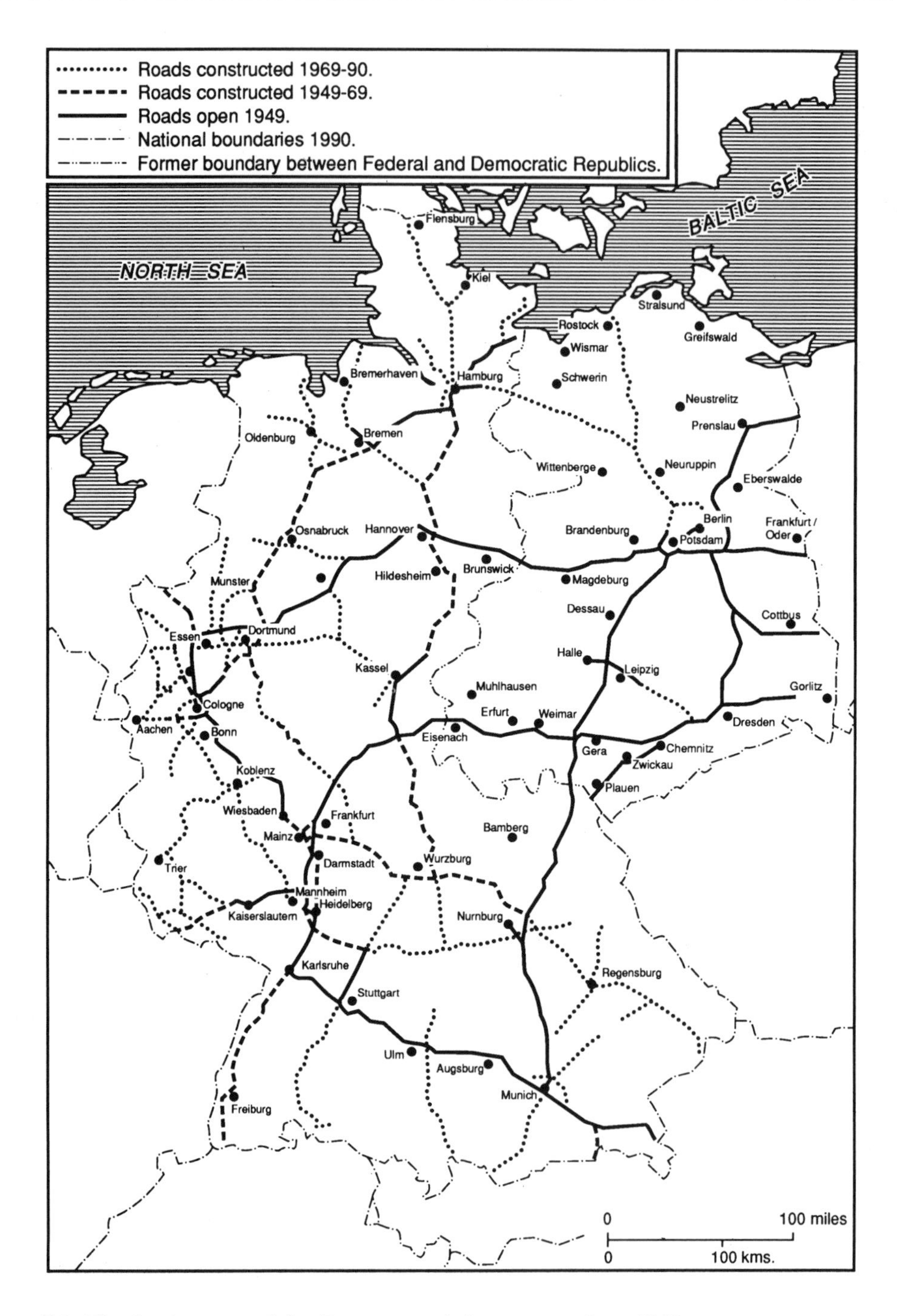

Figure 7.3 The development of the German autobahnen network to 1990.

Sources: Road maps published 1930–90.

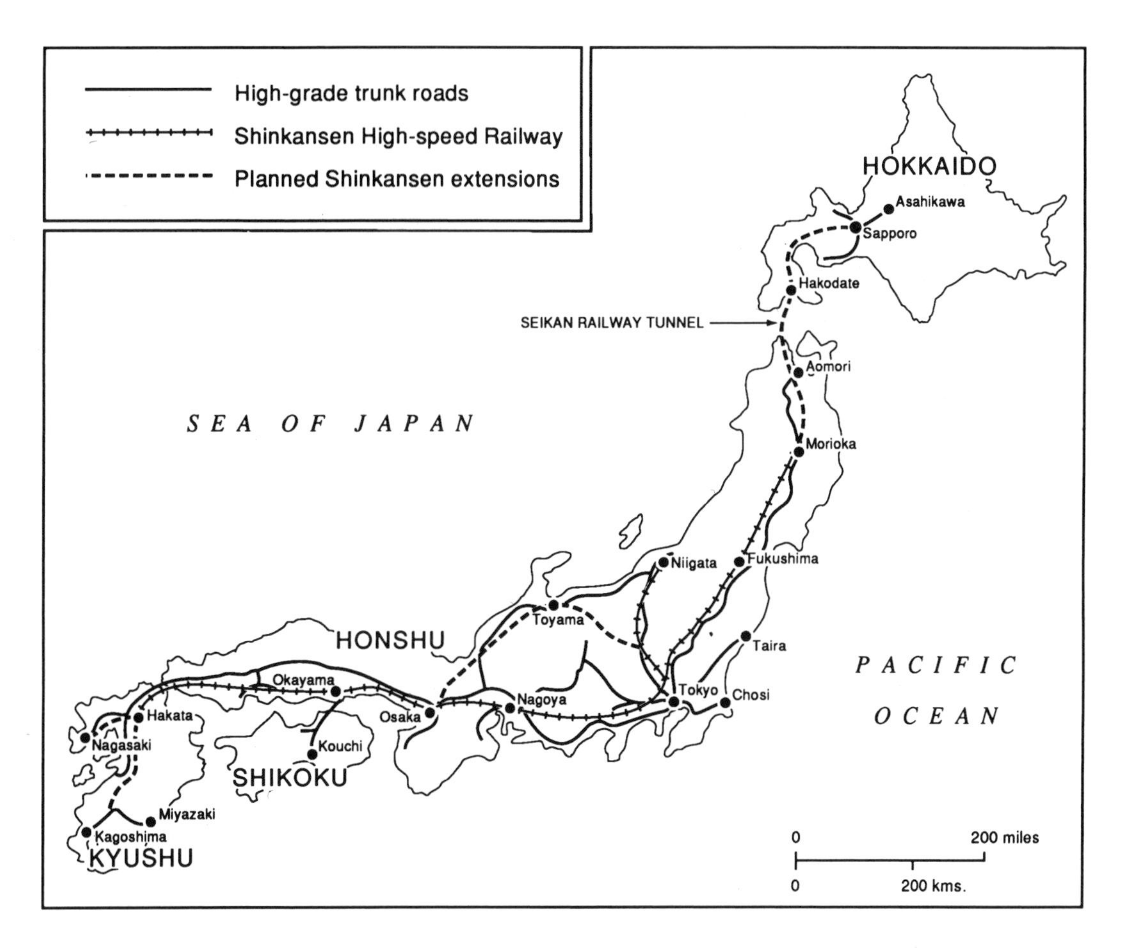

Figure 7.4 The Shinkansen rail and high-grade trunk-road networks in Japan.
Sources: Saito (1990); *Janes' World Railways* (1990).

services, while substantially reducing journey timings between selected cities, will not improve the conventional rail network and will probably benefit only a restricted range of business travellers (Whitelegg, 1988).

Japan

During the 1950s and 1960s the severe strains upon the Japanese inland-transport system caused by the rapid growth of the economy were remedied by ambitious motorway- and railway-improvement programmes (Figure 7.4). The national narrow-gauge rail network was steadily losing passenger traffic to both road and air transport. In 1964 the 500-km. high-speed standard-gauge Shinkansen route with electric traction was opened between Tokyo and Osaka to relieve existing lines and to counteract these growing challenges. In 1975 the new line was extended to Hakata on Kyushu and further routes linking Tokyo with Morioka and Niigata were built. These lines are restricted to passenger services at speeds of up to 220 kph but the initial expansion of traffic has not always been sustained, although traffic is still rising on the original Tokaido and Sanyo routes and speeds are to be increased to 260 km. per hour with the prospect of several new Shinkansen lines being built. However, despite these new lines and privatization of the railways in 1987, deficits have continued to accumulate and no national transport policy has been implemented (Saito, 1989; Ohta, 1989).

The United States

Although railways continue to carry a substantial share of inter-urban freight, their passenger traffic has been reduced to negligible proportions with the rapid expansion since the 1950s of domestic air services and private car traffic. In an effort to retain some share of the inter-urban market the bulk of the long-distance rail-passenger services were acquired in 1971 by the National Railroad Passenger Corporation (Amtrak), which then operated trains over the privately owned networks. During the 1980s over 44,000 kms of track carried inter-urban Amtrak services and traffic rose to 21.4 million passengers by 1988. Federal financial support has been progressively reduced and Amtrak now owns sections of the railway in the northeast corridor between Washington and Boston, which carries the heaviest traffic. The European initiative in high-speed rail travel, however, has found little application in the USA, where traffic levels are low and domestic air transport dominates long-distance travel – although a 250 km. per hour rail service is being introduced in Florida between Tampa and Miami, with tourist traffic offering the necessary potential (Nice, 1989).

The United States transcontinental motorway network now accounts for 15 per cent of the national trunk-road system and is the result of both state and federal construction programmes. It includes the 65,000 km. of the Interstate Highway system, which is largely a federal initiative designed to provide high-quality motorway facilities between most major cities. Heavily urbanized states, such as California and those of the north-east industrial region, have the densest networks but in regions such as the mid-west the system is less complex.

The USA provides one of the best examples of a well-established and complex pattern of inter-urban air services. Long distances between cities and the high living standards enjoyed by a large proportion of the population have favoured the expansion of internal air transport at the expense of rail services. By 1989 over 520 billion domestic passenger-km. were flown, an increase of 75 per cent on the 1979 total and the highest for any country. Services are provided by a large number of private undertakings concerned with either medium and long-haul routes, such as those linking east and west coast cities, or with more localized operations over shorter distances. Long hauls, defined as flights of over 1,280 km., usually connect cities with populations of over one million and distances range from 4,320 km. (Boston to San Francisco) to 1,330 km. (Portland to Los Angeles).

Before 1978 airlines were subject to considerable control from the Civil Aeronautics Board (CAB) on pricing policies and route operations but in that year the Airline Deregulation Act was passed, liberalizing many of the conditions relating to competition and enabling the CAB to permit new services seen as beneficial to the public. Where traffic levels were insufficient to cover costs, an airline could be required to continue the service for ten years but with a CAB subsidy to meet any losses incurred. Following the Act there was a large increase in passenger carriers and many airlines also extended their networks, providing many cities with a greater choice of inter-urban flights and more frequent daily departures. Many passengers travel by means of a hub-and-spoke pattern which can involve change of carrier, but with deregulation the same airline will be responsible for the overall flight. The significance of a hub within the national domestic-airways system can be defined in terms of the percentage of all US passengers that it handles or the percentage of all departures by a particular airline. Major hubs can be defined as those cities where this percentage is at least ten and during the 1980s most of the leading airlines have focused their flight patterns upon these centres. Between 1977 and 1984 all domestic departures rose by 24 per cent but the volume at the major hubs (such as Dallas/Fort Worth, Atlanta or Minneapolis/St Paul) almost doubled.

An analysis of traffic changes for 14 major carriers in the 1977–84 period showed that while total domestic departures fell by up to 46 per cent, increases of between 15 and 654 per cent were recorded in their traffic handled at the major hub airports. The growing importance since deregulation of the hub-and-spoke network pattern for inter-urban flights has produced a better quality of service overall, with increased numbers of departures between smaller centres where passenger demand is less, and with a greater percentage of all flights being handled by single carriers. It has also been claimed that the pattern of services that emerged in the 1980s is catering for a wider market than before where long-distance flights are involved and more middle-income and tourist travellers are being attracted by the lower rates. In contrast, for hauls of less than 320 km. rates are higher and business travellers are most likely to make use of them (Phillips, 1987).

China

In China transport policy has been designed to fulfil several requirements of a centrally-planned state established in 1949 and inheriting a road and rail infrastructure which was severely damaged during the war. Railway investment was initially concentrated upon the repair and improvement of existing inter-urban lines in the eastern coastal areas and the extension of this core network westwards to peripheral provinces such as Inner Mongolia and Yunnan. The Chinese achievement in expanding their railway system since the communist state was founded in 1949 is unparalleled anywhere else in the world. Of the 53,000 km. now in use, 57 per cent has been constructed since 1949 and current plans envisage a further 26,000 km. to be added by the end of the century. All provincial capitals apart from Lhasa are now connected to Beijing by rail, compared with only one-half of these centres in 1949; in order to secure this aim the rate of construction between 1949 and the early 1970s was on average 1,000 km. each year.

Before 1949 only one-fifth of the national network lay west of the main north-south Beijing to Guangzhou line but by the mid-1980s this share had risen to 45 per cent. Almost 80 per cent of the network is still single-track but in the east many lines have been double-tracked, including the trunk routes linking Beijing with Wuhan and Guangzhou and the Tientsin-Shanghai line. The electrification programme focuses on lines carrying bulk freight, such as that linking the Shanxi coalfield with the port of Qinhuangdou which has an annual capacity of 100 million tonnes. A further 1,500 km. of electrified track was due to be added to the present total of 6,500 km. by late 1991 (Leung, 1980; Howard, 1990).

Both economic and strategic motives were involved in this programme and although the phenomenal rates of railway construction recorded in the 1950–80 period are unlikely to be repeated, an ambitious programme of route-upgrading continues. The Chinese government now accepts that road transport could carry a much larger share of short-haul freight but an expansion of such activities is still severely hampered by a lack of suitable vehicles and, in some areas, poor road conditions.

Zimbabwe

National transport planning in Zimbabwe is closely related to the broader regional policies of the Southern African Development Coordination Conference established in 1980 by the independent states of southern Africa. These seek to reduce the reliance upon South Africa for access to seaports, and several elements of Zimbabwe's inter-urban transport policy are associated with the need to reinvigorate routes to alternative coastal outlets in Mozambique (Turton, 1989)

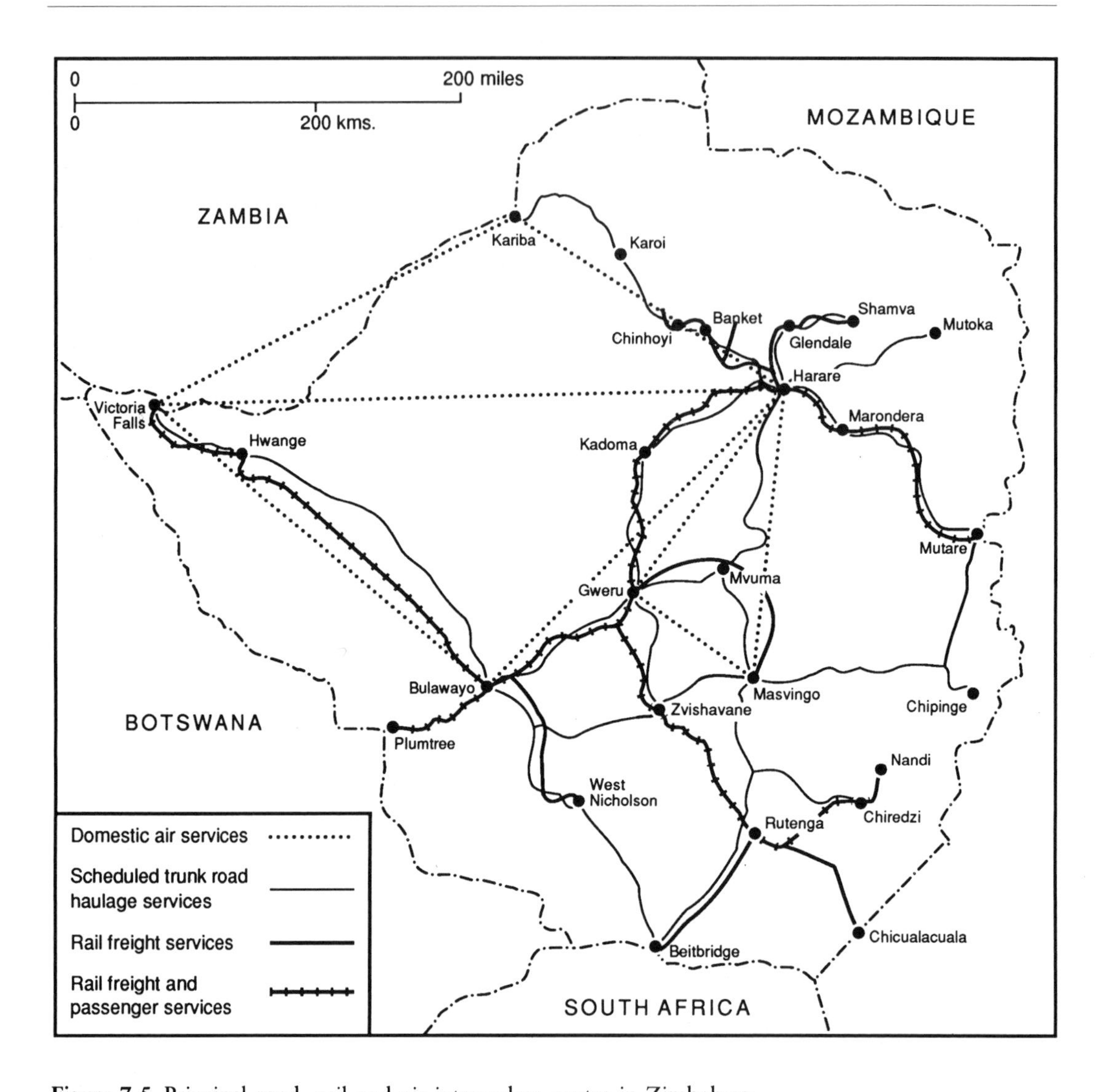

Figure 7.5 Principal road, rail and air inter-urban routes in Zimbabwe.
Sources: Current airline and rail timetables; information supplied by Swift Road Transport.

(Figure 7.5). The First National Development Plan of 1986 has the integration and coordination of internal transport facilities as one of its principal aims and the state-owned railway system is seen as a vital contributor to the economic growth programme. The railway deficits that have been accumulating since 1971 are compensated for by state subsidies and many of the freight-haulage rates are lower than actual operating costs, a situation accepted by a government that intends to make continued use of the railways for as much as possible of internal long-distance freight haulage (Zimbabwe, 1986).

The continuing importance of railways in Zimbabwe may be contrasted with the situation in Nigeria, where since independence the larger share of transport-development funds have been allocated to roads and many of the under-used inter-urban railways are now paralleled by a system of greatly improved trunk highways. During the four National Development Plans published between 1962

Table 7.7 Nigeria: investment in the transport sector 1962–1985

Development plan	*Percentage of total allocated to transport sector*	*Percentage of transport sector allocation*			
		Road	*Rail*	*Air*	*Water**
First 1962–8	19	58	10	7	25
Second 1970–4	12	67	9	11	13
Third 1975–80	22	71	9	10	10
Fourth 1981–5	15	60	25	6	9

*Ports, shipping and inland waterways.

Source: National Development Plans for Nigeria, 1962–1985.

and 1985 the Nigerian road system's share of funding varied between 54 and 72 per cent. That of the railways was much lower but some recognition of the latter's valuable role in hauling bulk mineral and agricultural goods eventually came in the 1981–85 Plan when one-quarter of transport investment was allocated to railways (Ezeife, 1984) (Table 7.7).

Conclusions

Inter-urban transport is provided by a variety of modes but road services have expanded to capture the greater share of total traffic in many countries. Railways, pipelines and waterways, however, continue to carry large volumes of bulk raw materials and manufactured goods. They are of vital importance in the economies of nations such as the USSR and China, where long distances separate urban centres in mining and industrial regions. In Western Europe railways are attempting to regain some of their lost traffic and in particular efforts are being made to sustain the rail-passenger market with the introduction of high-speed trains, which over distances of about 300 km. can effectively compete with air services.

Competition for traffic both between and within modes is still an important characteristic of many inter-urban transport systems but the effects of government plans and policies can also be detected in terms of investment, particularly with respect to railways and roads. Most major railway systems are now state-owned or controlled and the operation of many passenger services and the upgrading of routes is only possible with continued state financial support. Many developing countries have guided the expansion of their transport networks by the use of periodic development plans, allocating funds to those parts of the system judged to be the most effective contributors to economic growth.

The transport industry is a prodigious consumer of the world's petroleum and the most energy-efficient modes will receive more attention in the future in an effort to make the best use of available resources. This problem, coupled with that of minimizing the environmental impact of new transport initiatives, will be a major influence in the next few decades of inter-urban transport. A more widespread application of telecommunications and related developments in information transmission may also lead to a decrease in long-distance personal travel but this reduction could be countered by the expansion of tourism as the working week is shortened in many industrial nations.

References

Addus, A. A. (1989), 'Road transportation in Africa', *Transportation Quarterly* 43 (3), 421–33.

Bierman, D. E. and Rydzkowski, W. (1988), 'European waterways', *Transportation Quarterly* 42 (2), 289–306.

Chisholm, M. (1987), 'Regional variations in transport costs in Britain', *Transactions of the Institute of British Geographers*, 12 (3), 303–14.

Department of Transport (1977), *Transport policy* (London: HMSO).

Department of Transport (1989), *Roads for prosperity* (London: HMSO).

Department of Transport (1990), *Transport Statistics, Great Britain, 1988–89* (London: HMSO).

Dunn, J. A. (1981), *Miles to go: European and American transportation policies* (Cambridge,

Mass.: MIT Press).

Ezeife, P. C. (1984), 'The development of the Nigerian transport system', *Transport Reviews* 4 (4), 305–30.

Fowkes, A. S., Nash, C. A., Tweddle, G. and Whiteing, A. E. (1987), 'Forecasting freight mode choice in Great Britain', *PTRC Proceedings*, Seminar G, University of Bath, 43–54.

Gwilliam, K. and Mackie, S. (1975), *Economics and transport policy* (London: Allen and Unwin).

Haggett, P. and Chorley, R. (1969), *Network analysis in geography* (London: John Arnold).

Hay, A. (1973), *Transport for the space economy* (London: Macmillan).

Howard, A. (1990), 'Industry, energy and transport', *The geography of contemporary China*, ed. T. Cannon and A. Jenkins (London: Routledge), ch. 7.

Hoyle, B. S. (ed.) (1973), *Transport and development* (London: Macmillan).

Leung, C. K. (1980), 'China: railway patterns and national goals' (University of Chicago, Department of Geography, *Research Paper* 165).

McKinnon, A. (1989), *Physical distribution systems* (London: Routledge).

Nice, D. (1989), 'Stability of the Amtrak system', *Transportation Quarterly* 43 (4), 557–70.

OECD (1977), *The future of European passenger transport* (Paris: OECD).

Ohta, K. (1989), 'The development of Japanese transportation policies in the context of regional development', *Transportation Research* 23 A/1, 91–101.

Phillips, L. T. (1987), 'Air carrier activity at major hub airports', *Transportation Research* 21 A/3, 215–22.

Prideaux, J. (1988), 'InterCity: profits of change', *Transport* 9 (1), 27–30.

Quarmby, D. A. (1989), 'Developments in the retail market and their effects on freight distribution', *Journal of Transport Economics and Policy* 23 (1), 75–88.

Saito, C. (1989), 'Transportation coordination debate and the Japanese National Railways problem in postwar Japan', *Transportation Research* 23 A/1, 13–18.

Scott, D. (1985), 'The West German transport system', *Geography* 68 (3), 266–71.

Starkie, D. A. (1976), *Transportation planning, policy and analysis* (Oxford: Pergamon).

Starkie, D. A. (1982), *The motorway age* (Oxford: Pergamon).

Turton, B. J. (1989), 'Railways and the national economy of Zimbabwe', *Geographical Journal of Zimbabwe* 19, 47–57.

White, P. and Doganis, R. (1990), 'Long-distance travel within Britain' (Oxford, Transport Studies Unit, *Rees Jeffreys Discussion Paper 17*).

Whitelegg, J. (1988), 'High-speed railways and new investment in Germany', *Transport technology and spatial change*, ed. R. S. Tolley (Institute of British Geographers, Transport Geography Study Group).

Zimbabwe, Ministry of Finance (1986), *First National Development Plan* (Harare: Central Statistical Office).

Zimbabwe, Ministry of Transport (1984), *Study of the Zimbabwe road passenger transport industry, Vol. 1* (Harare: Central Statistical Office).

8
Rural Areas: the Accessibility Problem

Stephen Nutley

In rural areas the low density of population causes economic problems for public transport and hardship to people without cars. In developing countries the lack of rural roads severely retards development prospects. In rural areas of the UK, and in other countries, the central issue is lack of accessibility. Policies have failed to recognize this, although various alternative approaches are conceivable.

Introduction

The problematic nature of transport in rural areas stems directly from the inherent characteristics of the rural environment itself. It is always difficult to define rural areas and to demarcate them clearly from 'urban'; most social and economic indicators show increasingly complex patterns across urban and rural zones (Cloke and Edwards, 1986) and transport reflects this. Nevertheless, the crucial factors are low population density, a dispersed settlement pattern with low population totals at any point, a scattered pattern of small service outlets and a concentration of middle- and high-order facilities into widely separated urban nodes. Essentially, the 'problem' is defined in terms of *public* transport, and therefore bears upon the population dependent upon this sector. It also refers to the *local* scale of passenger demand, i.e. the ease or difficulty of carrying out everyday journeys such as shopping. It is overwhelmingly concerned with passenger rather than freight movements, and does not refer to major traffic flows which happen to pass through rural areas between one city and another.

Hence the 'demand surface' for transport is totally at variance with the optimum economic conditions for public transport operation. This means that, in terms of infrastructure, a high aggregate length of road or railway track is required to cover the spatial extent of a rural region which is disproportionate to the population or traffic potential of that region. Similarly with respect to services, to maintain a geographical coverage of all settlements in a rural region, even at low frequencies, means providing an aggregate vehicle capacity which is excessively high relative to the likely patronage. Transport supply, at the level of the individual bus or train, is indivisible, and cannot be broken down into very small units to match the scale of demand. There is a fundamental mismatch between the type and scale of transport provision and the nature of demand. This chapter reviews these issues, primarily in the context of the UK, but also with reference to other advanced and less-developed countries.

The 'rural transport problem' in developed countries

The term 'rural transport problem' originated as the title of a book by David St John Thomas published in 1963, the same year that the Beeching Report threatened the emasculation of railway lines in rural Britain to add to the effects of early cutbacks in country bus services. Thenceforth, the term has been widely understood by rural residents and policy-makers alike and has remained strongly in the public consciousness in the regions affected. In advanced countries the normal experience is that service costs are relatively high and passenger numbers and revenue low. Economic operation has been found only (a) historically, where demand for public transport was much higher, (b) in specialist market sectors such as schools traffic, hires and excursions, and perhaps (c) with radical innovations in operating practices, economic structures or even suitable new modes of transport.

Traditional public transport in rural areas has always been extremely vulnerable to competition from modes better suited to the dispersed pattern of demand. Other things being equal, by far the most appropriate mode for rural environments is the private motor car. It does not suffer from the fixed linear routes and timetables of the bus or train; it is personal and individualist, convenient and flexible; and it can carry heavy loads. Moreover, it does not suffer the problems experienced by cars in urban areas; in the countryside there are no congestion or parking problems (except in some tourist areas in summer) and pollution is rarely an issue. Car-ownership rates have always been higher in rural areas, ever since the first cars were acquired in the early 1900s by the landed gentry to run about their country estates. The main reason is *need*, not wealth. Precisely the same geographical, 'rural' factors that make public transport so difficult make car ownership so necessary and attractive. The enduring 'rural transport problem' is suffered primarily by those who for various reasons have no means of exploiting the car's advantages, and hence are dependent upon public modes.

Rising car ownership inevitably triggered off the initial decline in bus services as the public-sector market contracted. Bus companies cannot compete with the long-term process towards greater car ownership and use, and their possible responses are limited. Reduction of network length or service levels, and fare increases, either reduce the market further or risk consumer resistance. This produces a permanently unstable situation, with each adjustment failing to halt a downward trend of services and passenger numbers. Feedback relationships reinforce this process, in that perception of inadequate public transport encourages more people to acquire cars and to use them more. Hence the frequently cited 'vicious circle' of decline (Figure 8.1). Another available response in many cases is to seek external subsidy, but to maintain a stable service level then means a continuous increase in claims. Essentially the same process occurs on the railways, with decline taking the form of frequency reductions and infrastructure deterioration. Eventually, the 'final solution' of closure is proposed and, once implemented, is irrevocable.

Developing countries

Despite the hazards of generalizing enormous areas with different cultures, environments and levels of development, it is evident that 'rural transport' in the Third World has an entirely different meaning to that in the West and is frequently equated with agricultural development. Rural areas in the developing world are likely to hold a higher proportion of the national population and to play a bigger role in the national economy, but they are often neglected by governments whose resources are far below those necessary for even basic infrastructural improvements. Unlike the West, the rural economy in developing countries is overwhelmingly an agricultural one; villages are occupational communities. The purpose of rural transport

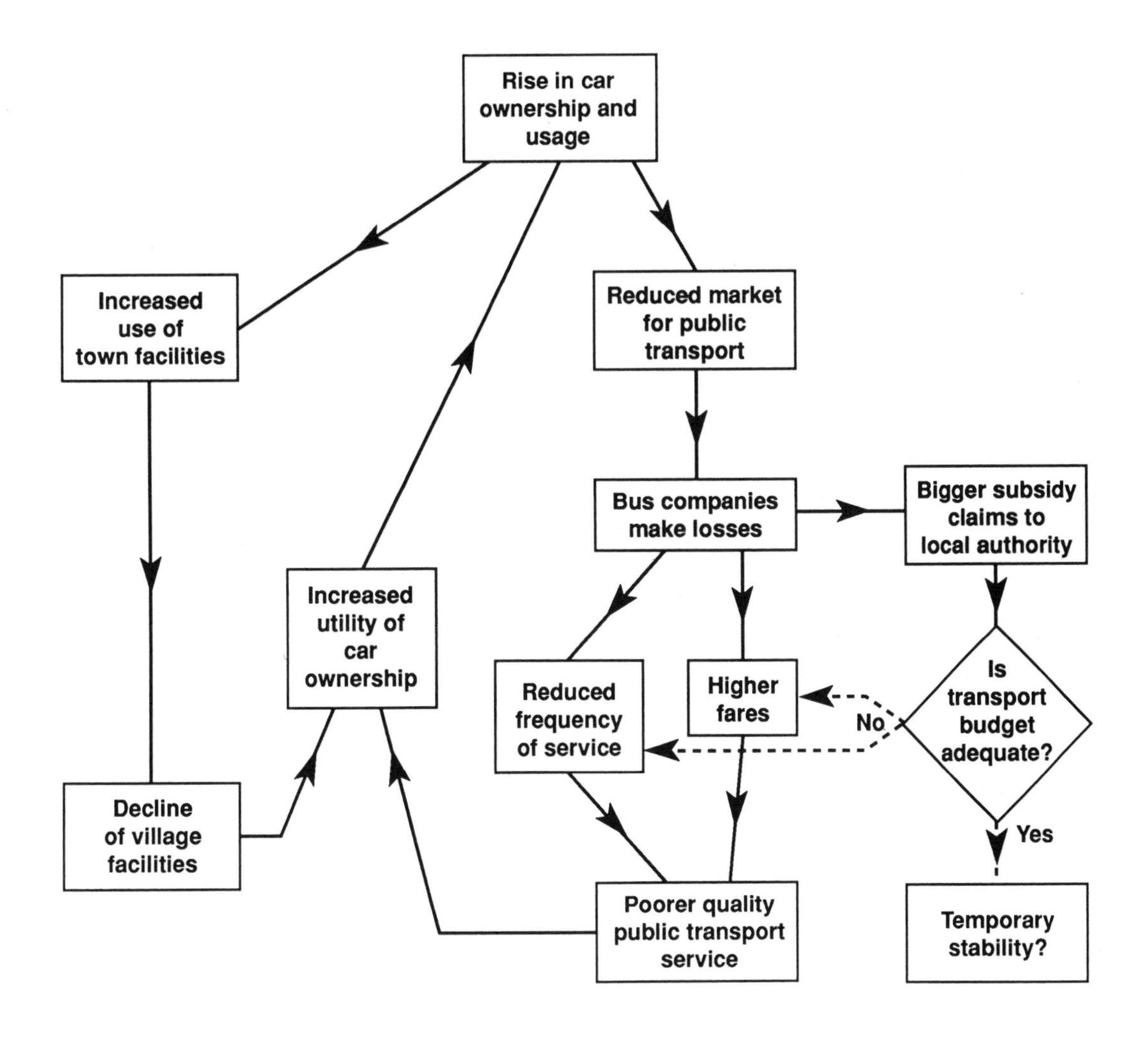

Figure 8.1 Rural transport 'vicious circles'.

is primarily to service agricultural demands, either at a subsistence level within villages or at a commercial level through local markets or regional export. Freight transport has a higher priority than passenger. Long journeys may be necessary for markets or basic commodities such as water and fuel, but otherwise travel patterns may be spatially restricted. The main issue here is not the provision of public passenger transport by the state or private sectors, but the development of appropriate technologies to reduce the transport cost of communal production and marketing.

Rural transport issues in the United Kingdom

These have attracted a great deal of attention: general works are Hillman et al. (1976), White (1978), Cresswell (1978), Moseley (1979), Halsall and Turton (1979) and Cloke (1985). Many local case-studies have been analysed in detail, e.g. Moseley et al. (1977), Nutley (1980a, 1983), Banister (1980), Stanley and Farrington (1981), Smith and Gant (1981), Kilvington and McKenzie (1985) and Moyes (1989).

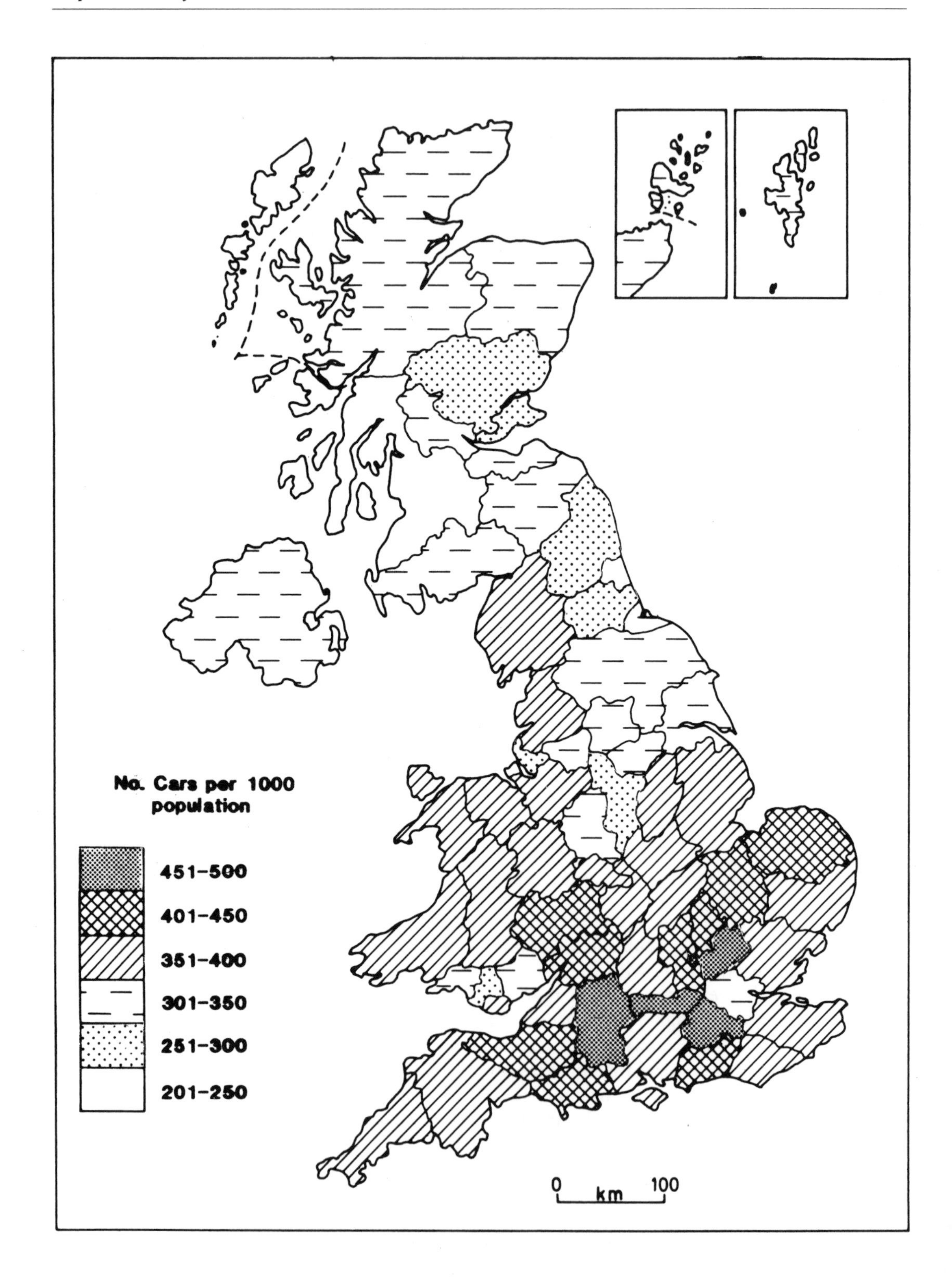

Figure 8.2 Car ownership, United Kingdom, 1989.

Car ownership

Car ownership began to increase dramatically in the early 1950s, with the ending of postwar restrictions and growth in the economy. For various reasons the costs of car purchase and use have risen by less than the overall cost of living, while public-transport fares have risen by more. Hence car acquisition has come within the means of successively lower income groups, and although a boon for rural people this has reinforced the public transport vicious circle of decline. Car ownership rates vary spatially and socially. The map pattern (Figure 8.2) displays an intriguing blend of two influences.

(a) *Rurality*. Shire counties and rural regions have significantly higher rates than cities, because of the dispersed location of facilities in the countryside, longer distances and the inadequacy of public transport. Many relatively poor rural families make great sacrifices to keep their car going.

(b) *Income*. Among the rural counties/regions there is a clear north-south transition. Areas with relatively more cars are both rural and affluent, with the highest rates in London's stockbroker belt.

Household car ownership rates in rural areas can be up to 80 per cent, but this may imply the whole family has constant access to the vehicle. Commonly, the main wage earner takes the car to work, leaving the rest of the family deprived during the day, during which time they may require a shopping journey and a school journey. Even two-car households may have some demand for public transport. Realistically, in rural areas 20–35 per cent of households and 45–65 per cent of adult individuals are still without a car. Government has belatedly recognized that the market for rural public transport is not just a 'declining minority', but there is an irreducible minimum population that will always need it. The young and the elderly may be unable to run a car because of age, infirmity and/or income; the disabled, those on low incomes or state benefits may be similarly deprived. Even wealthy retired people who have chosen to live in the countryside must accept that eventually they will have to give up their car and face an unfamiliar isolation. Some relevant data are given in Table 8.1. The carless are sometimes labelled 'transport-poor', suffering 'transport-induced deprivation' (Shaw, 1979; Nutley, 1980a), as their lack of mobility prohibits them from consuming other economic and social activities.

Bus issues

Nationally, the use of buses, measured by passenger journeys, has more than halved from its peak in 1955. It is rarely possible to find transport statistics relating to rural areas only, but here the decline was probably steeper. The end of postwar depression, rising car ownership and rural depopulation all contributed to the cycle of decline. Historic route maps show a remarkable density of services in earlier years, an abundance that had encouraged dependency and high expectations of public services among the rural population, now acutely aware of its loss. There is evidence (Nutley, 1982) that in the 1960s decline occurred mainly by route closures and in the 1970s by frequency reductions. Despite this, spatial coverage remained quite good, with no more than five per cent of the population of any rural region having no bus services at all (Nutley, 1980b). While the basic network remained intact, *frequencies* were the main concern. These normally conform to the route hierarchy, but in remote regions they may be consistently very low. A basic network of once-daily services prevails in the Scottish Highlands. Once a week services (usually market day) are not uncommon, e.g. in mid-Wales.

Beyond the more accessible lowland areas, Sunday services have virtually disappeared. Early morning and late evening workings have largely been cut out, and Saturdays (when the family car is usually available for shopping trips) may have thinner frequencies. Leisure trips by bus are extremely difficult. Monday-Friday peak services are

Table 8.1 Car ownership and driving-licence tenure

Urban/rural differences in car ownership (1985/6)

Number of cars in household	% of households Urban	Rural	% of people Urban	Rural
None	40	22	31	15
One	44	44	47	42
Two or more	16	34	21	42
Total	100	100	100	100

Household characteristics and car ownership (1985/6) [GB all areas]

Gross household income quarter	% hhs with car	cars/hh	Household structure	% hhs with car	cars/hh
Lowest	20	0.21	Single adult hhs:		
Second	56	0.63	Elderly	16	0.16
Third	80	0.99	Other	43	0.46
Highest	93	1.50	All single adult hhs	27	0.28
All	62	0.83	Hhs with 2+ adults	76	1.05
			All households	62	0.83

Driving-licence tenure (1985/6) [GB all areas]

Age	% M	F	All		%
17–19	31	25	28	Employers & managers	91
20–29	72	53	62	Professional	95
30–39	86	62	74	Intermediate non-manual	80
40–49	87	56	71	Junior non-manual	60
50–59	81	41	60	Skilled manual	76
60–69	72	24	46	Semi-skilled manual	47
70+	51	11	33	Unskilled manual	36
				Retired	38
All	74	41	57	Other (mainly inactive)	16
				All	57

Source: National Travel Survey, 1985/6 (Department of Transport, London, HMSO, 1988).

aimed at school rather than work journeys – rural commuters are overwhelmingly car owners. The former will most often support off-peak shopping services. A particular problem is making occasional but important journeys such as to hospitals. Compared with the national average, the rural bus market is dominated by education and shopping journey purposes (White, 1986). Apart from schoolchildren, the clientele consists predominantly of women and pensioners.

Since about 1960, however, the most persistent and intractable issue has been the *finance* of bus services in this unrewarding environment. The traditional expedient has been internal cross-subsidy, by which deficits on poorly patronized rural routes were compensated by profits from other sectors, such as town or inter-urban services, private hires and excursions, and school contracts. Over the course of time such profitable sectors have themselves been squeezed, putting further pressure on uneconomic routes unless external funding could be found. A vital source has been the local education authority in its payments to bus operators for the carriage of pupils to school. Especially in the remoter rural areas, school buses have a much denser network and carry many more passengers than public stage-carriage services. A bus and driver whose costs are covered by the school contract can easily be deployed on other duties in off-peak times (e.g. a shopping service) at marginal cost. The integration of public and schools traffic, by various means, offers major opportunities for cutting costs for both bus companies and local authorities, and has been implemented in most areas.

Despite this, external subsidy from public funds is normally regarded as indispensable to the survival of rural bus services. Demands were first made in the late 1950s. Subsidies ('revenue support') were introduced under the 1968 Transport Act, administered by local authorities for 'unremunerative' routes. Justification for this is the liberal 'social service' argument: it is unreasonable to expect rural services to pay their way yet local people depend on them for basic levels of mobility and access to facilities that are taken for granted in urban areas. An analogy is implied with health, education and the social services. The problem with subsidies is that once established they tend to get bigger. It must be emphasized, however, that despite the popular image of empty rural buses, their subsidy is lower per passenger than in the conurbations (White, 1986). Successive governments have tried to keep subsidies under control by new procedures, such as linking them to planning mechanisms under the Local Government Act 1972 and Transport Act 1978, and by deregulation under the Transport Act 1985.

There can be little doubt that the prime motives for deregulation were ideological and financial, with little commitment to upholding the level of service provided by the bus sector (see Chapter 3). Only one-quarter of the total pre-1985 subsidy was due to rural areas. Operators must either register as 'commercial' and run without subsidy or submit tenders to the local authority for financial support. Contrary to expectations, a high proportion of rural services were declared commercial and this allowed local authorities to finance remaining routes they thought to be important. Subsidy *has* therefore been significantly reduced. Network changes are highly variable and complex but, because of the lack of appropriate studies, it is uncertain whether the above comments on network integrity and spatial coverage still hold. Some rural areas have reported little change (Moyes, 1988) or a rough balance of gains and losses (Bell and Cloke, 1988; Forni et al., 1987). While there is much anecdotal evidence of service losses and personal hardship, this seems to make little impression on aggregate statistics (Cahm and Guiver, 1988).

Rail issues

Railways are more highly capitalized and less flexible than buses and are even more difficult to adapt to adverse circumstances. The current sparsity of rural lines is widely attributed to the Beeching Report of 1963 (Patmore, 1965). While the national railway network had remained at roughly its maximum extent throughout the period 1910–45, the passengers carried had declined steadily since 1920 and the system was plagued by financial crises both before and after nationalization in 1947. Beeching is not to blame for all rail closures and Patmore (1966) sees it as part of an historical process of innovation, competition and decline. The realization that 33 per cent of the mileage carried one per cent of the traffic brought home the effects of historical over-provision and inefficiency; the main culprits were rural 'stopping passenger trains'. By 1970 there was 30 per cent less route mileage and 57 per cent fewer stations than there had been in 1962. Some rural regions were decimated – particularly Wales (Figure 8.3), East Anglia, the South West and parts of the Midlands, although northern Scotland was largely reprieved. Despite enormous criticism at the time, it must be conceded that most of the lines closed were hopeless cases and major surgery was essential to bring the national railway system up-to-date. However, the motivation was overwhelmingly financial: no consideration was given to the social consequences on a line-by-line basis. Beeching's legacy is still apparent, not merely in terms of abandoned track and decaying structures, but also in the reduction in travel-mode choice of rural people and the narrowly

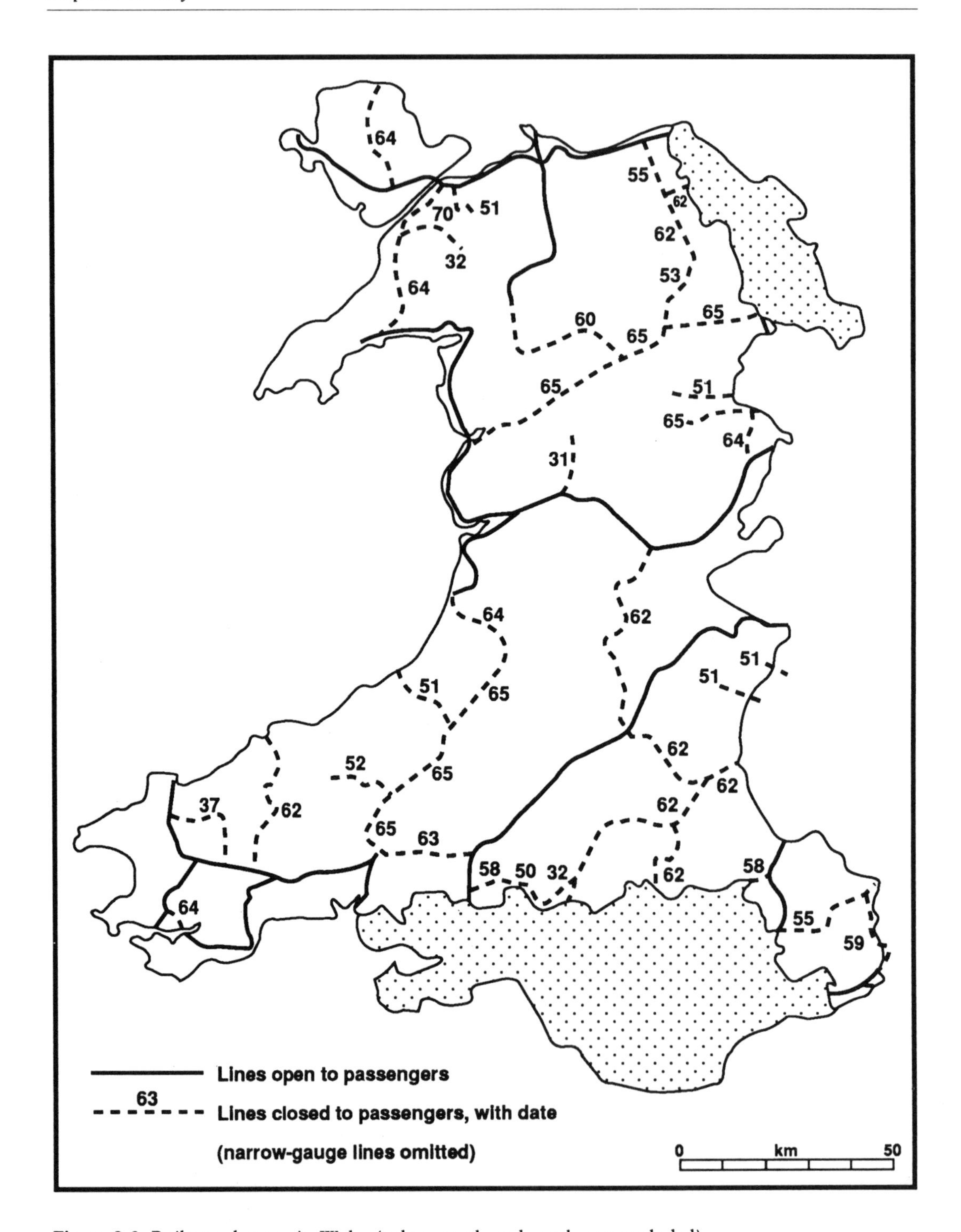

Figure 8.3 Railway closures in Wales (urban south and north-east excluded). Based on Nutley, 1982.

financial attitudes to further closure proposals.

There have in fact been very few line closures since 1970 and successive transport ministers have declared their opposition to substantial cuts in the network, but this has not inspired confidence. A persistent theme of the post-Beeching era has been a political climate of uncertainty which ensures that rural lines appear to be constantly under threat. The economic options are discussed by Kilvington (1985) and the social and political implications by Whitelegg (1987). Very briefly, the arguments against rural line closures are as follows:

(a) Closure causes serious hardship to former users. Finding alternative modes inadequate, they make fewer journeys in total. Some trips are stopped completely, others are made less often or diverted to other destinations. The carless, the elderly and the low paid suffer most (Hillman and Whalley, 1980).
(b) Substitute bus services have proved to be inadequate.
(c) The main network suffers disproportionately when a branch line is closed, as connecting passengers are lost to the main lines.
(d) Relevant issues could be assessed by a social cost-benefit analysis. These are sometimes done by academics (e.g. Richards, 1972), but there is no requirement to do this under the legal closure procedures and purely financial factors are paramount (Whitelegg, 1984).
(e) Marginal lines are deliberately run down by British Rail to a little-used and neglected condition – 'closure by stealth'.
(f) No attempts are made to invest in rural lines and actively promote them (e.g. for tourism).
(g) It was recently realized that very little money would be saved by closure.

A recent landmark was the reprieve of the Settle-Carlisle line in 1989, but this is unlikely to indicate a major change in official policy. Strategies for remaining lines depend on reducing costs. The 'low-cost rural railway' is based on simplification and basic minimum standards, such as single-line track, 'paytrains' with unstaffed stations and automatic signalling (Hodge, 1981). Associated with this is the opportunity to replace obsolete stock and infrastructure with simpler lightweight technologies. Ageing diesel-multiple-units have been partially replaced by 'railbuses', 'Pacers' and 'Sprinters' (Ford, 1986). Many new stations have been opened, although only a minority are truly rural. For an optimistic view, see Chapman (1988).

Rural transport issues in other countries

In continental Europe there is greater willingness than in the UK to provide state subsidies to rural transport. It is recognized that such services cannot be expected to run at a profit, and the basic 'welfare' argument prevails. Regional development, demographic and agricultural factors strongly influence rural support policies. Transport infrastructures are maintained as assets for future rail developments. National railway systems are politically favoured and marginal routes are much less likely to be closed than in the UK. Rail/bus integration is frequently impressive, with public and private sector bus companies enjoined to cooperate and not compete with railways. Rural bus systems are commonly run by the railways or the post office and often integrated with school transport. There is a general policy consensus about such matters in continental Europe, and transport is not the political football that it is in the UK.

Most countries of Western Europe have higher levels of state support to public transport than the UK. However, comparative statistics referring only to rural areas are impossible to obtain. As a crude indicator, ten out of 12 EEC countries have railway systems with a revenue/cost ratio of only 0.20 to 0.57, compared with 0.65 in Britain and Ireland (Whitelegg, 1988). Least subsidized in Europe are Switzerland (0.71) and Sweden (0.79). There are no suitable data for buses.

Scandinavia provides an excellent example of the conflict between service provision to low-density areas and modern economic realities (Fullerton, 1988). Across the region, a traditional concern for the maintenance of high standards of welfare and opportunity for all people, regardless of where they live, has compelled minimum standards for transport and other services, subsidized where necessary by the state. Strong environmental and regional lobbies, such as the imperative to stem remote area depopulation, have reinforced governments' willingness to support public infrastructure. The low density of population and long distances make this a high-cost policy. Road building, railways, buses, coastal shipping and ferries, and third-level air services have been state-funded with varying degrees of generosity. Such policies are contradicted, however, by the equally valued principle of consumer choice, which makes governments reluctant to protect public transport from road competition. Car ownership has grown to high levels. The public cost implications demanded a coherent policy.

Free competition obtains where traffic allows, as in road haulage, most coastal shipping and most air services, and intermodal competition exists on major routes and in less remote areas. Elsewhere, one approach is to exploit the complementarity of modes, such as road/ferry chains, although the trend in north and west Norway is to circumvent highly subsidized ferries by new roads. A general policy is to sub-divide large transport systems to achieve market segmentation and local accountability, such as Sweden's 'social railway'. Devolution of transport responsibilities to local authorities enables funding allocations to be spent on the basis of overall community benefit. In remote areas school- and post-buses, taxis, dial-a-rides, ferries and small scale air services are all vital. Strict approval is needed for service closures.

In the densely populated Netherlands, by contrast, the main policy objective is to ensure that rural inhabitants have access to urban jobs and facilities (Eckmann, 1981). Rigid regulation by the state is most unhelpful in a rural context where flexibility and local sensitivity are needed, but this is the situation in the Irish Republic: ironically, for one of the most rural countries in Europe. For historical-political reasons, the state-owned transport corporation, CIE, has a virtual monopoly of all bus and rail services, and operations are tightly regulated (Barrett, 1982). There is no scope for competition and very little for innovation: the only challenge comes from charter coach services of doubtful legality. Deregulation was proposed in 1989 (Barrett et al., 1990). Subsidies are not high by European standards, and there is no clear evidence whether or not accessibility is worse than elsewhere.

Public transport is dominant in Eastern Europe and the former USSR as low incomes, shortage of vehicles and official discouragement keep car ownership low. Unlike the West, car ownership is lower in rural areas than urban; it is increasing significantly in Eastern Europe but in rural areas of the former USSR it is still less than ten per cent by household. Local transport is provided by heavily subsidized state-run bus services, using old vehicles with high maintenance costs. In Eastern Europe buses provide a high level of service in the countryside and are well-used, but in the former USSR they are sparse, unreliable and severely deterred by inadequate rural roads. There is an extraordinary dependence on railways, which are under-capitalized and overloaded. The most urgent problem in the countryside is the lack of surfaced roads, which is a severe impediment to the agricultural economy – 25 per cent of farms in Russia have no roads (Crouch, 1985). People get lifts on trucks and tractors, and vehicles are frequently driven across fields. Analogies with the Third World are strongly suggested.

At the other extreme, it might be thought that rural accessibility is not a problem in affluent countries with high mobility. The US example shows that even with rural car-ownership rates of 85 per cent by household, there are considerable problems of mobility deprivation and lack of accessibility (Briggs, 1981). In 1980, 45 per cent of the rural elderly and 57 per cent of the rural poor had no car. Briggs and McKelvey (1975) identified the 'transportation disadvantaged' as social groups similar to those in the UK suffering in a similar way from local-facility and public-transport decline, in a society which takes for granted car-borne mobility. In the next decade the situation was unchanged: Rucker (1984) calculated a needs index showing higher scores for rural than urban America, higher scores for blacks and a deterioration between 1970 and 1980. A particular problem in North America is that bus services are limited to long-distance intercity routes, have been withdrawing from smaller towns and do not penetrate interstitial rural areas. Less than two per cent of towns under 50,000 population have regular scheduled public transport (Burkhardt, 1981). Since bus deregulation in 1982 this has got even worse (Kihl, 1988). Over 60 per cent of transit vehicles in non-metropolitan areas belong to 'human service agencies', i.e. welfare or voluntary sectors (Rucker, 1984). Under one per cent of the rural workforce use public transport. Railways, mainly privately owned, have suffered neglect and decline and are not orientated to local passenger movements. The subsidy of small-scale air services is also a valid issue.

Conditions similar to these are likely to be experienced in other affluent, low-density countries such as Australia (Lonsdale and Holmes, 1981), Canada and New Zealand. In developing countries, however, public transport is heavily used, although most often based on old, unreliable vehicles and infrastructure. In the better-off countries where rural roads are adequate, buses, taxis and trucks are able to provide passenger transport to much of the countryside. Government policies are usually preoccupied with urban areas, freight transport and ports; rural areas receive low priority. In the poorest countries the lack of roads over vast areas can act as a total block on any kind of development (Owen, 1987). Large populations are prohibited from entering the market economy and condemned to subsistence agriculture. Any kind of improvement requires access for delivery of fertilizers, seed, fuel, water-supply equipment, building materials and for export of even basic products such as milk. Provision of education and medical services are also deterred by the lack of passable roads. Non-agricultural employment in other places cannot be reached. In such circumstances goods are carried by headloading, pack animals or animal-drawn carts; sometimes bicycles are feasible. Whether measured in terms of human effort, energy or time, the costs are enormous and efficiency very low (McCall, 1985).

It is increasingly being realized that development agencies have been too concerned with big infrastructural projects, tarmac roads and motor transport. A focus instead on local rural roads and transport systems at a simpler 'appropriate' level of technology would be more likely to benefit the poor and also spread resources more widely (Howe and Richards, 1984; Barwell et al., 1985). Roads could then be upgraded in stages corresponding to the growth in traffic. It is also preferable to provide village access as part of an integrated development programme, which might also include school building, irrigation works or power supply. There is a desperate shortage of simple, robust trucks, but for non-motorable roads the best policy might be to develop more efficient bullock carts or combined freight/passenger wagons.

Policy and alternatives

Rural Britain – policy or the lack of it

In the UK there has never been any distinctive or coherent government policy exclusively for *rural* public transport. There has been no shortage of *national* policies, which are always strongly influenced by the urban and intercity sectors, and rural operations have to abide by the same system as the rest of the country. To compound this problem, national policies have fluctuated dramatically with every change of government, with the result that rural and urban systems have suffered from the absence of any long-term consistency, stability or economic security. Macro-policy has alternated between nationalization and privatization, regulation and deregulation, intervention and *laissez-faire*, planning/coordination and free market/competition (see Chapter 3). Perversely, it is by no means obvious which of these policies has been most successful, for rural consumers or for anyone else.

Some legislation does show recognition of the problem of uneconomic but socially important services. The 1978 Transport Act required local councils to relate subsidy to the fulfilment of social 'need' for transport. The 1985 Transport Act instituted a tendering procedure for 'non-commercial' services. But it seems the only truly *rural* provisions have been the setting up in 1977 of the Rural Transport Experiments (RUTEX) to encourage the development of 'unconventional modes' and 'community transport', and the creation, under the 1985 Act, of the Rural Transport Innovation Grant to assist new services of an 'innovative' nature. Despite these, it must be emphasized that there is little sign of relating policy to a scientific analysis of the problem.

Different approaches

This brief descriptive summary of the main trends and issues in rural transport has assumed the traditional interpretation that the problem is essentially *economic*. Until the mid-1970s discussion was largely confined to the economic problems of bus companies or branch lines. This is still the main preoccupation of law-makers and decision-makers, who are accustomed to identifying much more with service providers than with consumers. The 'problem' however is that of *rurality* and its implications; transport is but one component of a wider scenario.

In general, transport cannot be divorced from its socioeconomic and cultural setting, and it is the latter that generates the 'problem'. It is essential to recognize the influence of variation and change in rural regions. Demographic and social changes, such as the population 'turnaround', counterurbanization, commuter settlements and retirement homes, all affect mobility patterns and demand for transport in the countryside (Moseley, 1979). Other influences are the closure and centralization of local facilities (shops, doctors, etc.), rural industrialization and the growth in outdoor recreation. Great variation among rural regions must be acknowledged. A distinction is often made in Britain between 'accessible' and 'remote' zones (Cloke, 1979). The former are within commuting range of major cities, containing larger settlements and under severe pressure for building development, while the latter retain many elements of the traditional rural economy but with a weaker service infrastructure and a growing dependence on tourism.

As an example, the most distinctive rural region in the UK is the Scottish Highlands and Islands, where one quarter of the population lives on islands and hence sea and air communications are of vital importance. Here 'rural transport' means a four-mode situation instead of the usual two. Scotland is often compared with Norway, where topography is similar and distances even more extreme. The crucial difference is that in Norway, as in the whole of Scandinavia, transport is supported by the state as a vital element of a strong regional development policy. It is accepted as a state responsibility to ensure that remote populations enjoy the same service standards and opportunities as those in central zones (Fullerton, 1988). In Norway, Sweden and Finland subsidies are disbursed on a geographical basis, increasing with remoteness. Norwegian ferries are subsidized on the basis that tariffs should not exceed the cost of an equivalent road journey (Knowles, 1979). British governments have been unable to face the public cost implications of such policies, and reluctantly concede that ferries in Scotland have to be

Photograph 8.1 Traditional rural public transport. Taken in the late 1970s, a picture of public and private sector buses serving a small market town (Lampeter, Dyfed).

supported as a 'lifeline' to the islands.

Another vital concept is that transport operations and planning should not be 'demand-led' as at present, but 'needs-based'. Current attitudes see transport – whether private or public sector – as a commercial business with a duty to make a profit, not as a piece of social infrastructure to provide a service to the community.

Breaking down the problem

Several key concepts explain the true nature of the 'rural transport problem'. In particular, the concept of 'accessibility' has been developed and emphasized significantly from the mid-1970s. Although the concepts are general ones, they have succeeded in highlighting the distinctiveness of the rural situation.

It is essential to appreciate that the purpose of transport is to provide *accessibility*, or the ability to make a journey for a specific purpose. Transport is not consumed for its own sake, but is merely a means to an end (a derived demand). Hence residents in location A seek *access* to location B in order to acquire goods or services or partake in activities that are not available at A. If A and B are beyond walking distance apart, then transport is needed to overcome the distance barrier that separates them. Commonly, A is a rural village and B a country town. Compared with urban areas, a greater proportion of desired facilities (shops, doctors, work and leisure places, etc.) are likely to be at distant locations. Hence there is a *greater need for transport* in the countryside, in an environment where (except for car users) it is less likely to be available. The assumption is made that

Table 8.2 Aspects of mobility in rural areas

Trips/person/week (1978/9)

Mode	Urban areas (pop > 25000) households with cars	households without cars	Rural areas (pop < 3000) households with cars	households without cars
Car	12.7	1.7	13.5	2.4
Public transport	2.3	5.6	1.4	3.4
Walk	7.5	9.8	5.7	7.8
Other	0.8	0.8	1.2	1.6
Total	23.2	17.9	21.8	15.2

Journey purpose and mode – Cotswolds (1979)

% of trips

Mode	Work	Groceries	Clothing	Post Office	Chemist	Social
Walk/Cycle	29	24	2	56	27	54
Car	62	63	70	40	58	42
Bus	7	12	17	4	13	3
Other	2	1	1	–	2	1
Total	100	100	100	100	100	100

Journey purpose and mode – County Londonderry (1988)

% of trips

Mode	Sec. School	Groceries	Luxuries	Bank	Doctor	Visit Parents
Walk	–	35	–	3	13	30
Car (own)	–	48	75	79	58	64
Car (lift)	–	12	16	11	22	3
Bus (stage)	27	2	6	4	3	1
School Bus	73	–	–	–	–	–
Taxi	–	3	4	4	5	–
Other	–	1	–	–	–	1
Total	100	100	100	100	100	100

Sources: (top) *National Travel Survey 1978/9* (Department of Transport, 1983, HMSO, London); (middle) Smith and Gant, 1981; (bottom) author's unpublished research.

access to a basic minimum range of facilities/activities is economically and socially necessary for the pursuit of a normal way of life; rural dwellers are entitled to the same standards as their urban counterparts. The rural 'problem' therefore is *lack of accessibility* – that current levels of public transport are inadequate to provide these standards for the non-car population.

This interpretation is at variance with the viewpoint normally taken by government and transport operators which focuses on actual trip-making behaviour. *Mobility* is simply the ability to move around, without any particular destination. *Actual* mobility, measured in terms of traffic levels or trip rates (trips/person/week), has always been the dominant influence on UK government transport policies, demonstrated in a permanent bias towards roads and the private car. Selected data on rural mobility are given in Table 8.2. That a low proportion of total trips in rural areas is made by public transport, and that non-car owners have lower trip rates, are often seen by officialdom as evidence that there is little demand for bus or train services and hence they are not worth supporting.

Photograph 8.2 The fate of many rural railway lines. The old station building is now a national-park information centre (Hawes, North Yorkshire).

The crucial error in this reasoning is to assume that low patronage of rural transport services is an accurate expression of demand or need for them. The reality is that service provision has declined to such low levels that they now act as a severe constraint on the trip opportunities available to consumers. Many travel needs are frustrated because of the absence of suitable transport services (*suppressed* or *latent* demand). Merely to equate low bus patronage with low demand is to argue the problem out of existence – the whole point is that total demand exceeds transport supply.

A serious gap now exists between the degree of accessibility needed by the local population and that which can be provided under the prevailing land-use/transport system. This is the crux of the problem. While understandable in general terms, measurement requires a definition of *need*, i.e. which local facilities the population needs access to, and under what conditions, e.g. how often? A certain proportion of needs will be satisfiable by the facilities and transport in any area, but the residue of *unfulfilled needs* can be taken as a measure of the scale of the 'problem'. In this context, needs are usually assessed by normative methods, where a set of desired facilities and minimum conditions are specified externally. This has the advantage of consistency and allows comparisons among areas and population groups.

Planning for solving problems should therefore be *needs-based* and not demand-led. The only UK legislation to recognize this was the 1978 Transport Act, which required local authorities to define the needs of communities in their area and to compile a

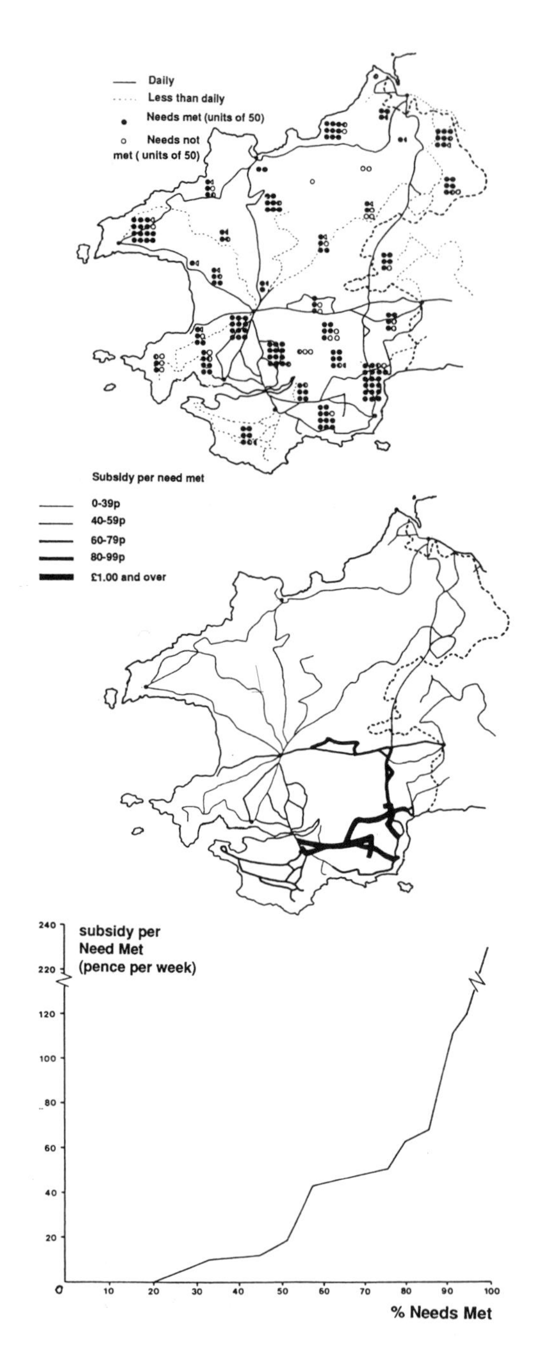

Figure 8.4 Public-transport needs and subsidy, West Dyfed, 1981. Top: Distribution of assessed needs. Middle: Bus network support costs. Bottom: Cost of meeting needs.
Based on Dyfed County Council, *Public transport plan, 1981–6.*

Public Transport Plan aimed at meeting those needs. This was a means of ensuring that subsidies were well-targeted. One county's attempt is illustrated in Figure 8.4, which also demonstrates the cost implications. One methodology for assessing needs is given by Moyes (1989), while techniques for estimating the proportion satisfiable by the transport system are explained below.

Travel behaviour in rural environments is a function of needs, under the influence of constraints. An alternative way of exposing likely problem areas is to map the constraints (Stanley and Farrington, 1981). (1) The social and economic characteristics of the population, such as income, car ownership and the proportion of elderly people are indicators of *potential* (as opposed to actual) mobility. (2) The numbers and relative location of local facilities constrain the length, cost and choice of possible journeys. (3) Public transport provision is qualified by its frequency, timing and fares levels.

Analysis is undertaken at the local scale, the level at which problems are experienced and policy is implemented. Traditional accessibility measures were based on places within a network or the number of facilities within certain time bands, which were less specific and more suited to the regional scale. Also, it is essential that the population is disaggregated into social groups such as the elderly, working people, housewives and schoolchildren, as each has different needs and personal circumstances.

It is possible to link the three constraints of population characteristics, facility distribution and public-transport services in order to test at a local scale whether a specific social group in a specific village has or has not adequate access to particular facilities. This is usually depicted as in Figure 8.5a (Moseley et al., 1977). No scale is attached to the 'link', which could be expressed as distance, frequency, time or cost. It is a situation of spatial supply and demand, but in a rural environment the question is whether the connection can be made between people and facilities. Testing this is an issue of time-space coordination. Transport routeing has to

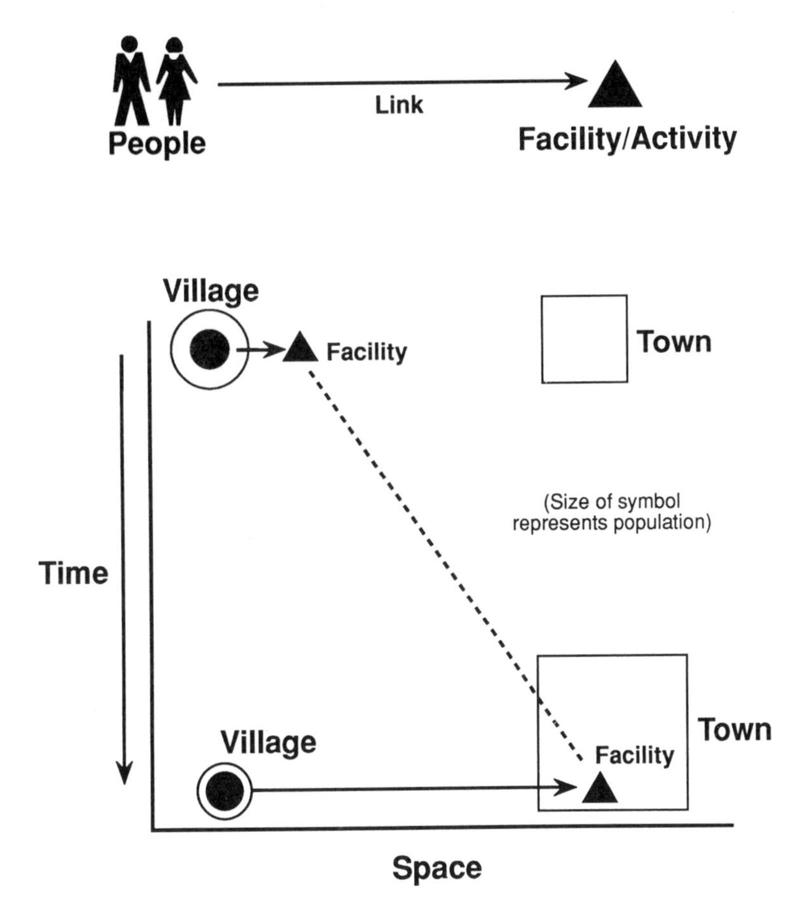

Figure 8.5a The basic components of accessibility.
Figure 8.5b Accessibility over time, assuming the decline of local facilities and rural depopulation.
Based on Nutley, 1983.

connect the locations, and its timing and frequency has to coincide with the times when the population is free to travel and the facility is available for consumption. Normative assumptions have to be made about time budgets, desirable frequencies and other conditions. Car owners can be dealt with under a simpler procedure. Accessibility can then be calculated for any area, village or social group, as the percentage of original needs able to be satisfied.

Figure 8.5b exemplifies the interdependency between facility location and transport. A widespread phenomenon is the 'decline of rural services' – the closure because of high unit costs of small shops, post offices, banks, etc., and the rationalization of doctors' practices, hospitals and schools into more distant larger units (Shaw, 1979; Phillips and Williams, 1984). This may or may not be associated with rural depopulation. The result is that one's nearest outlet becomes further away, at the same time as the ability to make that journey by public transport is being diminished.

In principle the concept of accessibility is equally valid, with appropriate modifications, in any culture and environment worldwide. The Moseley model has been adapted to urban areas, but the crucial difference is that in the latter more facilities are within walking distance and there is

Photograph 8.3 Mobile services may replace facilities absent from villages or shops previously closed. Butchers' vans are very rare (Carrbridge, Highland Region).

usually a choice of modes and routes. In rural areas the issue is much starker and the main question is whether facilities are accessible *at all*.

Applications and examples

The work of Moseley et al. (1977) in East Anglia has proved extremely influential and the local, personal scale and the time-space approach have been adapted by others for different circumstances. Applications can be categorized as follows:

(a) Regional scale — To apply at a non-local scale a simplified methodology is required. This was done at a parish level for the whole of rural Wales by Nutley (1980a). Various tests for accessibility were applied. The most severe result was that 14.5 per cent of the study area population had no public-transport access to employment centres. Results of all tests can be combined by a simple weighting scheme, to produce the map shown in Figure 8.6. The two lowest categories contained 8.8 per cent of the population. An advantage of the regional scale is that it identifies 'problem areas' for detailed analysis.

(b) Local scale — Access problems can be identified at the level of individual experience, for specific social groups, villages or destination facilities. Following the applications by Moseley et al. (1977) in Norfolk, similar techniques have been used by Nutley (1983) in Wales and by Kilvington and McKenzie (1985), again in Norfolk. Potential data output from such exercises is

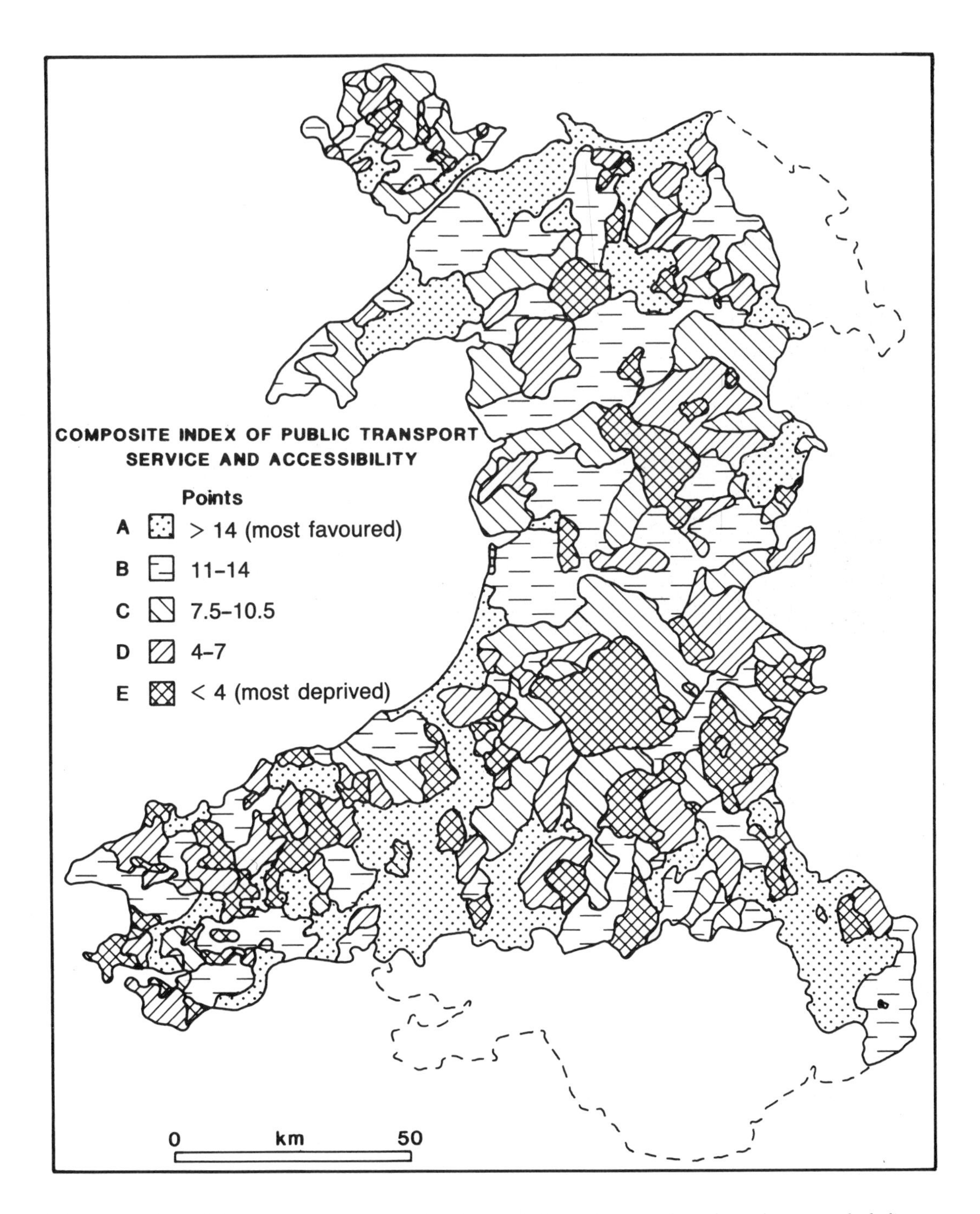

Figure 8.6 Accessibility at the regional scale, Wales, 1979 (urban south and north-east excluded). Based on Nutley, 1980a.

Photograph 8.4 One way of serving pockets of low demand is to use cars instead of buses. The 'Countrycar' is licensed as a public-service vehicle (Halstead, Essex).

enormous and cannot be summarized here: a selection from the three studies is given in Table 8.3. The social dimension is represented in Table 8.4 and Figure 8.7. While Moseley's work stressed the problems of the elderly, Nutley's case-studies also exposed problems for schoolchildren and working people without cars. A result of the time-space approach was to reveal the 'time-clash problem', whereby certain groups are denied access to facilities through being occupied elsewhere (usually work or school) when the facilities are open. This requires a 'non-transport solution' such as extended opening hours.

(c) Historic change — The method is easily extended to compare present access levels with those of the past, although for data reasons the earliest realistic starting date is 1950. Owens (1978) used a variety of data sources for a longitudinal study of two Norfolk villages, and found widening disparity in access levels among different social groups. Nutley (1983) compared accessibility in a part of Dyfed between the early 1950s and the late 1970s. Ironically, overall accessibility had actually *improved* over the period, from 51.6 to 59.1 per cent of total needs, due entirely to a four-fold increase in car ownership. But 'car' and 'non-car' groups, taken separately, both showed *declines* due to deteriorating local services. In both the studies cited, the closure of local facilities had a greater impact than cuts in public transport.

(d) Policy appraisal — By taking a 'before and after' approach, we can easily compare the access levels before and after the

Table 8.3 Accessibility to rural facilities: selected results from three studies (not strictly comparable due to different methods)

Facility, activity	*N. Walsham* (1975) % with access Working males	Retired	*Llandovery* (1977) % with access Car owners	Non-car owners	*Breckland* (1983) % parishes with access by foot or public transport	% pop. with access problem
Work – large town	74	n/r	100	0	38.8	0.4
Work – small town	78	n/r	100	66	38.8	0.4
Shopping – regional centre			93	55	69.9	0.1
Shopping – large town	74	30	100	73.5		
Shopping – small town	80	47	100	80.5	88.5	1.0
Local store	95	85	32.5	59	84.7	0.1
Doctor	83	59	32.5	34	31.2	6.8
Dentist	81	52	32.5	34	40.6	6.2
Chemist	76	56	32.5	34	57.4	1.5
Hospital – appointments			32.5	0	8.7	1.7
Hospital – visiting	73	30	32.5	0	33.3	2.3
Post Office	96	88	100	94		
Bank/Building Soc.	–	50	100	80.5	81.8	0.1
Employment Exchange/ Jobcentre	77	n/r	27.5	37	53.1	0.1
DHSS	–	29	32.5	47		
Council Offices	–	33	32.5	47	58.3	0.1
Leisure – regional centre			93	58	Leisure – evening	
Leisure – large town	73	28	100	0	7.8	5.1
Leisure – small town	78	45	100	0	Leisure – weekend	
Library	–	52	100	73.5	75.0	0.5
Pub	88	73	100	95		
After-school activities	n/r	n/r	n/r	64	18.0	2.1
Village hall			76	66.5		
Evening classes	75	30				

n/r = not relevant

Sources: N. Walsham, 1975, from Moseley et al., 1977; Llandovery, 1977, from Nutley, 1983; Breckland, 1983, from Kilvington and McKenzie, 1985.

Table 8.4 Rural accessibility by social group

Social group	*Accessibility by social group – rural Wales (1977–81)* % total needs that are accessible (lowest–highest, of six case study areas) Car owners	Non-car owners	Total popn.
Working males	62–65	20.5–33	49.5–57
Working females	62–65	20.5–33	43–50
Inactive (housewives et al.)	96.5–100	48–61.5	71–79
Elderly	75–91	29.5–51	40–59.5
Schoolchildren	n/r	29.5–41	29.5–41
Total popn.	72–74.5	32–46	48.5–59

n/r = not relevant

Source: Nutley, 1983.

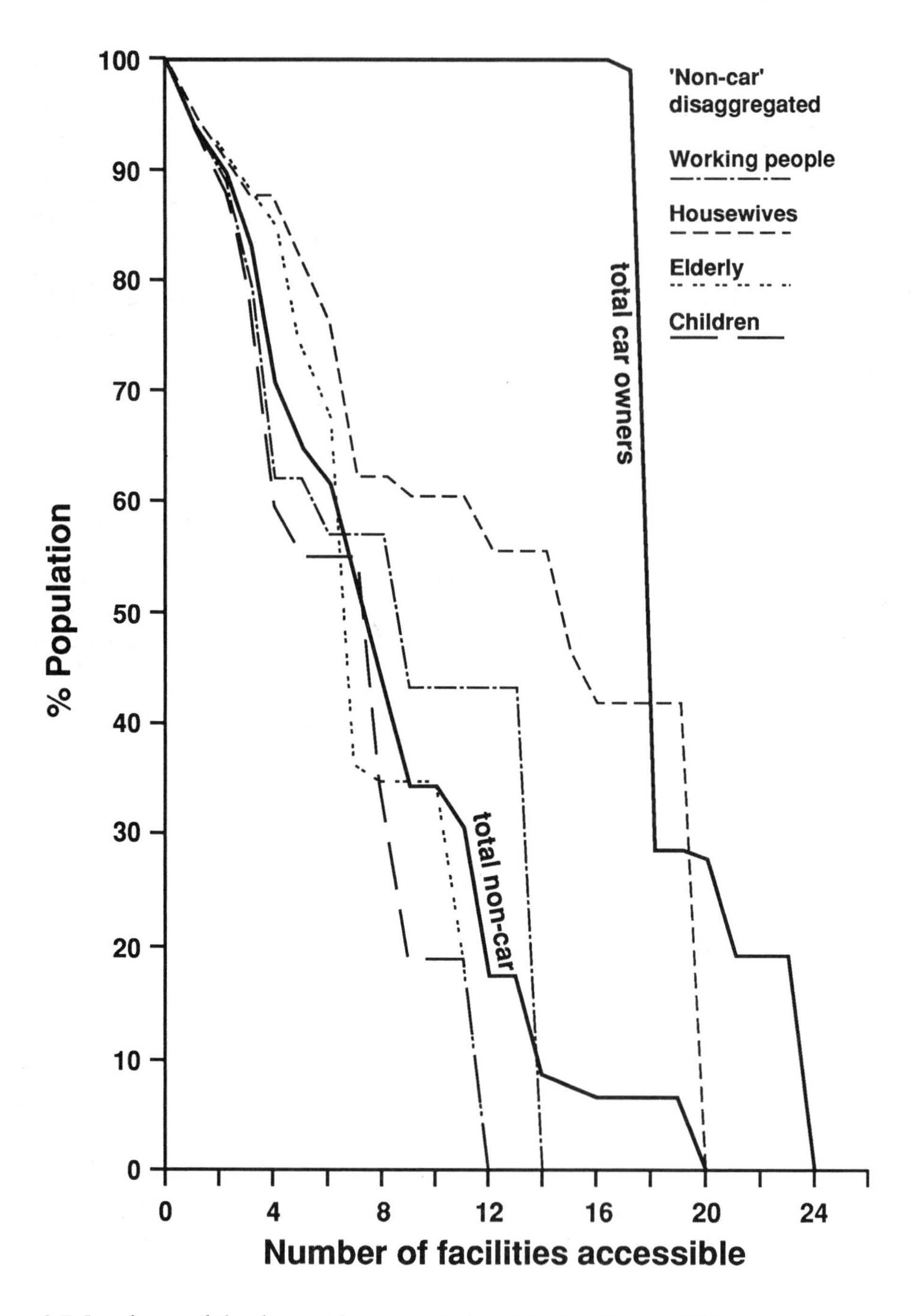

Figure 8.7 Local accessibility by social groups, Radnor District, Powys, 1981. Based on Nutley, 1985.

Photograph 8.5 An unconventional mode. The postbus waits for passengers outside the railway station (Llandrindod Wells, Powys).

introduction of any specific policy and hence deduce the benefits of that policy in accessibility terms. We can also discover which groups of people benefit the most and which the least. Moseley et al. (1977) assessed the impact of the Norfolk Structure Plan. Nutley (1983) evaluated the effects of the 1977–9 Rural Transport Experiments in part of Dyfed. Here the innovations – postbuses, social car schemes and school buses – could be separately analysed and costed. This provides a new form of cost-benefit analysis more appropriate to rural areas. While the access gains produced by RUTEX appeared relatively low, they were significantly greater for some groups of people. The social car scheme was the most cost-effective innovation.

(e) Planning — The same techniques can be applied to hypothetical future states of the system. Examining a part of Powys with very low non-car accessibility, Nutley (1985) designed 40 alternative planning options for raising standards. These included more conventional bus services, various 'unconventional modes' (postbuses, social car schemes, community bus, etc.), and 'non-transport solutions' (more village facilities, mobile services, extended opening hours). Accessibility benefits and costs were calculated for each. The general conclusions were that simply increasing bus-service levels brings diminishing returns, while some unconventional modes are promising if modest gains are acceptable. The best is probably a basic service by conventional buses, with a volunteer-run demand-responsive mode to fill in the gaps.

Techniques of this kind provide a tool for

evaluation and planning which recognizes the true nature of the problem, measures the social impact of policy and provides a realistic benefit/cost assessment. Unfortunately, by 1990 the political climate was such that any commitment to meeting social needs had largely evaporated.

Future scenarios

Continuing the wider view of the rural accessibility problem, which sees transport as but a component of a spatial population/land-use system, we can now present a general framework for communications in geographically extensive, low-density environments. The classification advanced by Moseley (1979) has become well-known, in which the alternatives are described as (a) the 'transport option', where people travel to facilities; (b) the 'mobile services option', where goods and services are brought to the people; (c) the 'fragmented service option', where facilities are small-scale and dispersed among the population; and (d) the 'key village option', where the population is concentrated into more economic units. To these might be added a 'telecommunications option'.

The transport option

This has long been considered the 'normal' response in developed countries, almost to the complete exclusion of anything else. Future progress here seems to depend on *either* a radical reform of the institutional, legal and economic system governing the public-transport sector in order to break the cycle of decline, *or* the introduction of new modes of transport, through technological or operational change, that are better suited to rural demand patterns. It is still impossible to say whether deregulation has succeeded in fulfilling the first of these. An urgent need is for local accessibility analyses to be done, comparing the situations before and after deregulation.

'New modes', however, are available. Given that conventional public transport is severely hampered in rural areas by its rigid mode of operation, fixed routes and timetables and large vehicles, then greater success might be achieved by developing more flexible modal types. 'Unconventional modes' (UCMs) have been emerging steadily and unobtrusively since the late 1960s, and now constitute a valid policy option, albeit for specialized niches in the market. Unconventional features are: flexible routeing and timing (perhaps demand-responsive); multipurpose operation, by combining different types of business on the same vehicle; management and/or operation by local volunteer labour, and the use of private vehicles; restricted eligibility to specific client groups and/or destinations; alternative sources of funding; and the use of new design and technology. Modes using volunteer labour are also known as 'community transport' (CT), an extension of traditional charity or welfare services. Figure 8.8 classifies 18 types, although this list is not definitive. Operational details of each type and examples of services are described in Nutley (1988, 1990).

What is 'unconventional' in one country may be normal in another, and international comparisons can be instructive. There are many UCM systems in North America and the Third World (known as 'paratransit'), but these are mainly urban. In many countries of Western Europe the rural transport network is based on postbuses (Holding, 1983). The Netherlands has a well-developed system of community buses (Banister, 1982). The West could look to the Third World for promising examples of appropriate technologies.

UCMs and CT have been encouraged in Britain by the government-sponsored RUTEX programme in 1977–9 (Transport and Road Research Laboratory, 1980) and a sequence of permissive legislation between 1977 and 1985. Further impetus has come from public-expenditure cuts forcing local authorities to seek 'low-cost solutions', and the growth of the 'self-help' movement

MODE \ UNCONVENTIONAL FEATURES		Restricted eligibility	Flexible routeing & timing (or D/R)	Multi-purpose transport (vehicles or operations)	Community management and/or operation	Alternative finance	New technology
	D/R diversion (Flexibus)	-	✓	-	-	-	-
	Multiple service bus	-	♦	✓	-	-	-
	Contract bus	-	-	-	♦	✓	-
	Subscription bus	-	-	-	✓	✓	-
	Free shoppers bus	-	-	-	-	✓	-
	School bus	-	-	✓	-	-	-
	Postbus	-	-	✓	-	-	-
CT	Community bus	-	✓	✓	✓	♦	-
	PSV dial-a-ride	-	✓	-	-	-	♦
CT	'Welfare' dial-a-bus'	✓	✓	-	♦	♦	♦
CT	Social car	♦	✓	-	✓	♦	-
CT	Hospital car	✓	✓	-	♦	♦	-
CT	Lift-giving scheme	-	✓	-	✓	♦	-
CT	Car pooling	-	✓	-	✓	♦	-
	Shared taxi/hire car	-	✓	♦	-	-	-
	Passenger/freight service (Courier)	-	♦	✓	-	-	-
	Demountable vehicle	-	-	✓	-	-	♦
	Railbus	-	-	-	-	-	✓

CT 'Community transport' mode
✓ Primary unconventional feature
♦ Secondary unconventional feature or important in some cases only

Figure 8.8 A classification of unconventional modes.
Based on Nutley, 1988.

Photograph 8.6 An unconventional mode that also ranks as 'community transport'. A volunteer-run community bus at Docking, Norfolk.

encouraging community action (Wilmers et al., 1981). At first sight, the UCM/CT sector has a specialized role to play, but it is possible that a large proportion of conventional buses could adopt 'unconventional' characteristics for the mutual benefit of bus companies and rural consumers (Nutley, 1990).

The mobile services option

'Bringing the service to the people' is a marginal mode in the contemporary UK, mainly associated with mobile shops and libraries in some areas. Lower order delivery services such as mail, milk, bread and meals-on-wheels are widespread but often overlooked. Others, such as mobile banks and playbuses, are rare. A detailed survey of such services has been done by Moseley and Packman (1983). These are the descendants of the 'village carrier', which until the early twentieth century serviced villages on a peripatetic basis with a variety of goods from a donkey and cart. Peripatetic services are also common in the Third World. It is difficult to conceive of a revival of such modes in developed nations. Nutley's (1985) comparative accessibility study found them not very cost-effective. While investment in a mobile mode might provide at most two or three facilities, the same investment in a bus service would give access to all the facilities of a town.

The fragmented services option

Again, this harks back to an earlier era when

rural communities were much more self-sufficient in local services, and also has Third World analogies. A dispersed pattern of small-scale facilities runs counter to all contemporary trends towards rationalization and centralization, for both public and private sectors. With the partial exception of self-help initiatives (Wilmers et al., 1981), there seems to be no mechanism for reversing current trends. Like other 'non-transport solutions' (mobiles, extended opening hours), the additional accessibility is unlikely to justify the costs, in comparison with subsidizing a bus service (Nutley, 1985).

The key village option

Another means of reducing the need to travel is to concentrate population into larger settlements where basic facilities could be more efficiently located. 'Key villages' have been a common planning strategy since the 1950s, in the belief that it would eventually produce a more economic settlement pattern (Cloke, 1979). Such policies, however, have been strongly criticized on the grounds that 'non-key' settlements are discriminated against and would experience greater isolation. Nevertheless it is true that most planning authorities have adopted discriminatory strategies for rural settlement – calling them key villages or growth points, tiers or hierarchies (Cloke and Shaw, 1983). Village rationalization and resettlement schemes have been applied in various parts of the Third World and the USSR, although transport is not the main criterion. Theoretically there is a range of alternatives from complete dispersal of population and facilities through to high concentration (Cloke, 1979), but each has transport-cost implications. Even with consistent political support, settlement concentration can only be achieved in the long term. The main problem is how to accomplish the transition from the present to the future state while continuing to maintain accessibility.

The telecommunications option

Advanced telecommunications have the potential to replace journeys for social, information and financial purposes, and in some cases allow working from home. The technical possibilities have been discussed spasmodically since the 1960s. Telephone use is dominated by the better-off and mobile; it may actually generate *more* trips (Clark and Unwin, 1981). Computer-based information systems could be useful in rural communities, although there would be practical problems of finance, implementation and social acceptability. However, the apparent success of electronic and satellite-based information systems for educational purposes in rural India and Malaysia shows what can be done. 'Telecommuting' from the countryside is another possibility, but despite some publicized exceptions (*Independent*, 1990) appears to be insignificant. This is a field where technological potential is far ahead of practical reality.

Conclusions

Rural society in Britain is steadily polarizing. The population 'turnaround' in all rural areas, including the remotest, since 1975 reflects the freedom of choice of affluent car-owning groups – commuters, second-home owners, the newly retired. It has *not*, however, led to any revival of rural facilities or public transport. Recent social and economic trends can only further undermine the opportunities available to the disadvantaged non-car population. Car ownership will never be universal. The vulnerable, public-transport dependent groups will always exist. The problem will not go away.

The above options are all *spatial* solutions, recognizing the essentially geographical nature of the accessibility gap. Traditional attempts at 'solutions' aim at economic, legal or institutional change. An approach along the lines suggested above needs a commitment to the concepts of need and accessibility – and a wider vision, able to integrate the numerous policy measures necessary to make it happen. With the partial exception of the 1978 Transport Act, there has never been any sign of this in the UK at national level. Only the 'transport option' seems realistic in the foreseeable future, but even here, after deregulation there is only *laissez-faire* with a modest safety net for uneconomic (many rural) services. The only other new element is the unconventional and voluntary sectors, but there is no clear guidance on their role.

A major impediment is the long-standing habit of seeing problems and making policy in sectoral terms, with 'transport' distinct from land-use planning, employment and housing, etc. It is increasingly being argued that policy should be made on regional or environmental, rather than sectoral, grounds (Blunden and Curry, 1988). The basic problem is not a transport one but a rural one. In this respect other parts of the world such as Scandinavia and even many developing countries are ahead of the UK. Sectoral thinking together with the free market produces an arbitrary and damaging division between the affluent and mobile and the isolated underprivileged. An integrated 'rural policy' is the best solution.

Acknowledgements

I am grateful for the assistance given through discussions with Tony Moyes of UCW Aberystwyth and Dr Colin Thomas of my own department, and also to Nigel McDowell and Kilian McDaid for photography and illustrations.

References

Banister, D. (1980), 'Transport mobility in inter-urban areas: a case-study approach in South Oxfordshire', *Regional Studies* 14, 285–96.

Banister, D. (1982), 'Community transport for rural areas: panacea or palliative?', *Built Environment* 8, 184–9.

Barrett, S. (1982), *Transport policy in Ireland* (Dublin: Irish Management Institute).

Barrett, S., Keenan, M. et al. (1990), 'Irish Republic special report', *Transport* 11(4), 103–15.

Barwell, I., Edmonds, G., Howe, J. and De Veen, J. (1985), *Rural transport in developing countries* (London: Intermediate Technology Publications).

Bell, P. and Cloke, P. (1988), 'Bus deregulation in the Powys/Clwyd study area: an interim report', *Contractor Report 75* (Crowthorne: Transport and Road Research Laboratory).

Blunden, J. and Curry, N. (1988), *A future for our countryside* (Oxford: Blackwell), ch. 8.

Briggs, R. (1981), 'Federal policy in the US and the transportation problems of low-density areas', *Settlement systems in sparsely populated regions: the United States and Australia*, ed. R. Lonsdale and J. Holmes (New York: Pergamon), ch. 11.

Briggs, R. and McKelvey, D. (1975), 'Rural public transportation and the disadvantaged', *Antipode* 7(3), 31–6.

Burkhardt, J. (1981), 'Rise and fall of rural public transportation', *Transportation Research Record* 831, 2–5.

Cahm, C. and Guiver, J. (1988), *A passenger view of bus deregulation* (London: Buswatch, National Consumer Council).

Chapman, R. (1988), 'Rural rides back on the right lines', *Geographical Magazine* 60(5), 44–9.

Clark, D. and Unwin, K. (1981), 'Telecommunications and travel: potential impact in rural areas', *Regional Studies* 15, 47–56.

Cloke, P. (1979), *Key settlements in rural areas* (London: Methuen).

Cloke, P. (ed.) (1985), *Rural accessibility and mobility* (Lampeter: Institute of British Geographers Rural Geography Study Group).

Cloke, P. and Edwards, G. (1986), 'Rurality in England and Wales 1981: replication of the 1971 index', *Regional Studies* 20, 289–306.

Cloke, P. and Shaw, D. (1983), 'Rural settlement policies in structure plans', *Town Planning Review* 54, 338–54.

Cresswell, R. (ed.) (1978), *Rural transport and country planning* (Glasgow: Leonard Hill).

Crouch, M. (1985), 'Road transport and the Soviet economy', *Soviet and East European transport problems*, ed. J. Ambler, D. Shaw and L. Symons (London: Croom Helm), ch. 6.

Eckmann, A. (1981), 'Rural passenger transportation in The Netherlands', *Transportation Research Record* 831, 69–73.

Ford, R. (1986), 'The DMU replacement story', *Modern Railways* 43, 315–20.

Forni, J., Cross, A. and Oxley, P. (1987), 'Monitoring the effect of the Transport Act

1985 in the western Wiltshire area', *Contractor Report 67* (Crowthorne: Transport and Road Research Laboratory).
Fullerton, B. (1988), 'Scandinavia adopts the new realism in transport policy', *Department of Geography Research Series* 15 (Newcastle upon Tyne: The University).
Halsall, D. and Turton, B. (eds) (1979), *Rural transport problems in Britain: papers and discussion* (Keele: Institute of British Geographers Transport Geography Study Group).
Hillman, M., Henderson, I. and Whalley, A. (1976), *Transport realities and planning policy*, Report 567 (London: Political and Economic Planning).
Hillman, M. and Whalley, A. (1980), *The social consequences of rail closures*, Report 587 (London: Policy Studies Institute).
Hodge, P. (1981), 'Low-cost rural railways and the BR-Leyland railbus', *Tenth Annual Seminar on Rural Public Transport*, ed. P. White (London: Polytechnic of Central London), 7–20.
Holding, D. (1983), 'The Swiss postal bus system: a comparison with British rural bus operation', *Eleventh Annual Seminar on Rural Public Transport*, ed. P. White (London: Polytechnic of Central London), 44–74.
Howe, J. and Richards, P. (1984), *Rural roads and poverty alleviation* (London: Intermediate Technology Publications).
Independent, The (1990), 'Joys of a city job with a lochside view', 9 June.
Kihl, M. (1988), 'The impacts of deregulation on passenger transportation in small towns', *Transportation Quarterly* 42, 243–68.
Kilvington, R. (1985), 'Railways in rural areas', *International railway economics*, ed. K. Button and D. Pitfield (Aldershot: Gower), ch. 11.
Kilvington, R. and McKenzie, R. (1985), 'A technique for assessing accessibility problems in rural areas', *Contractor Report 11* (Crowthorne: Transport and Road Research Laboratory).
Knowles, R. (1979), 'Road Equivalent Tariffs: the Scandinavian experience and the Scottish prospects', *Rural transport problems in Britain: papers and discussion*, ed. D. Halsall and B. Turton (Keele: Institute of British Geographers Transport Geography Study Group), 113–26.
Lonsdale, R. and Holmes, J. (1981), *Settlement systems in sparsely populated regions: the United States and Australia* (New York: Pergamon).
McCall, M. (1985), 'Accessibility and mobility in peasant agriculture in tropical Africa', *Rural accessibility and mobility*, ed. P. Cloke (Lampeter: Institute of British Geographers Rural Geography Study Group), 42–63.
Moseley, M. (1979), *Accessibility: the rural challenge* (London: Methuen).
Moseley, M., Harman, R., Coles, O. and Spencer, M. (1977), *Rural transport and accessibility* (Norwich: University of East Anglia).
Moseley, M. and Packman, J. (1983), *Mobile services in rural areas* (Norwich: University of East Anglia).
Moyes, A. (1988), 'Travellers' tales in rural Wales', *Geographical Magazine* 60(6), 40–1.
Moyes, A. (1989), 'The need for public transport in mid-Wales: normative approaches and their implications', *Rural Surveys Research Unit Monograph 2* (Aberystwyth: University College of Wales).
Nutley, S. (1980a), 'Accessibility, mobility and transport-related welfare: the case of rural Wales', *Geoforum* 11, 335–52.
Nutley, S. (1980b), 'Rural public transport coverage: problems of measurement and inter-regional comparisons', *Ninth Annual Seminar on Rural Public Transport*, ed. P. White (London: Polytechnic of Central London) 61–70.
Nutley, S. (1982), 'The extent of public transport decline in rural Wales', *Cambria* 9, 27–48.
Nutley, S. (1983), *Transport policy appraisal and personal accessibility in rural Wales* (Norwich: Geo Books).
Nutley, S. (1985), 'Planning options for the improvement of rural accessibility: use of the time-space approach', *Regional Studies* 19, 37–50.
Nutley, S. (1988), '"Unconventional modes" of transport in rural Britain: progress to 1985', *Journal of Rural Studies* 4, 73–86.
Nutley, S. (1990), *Unconventional and community transport in the United Kingdom* (London: Gordon & Breach).
Owen, W. (1987), *Transportation and world development* (London: Hutchinson).
Owens, S. (1978), 'Changing accessibility in two North Norfolk villages, Trunch and Southrepps, from the 1950s to the 1970s', *Social issues in rural Norfolk*, ed. M. Moseley (Norwich: University of East Anglia), ch. 2.
Patmore, J. A. (1965), 'The British railway network in the Beeching era', *Economic Geography* 41, 71–81.
Patmore, J. A. (1966), 'The contraction of the network of railway passenger services in England and Wales 1836–1962', *Transactions of the Institute of British Geographers* 38, 105–18.

Phillips, D. and Williams, A. (1984), *Rural Britain, a social geography* (Oxford: Blackwell), chs. 6, 8.

Richards, K. (1972), 'The economics of the Cambrian Coast Line', *Journal of Transport Economics and Policy* 6, 308–20.

Rucker, G. (1984), 'Public transportation: another gap in rural America', *Transportation Quarterly* 38, 419–32.

Shaw, J. M. (ed.) (1979), *Rural deprivation and planning* (Norwich: Geo Books).

Smith, J. and Gant, R. (1981), 'Transport provision and rural change: a case-study from the Cotswolds', *The spirit and purpose of transport geography*, ed. J. Whitelegg (Lancaster: Institute of British Geographers Transport Geography Study Group), 97–114.

Stanley, P. and Farrington, J. (1981), 'The need for rural public transport: a constraints-based case-study', *Tijdschrift voor Economische en Sociale Geografie* 72, 62–80.

Thomas, D. St J. (1963), *The rural transport problem* (London: Routledge).

Transport and Road Research Laboratory (1980), 'The Rural Transport Experiments', *Supplementary Report 584* (Crowthorne: Transport and Road Research Laboratory).

White, P. (ed.) (1978), *Rural public transport. A selection of papers from Rural Public Transport Seminars held at the Polytechnic of Central London 1972–76 with bibliography* (London: Polytechnic of Central London).

White, P. (1986), *Public transport: its planning, management and operation* (London: Hutchinson), ch. 8.

Whitelegg, J. (1984), 'Closure of the Settle-Carlisle railway line. The case for a social cost-benefit analysis', *Land Use Policy* 1, 283–98.

Whitelegg, J. (1987), 'Rural railways and disinvestment in rural areas', *Regional Studies* 21, 55–63.

Whitelegg, J. (1988), *Transport policy in the EEC* (London: Routledge).

Wilmers, P. et al. (1981), 'Self-help', *The Planner* 67, 59–74.

9

Transport for Tourism and Recreation

David Halsall

Considerable changes in mobility have profound consequences upon recreational opportunities. Transport is an integral part of much recreational and intra-national tourist behaviour. Approaches to recreational travel patterns are reviewed through the consideration of trends of use of differing transport modes at varying scales. The attraction of a transport heritage is assessed. Examples are drawn from contrasting modes and locations.

The relationship between transport, leisure, recreation and tourism

Patmore (1983, 5–6) identifies three basic contexts of leisure: of *time* not required for work or basic functions such as eating and sleeping, of *activities*, or *recreation* within leisure time, and of an *attitude of mind* based upon a perception of pleasure and enjoyment, recognizing that there may be blurring within and between these areas. While leisure time has become more of a routine based upon the growing availability of daily, weekly and annual periods of non-work time, this time is not entirely leisure 'since especially for women, "leisure" means "work"', (Urry, 1990, 154) or for the retired, leisure may be the remainder of their lives. Recreation is dependent upon a mix of social and economic variables operating within different environments.

Tourism is also an activity within leisure time, 'one manifestation of how work and leisure are organized as separate and regulated spheres of social practice in "modern" societies', (Urry, 1990, 2) and recreation is an intrinsic part of much tourism: 'recreational environments . . . are important tourist destinations' (Smith and Godbey, 1991, 97). While leisure, recreation and tourism are each difficult to define precisely, their interaction should form the basis of concepts and research (Fedler, 1987). Many recreational and associated transport resources cater for local groups as well as visitors, although in developing countries infrastructure and accommodation often represent investment for tourists with leisure contrasting with a local population lacking similar opportunities. Overall, tourists are distinguished by their journeys to places to *stay* for temporary leisure periods away from their work and residential locations. International journeys by tourists are considered in Chapters 12 and 13.

The growth of overall mobility levels has considerable impact upon the use of leisure time:

> Transport is an integral part of much recreational behaviour, both as an aid to access to recreational opportunities, and as a recreational activity in its own right . . . Progressive reductions in the relative costs of travel, and in the frictional effects

of distance have dramatically increased the demand for recreational trips. In particular, the growth of car ownership has extended both the distances travelled, and the range of recreational foci. (Halsall, 1982a, ii)

Studies of access and mobility to recreational opportunities

Access is a crucial element in potential conflict between recreation use and resource management. Access includes two key elements: legal and social variables including the laws of trespass and the implications of sale of land such as the privatization of Forestry Commission land in Britain, and physical aspects including the maintenance of the infrastructure and people's mobility. Geographical approaches to the study, measurement and analysis of access have developed from Ullman's bases of spatial interaction (1954/1974): *complementarity* between areas of demand and supply which influences the development of transport routes, *intervening opportunity*, or alternative source/location of supply, and *distance*, which constrains interaction through temporal and financial costs. 'The factor of *intervening opportunity* results in a *substitution of areas* and the factor of *distance* results in a *substitution of products*' (Ullman, 1974, 33). These attributes relate appropriately to recreational opportunities. The provision of recreational resources within urban fringe belts provides alternatives between inner-city locations and recreational sites further afield (Bryant et al., 1982, 134–6). The establishment of British 'Country Parks' from 1966 aimed to increase outdoor leisure opportunities for urban populations without undue increases in traffic congestion, journey length and damage to the countryside, and to reduce stress upon remote areas. Recreational trip-making has generated a vast quantity of research. Elson (1979) suggests four emphases of approach linking travel problems to the spatial distribution and management of recreational facilities and decisions to solve these distribution problems:

1 Spatial interaction — includes four broad types of model: trip generation (concerned with the number of trips generated from zones of origin); trip attraction (the ability of each possible destination to attract trips); distribution (where the trips go); and assignment models (which trips are chosen) (Elson, 1979, 6). Relationships between an area's land-use and its socioeconomic characteristics, its patterns of movement (trip generation) and the pattern and distribution of trips within and beyond the area is considered through the use of several types of mathematical model and process: gravity models, models of intervening and competing opportunities, electrostatic field models, multiple linear regression analysis and linear programming (Bruton, 1985). Thompson (1979) uses such models to explain why campers in Ontario use certain parks: the popularity of Canadian Shield sites, site characteristics, size and relative location, together with the size of the origin city, appear to be key variables. Smith (1983, 121–51) and Baxter and Ewing (1981) provide further general reviews. In the UK, distance decay is important, but travel time may influence longer journeys more than short trips. Pigram (1983, 31) confirms that 'the strength of interaction declines as distance increases . . . this means that recreation sites at a greater distance, or for which the journey is perceived as involving more time, effort or cost, are patronised less . . . for some people and some occasions, e.g. ocean cruises, travel becomes so stimulating as an integral part of the recreational experience, that the further the distance, the greater the desire to prolong it'.

Luxury rail trips such as the 'Royal Canadian' operated by the Trans-Canadian Railroad Co. Ltd. from Toronto to Vancouver (4,300 km.) demonstrate this theme. In the UK, luxury trains are operated over shorter distances. Charter trains include the privately owned 'Royal Scotsman' and 'Orient Express'. British Rail Inter City Charter operated 22 'Luxury Days Out' and 27 'Land Cruises' in the summer of 1990. The latter provide a high quality of service for a three-day journey including three nights accommodation, full board, courier service and routes passing through magnificent scenery in Scotland and south-west England.

Huff (1960) suggests that distance is modified by the socioeconomic profiles and tastes of consumers, their preferences for differing commodities and ability to travel. Trips and their length are a response to environmental stimuli filtered by socioeconomic and perceptual characteristics of the individual. Furthermore, many journeys are often looped or circular: 'the journey itself is part of the whole leisure experience' (Duffield, 1975, 30). Senior, too, (1979) asserts that gravity models require adjustment towards socioeconomic variables and the relationships between group and individual interaction. Bruton (1985, 165) warns that analysis of trip distribution remains a dubious process, a view shared by Elson (1979) with reference to recreational trip behaviour.

2 Constraints upon recreational trip making: time budgets — Recreational trip-making is affected by a variety of socioeconomic constraints which may be viewed through *time budgets*, personal allocations of 'uncommitted' time; in simple terms, that time outside basic functions such as eating and sleeping, family and social ties, work, education and commuting. In practice leisure time is more complex, overlapping other functions. These ideas are founded within studies of human-activity systems with constraints upon individuals' uncommitted time varying with age, gender, work and family status. Income affects the use but not the quantity of discretionary time. Individual use of outdoor recreational time can be viewed at diurnal, weekend, weekly and annual scales within time budgets, including the need for travel time: constraints to outdoor pursuits include seasonal change and weather, exemplified by visitors to East Lothian coastal sites in Scotland, 1968–71 (Coppock and Duffield, 1975, 8–23).

Time

The distribution of periods of time is a crucial element of recreational trip-making. Patmore (1971, 70) comments that the 'journey to play' lacks 'the studied monotony and regularity of the daily journey to work, but in weekly or seasonal rhythms it can reach even greater intensity and cause even more widespread congestion'. The southbound carriageway of the M6 motorway between Lancaster and the M61 junction south of Preston bears testimony to this during late Sunday afternoon periods. Recreation is characterized by often intense use of facilities at particular places during particular, generally brief, time periods, especially extreme in the summer season. In urban areas, seasonal variations are less than in rural and coastal areas, but traffic flows are still greater in summer months, especially during weekends (Daniels and Warnes, 1980). Increasing overlaps between recreation and other functions, especially going out for a meal and shopping (Martin and Mason, 1987), and demonstrated by leisure functions and the idea of 'a day out shopping' in Gateshead Metro Centre, Sheffield Meadowhall Centre and other out-of-town shopping centres in the UK, do not detract from the importance of the evening for recreational pursuits and travel on weekdays in towns. In the UK this specifically timed need for travel is not always met by public transport services. Moseley (1979, 61) demonstrates the paucity of bus services returning from urban entertainment centres to rural areas of Derbyshire after 10.00pm and the continuing inability of bus operators to achieve viability in evening services in the face of rising costs and increasing levels of car ownership. Deregulation of bus services in the UK has resulted in further reductions in such off-peak services. Temporal variations are also reflected strongly in public transport provision in areas of strong seasonal recreational/tourist traffic. Figure 9.1 shows the enhancement of summer Saturday railway passenger services in North Wales in 1939. Improvements to the infrastructure and

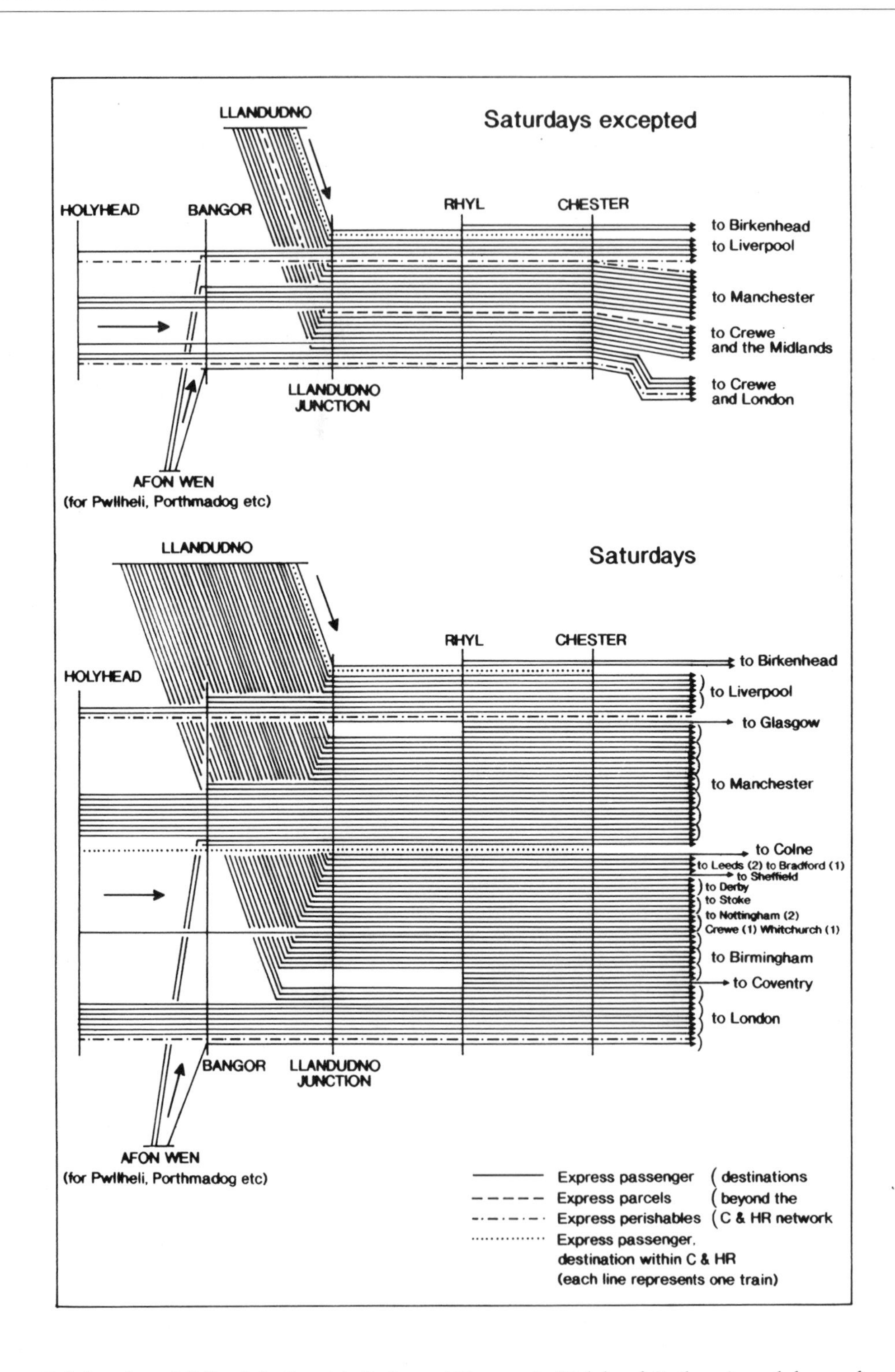

Figure 9.1 London, Midland & Scottish Railway (Chester & Holyhead Railway) weekday and Saturday west–east express train services, August 1939.

Source: LMSR timetables, 1939.

technical development of vehicles can much reduce journey time and thus make more distant sites feasible possibilities, although many roads suffer short duration peak congestion of heavy recreational traffic, but at other times show little justification for investment in increased capacity.

Opportunities and constraints

Moseley (1979, 67) adopts Hägerstrand's development of time-space budgets: 'measuring *opportunities* rather than predicted behaviour'. The individual is affected by two sets of constraints: those concerning him/herself – time constraints and the inability to be in more than one place at a time – and those concerning his/her environment, including the technology and means of transport and characteristics of the activities to which access is required, especially 'location and temporal availability' (Moseley, 1979, 69). These may be broadly classified as capability constraints – socioeconomic variables which affect behavioural disposition (e.g. age, income and nature of household) – and coupling constraints (how individual trip-making fits in with other members of the household or family group).

Those most constrained, and as a result, mobility deprived, include women lacking access to cars, children and the elderly. The constraints approach does not offer full explanation, but draws attention to factors which negate full choice, to the importance of time-space relationships, and focuses upon accessibility, drawing attention to mobility-deprived sub-groups of the population (Hillman et al., 1976; Moseley, 1979).

Gender and age

Women's and men's daily and weekly time budgets are unequal in uncommitted time available for leisure. Even on Sundays, 'housewives reduce their workload by only 30 per cent' compared with men 'whose work duties account for only a third of their average weekday workload' (Coppock and Duffield, 1975, 14–16). The dualism of women's role in modern Western society, as both waged and domestic workers, intensifies this situation. Moseley (1979) assesses the difficulties of accessibility for female rural dwellers, especially housewives, based upon time-space realms. Women are travel disadvantaged, 'more reliant than men on public transport to meet their travel requirements' which are in turn 'increasing as their dual role develops' (Pickup, 1988, 98, 100). Mobility differences are accentuated by patterns of men's ownership and usage of the 'family' car, and intrinsically linked to the gender role through time-based, spatial and transport-availability constraints.

The gender role constraint is especially marked where women have young children. Such women are restricted to 'leisure' pursuits such as playground/park visits with their children which are themselves part of the domestic, child-care role; meeting other mothers in these locations and visiting friends and relations are important social/leisure activities 'because of the relative inability of such women to take part in other outside activities' (Tivers, 1988, 93). Women's inequality in leisure is also demonstrated by working-class women's leisure opportunities in Armley, Leeds, constrained by 'lack of time, money, resources, of family commitments and expectations' (Dixey, 1988, 121). Leisure opportunities at home such as TV and video further blur the distinction between 'work' and 'leisure' time, and the decentralization of services away from such inner-city areas further disadvantage these women. Lack of mobility is a major constraint; most journeys are made by walking. Thus leisure activities are local, community based – playing 'bingo' is often 'the most frequent leisure activity outside the home, apart from going to the pub' (Dixey, 1988, 125). Constraints on women's leisure activities require further research: 'with few exceptions feminist academics in Britain have been slow to recognize the importance of inequality in leisure and its relation to other areas of women's oppression' (Dixey, 1988, 117).

The elderly are similarly disadvantaged. Old people walk most journeys and thus 'the spread of out of town leisure centres and hypermarkets . . . accessible only to car travellers further restricts their access' (Robson, 1982, 267). Ageing emphasizes class and gender differences in car-ownership patterns, and declining health and personal capability, decreasing income with retirement and the effects of widowhood contribute to reduced mobility, a situation aggravated by declining public-transport provision (Smith and Gant, 1982). Despite the importance of recreation in an individual's quality of life, recreational journeys are less vital than those for shopping, medical and other necessary services; and given the elderly's different use of time, often with increasing frailty, recreational travel may be curtailed, particularly so for those in residential care (Peace, 1982). Many elderly people, however, do spend much time in recreational activities outside the home; class and gender are important determinants - for example, many women remain full-time housewives - and transport operators provide varying levels of reduced fares, supported in certain areas by local authorities' concessionary rates (Skelton, 1982).

Further work is also needed on the plight of the disabled who also suffer from inaccessible transportation and failures of national and international policies to combat fully these problems, thus maintaining obstacles to local and distant leisure facilities (Smith, 1987).

3 Cognition, behaviour and spatial choice — Elson (1979, 25) suggests that concentrating on constraints assumes that people are 'passive rather than purposeful' and that a decision-making approach is preferable. This is measured through cognitive maps, representations of individuals' action spaces determined by a set of stimuli, the characteristics of the individual and the intention of the response. Often knowledge of recreational opportunities is restricted to the local residential area and narrow areas alongside routeways. It is affected, too, by factors such as an individual's values, length of residence and inclination to spatial search. Many recreational journeys, and especially stops at sites, are decided on impulse. The underlying problems are those of the validity and reliability of cognitive maps, compounded by many recreational trips resulting from recommendations rather than from direct experience and knowledge.

Smith (1983, 73–101) analyses the links between push and pull forces - socioeconomic and psychological - which influence motivation for recreational trips, thereby affecting recreational action space, a view supported by Mercer (1971, 66). Clawson's model of the recreational experience demonstrates the integral nature of recreational travel in the decision-making processes of recreational activity (Pigram, 1983, 30–3) (Figure 9.2).

4 Trip-making, leisure and the family — Individual decisions are important foci, especially where aggregated data mask differences of gender, age and so on. However, they should be viewed within their social context: leisure trips link with the family situation as well as work and social networks. Elson (1979, 34–5) suggests that the family is a 'conditioning, learning and decision-making environment'; countryside trips in particular are a family activity. Moreover, life-cycle stages may underlie varying interests and activities through time, and thus people's needs for leisure - the desire to be free from daily chores and responsibilities, to relax, to 'get away from it all' - can be developed in this context. Mercer (1971) suggests a developmental sequence in which a child's initial learning of leisure activities and preferences underpins changing patterns of recreational behaviour through adulthood. Elson (1979, 39) advocates an holistic approach.

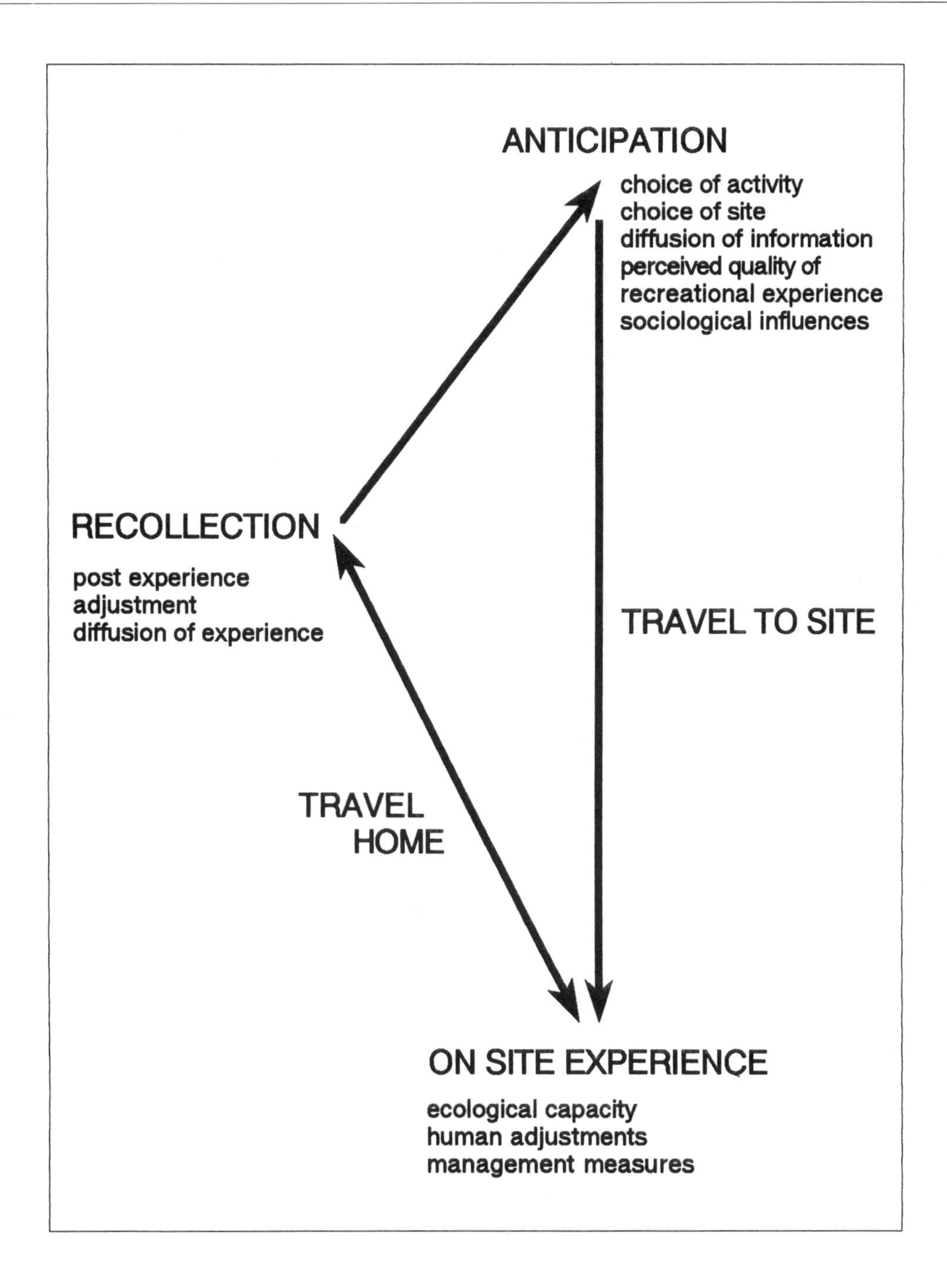

Figure 9.2 The Clawson 5-phase model of the recreation experience.
Source: Pigram, 1983, 22.

Changes in demand, supply and transport mode

An historical basis

Historical geographical studies are an important focus of study for their own sake, and for understanding the links between the legacy of investment in inherited infrastructure and related activities which much affect contemporary possibilities in transport planning and in recreational activities; in a special sense, the heritage element is of particular significance in recreation. Links between transport change and recreation are illuminated by Patmore (1970, 21–42; 1983, 30–53) in his discussions of the evolution of the demand for recreation. Two case-studies demonstrate some major aspects.

North Wales illustrates the developing relationship between increasing access, changing social and economic opportunities and the growth of the seasonal tourist industry in the UK. The popularity of Wales, especially Snowdonia, grew from the 1770s within the context of a fashionable cult of mountains, at a time when the area's isolation, lack of roads and of hotels made travel exceedingly difficult and the Napoleonic wars made continental travel impossible. In 1825, Miss Weeton (Figure 9.3), explored Snowdonia mainly by foot, with some coach and ferry travel after arrival by coastal steamer from Liverpool (Hall, 1969).

The railways consolidated the demand developed by the steam boat, expanding the popularity of seaside holidays at the growing coastal resorts and aiding access to the mountains. By 1850, rail travel was an accepted part of tourism. The range of railway tourist tickets, hotel lists and guides grew; circular road and rail tours operated jointly by local railway companies and coach operators covered a variety of routes in Snowdonia (Figure 9.4), all of which remain as important tourist flows by road in the late twentieth century. By the late nineteenth century there was a clearly discernible difference between augmented traffic flows on the Chester and Holyhead Railway in summer, especially Saturdays, and those of the other seasons (Figure 9.2). This persists to the present day, although reduced in magnitude because of competition from cars and coaches.

In contrast, Egypt's tourist industry shows more modern development (Gamble, 1989). Despite the threat of political instability in the Middle East, its heritage of well-preserved monuments of ancient cultures, particularly the Egyptian remains of the Nile valley and biblical associations of Sinai, together with the developing sun-based resort holidays, diving in the Red Sea and dramatic mountain scenery in Sinai are foundations of developing tourism (Figure 9.5). Heritage elements attract Europeans and Americans; Arabs appreciate the resort facilities. Egypt's 'cultural tourism' is based upon the Nile valley, competing with 90 per cent of the country's population and attendant congested land-uses. Transport for tourism is primarily along the Nile, with short trips on the traditional sailing boats (*feluccas*), and longer excursions (up to 14 nights) in luxurious floating hotels/motor cruisers. By 1991 there were over 140 cruise boats on the Nile (compared with seven in 1977 and 82 in 1986), embarking from Aswan, Luxor or Cairo. Over 7,000 cruise departures per year cover from about 225 km. to 950 km. each, increasing pressure on berths and river congestion in addition to other commercial traffic such as the Sudanese Railway steamer service from Aswan to Wadi Haifa. Legislation to allow 'tax concessions, duty-free imports and the repatriation of profits' (Gamble, 1989, 30) have encouraged foreign investment in the tourist transport infrastructure. American investment (by Sheraton and Hilton hotel groups, for example) in Nile cruises, and Belgian (*Wagons-Lit*) investment in the Cairo-Aswan railway sleeping-car service demonstrate Western confidence in Egypt's tourist industry.

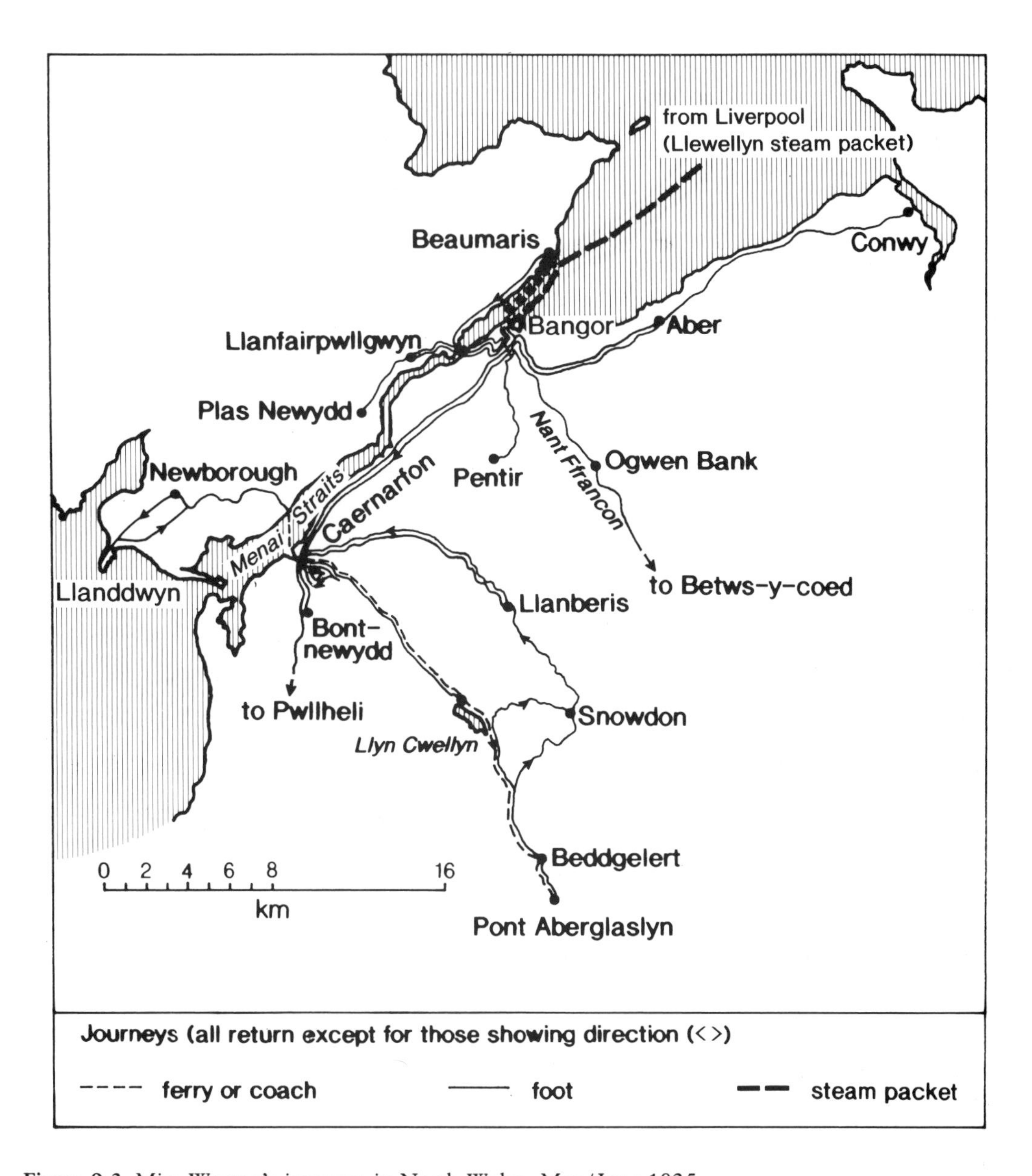

Figure 9.3 Miss Weeton's journeys in North Wales, May/June 1825.

Modal change and services: land and water

Differing transport modes and routes serve recreational demands and affect them with varying implications. Older modes have suffered considerable contraction and/or become recreational/heritage attractions in their own right, while modern services include attributes aimed specifically at recreational traffic. In Britain, the primary purpose of many canals is as routeways for pleasure traffic; certain nodes preserve and/or adapt canalside industry and warehouses as at Wigan Pier (Leeds and Liverpool Canal) as heritage sites.

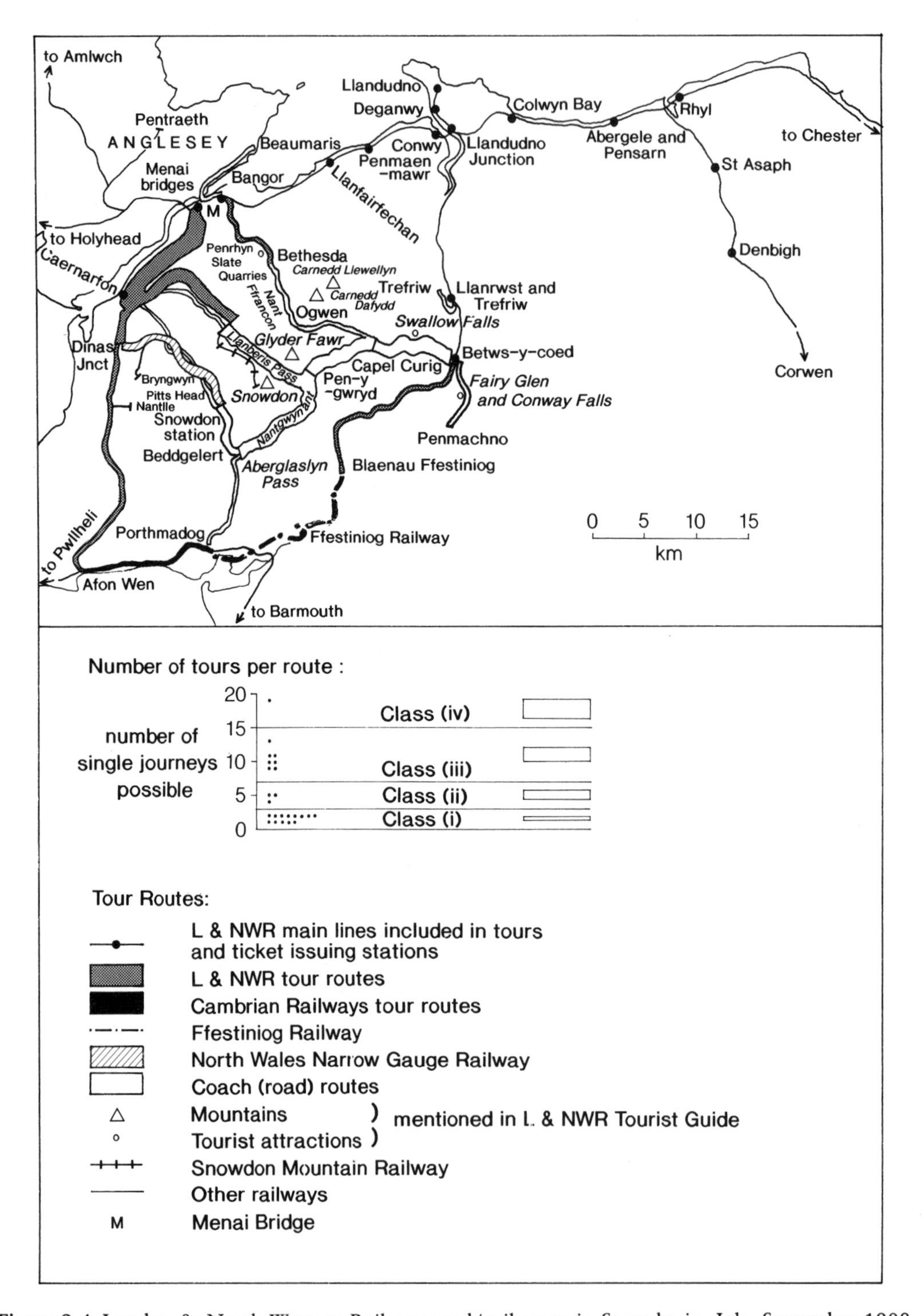

Figure 9.4 London & North Western Railway road/rail tours in Snowdonia, July–September 1908. *Source:* LNWR Tourist Guide, 1908.

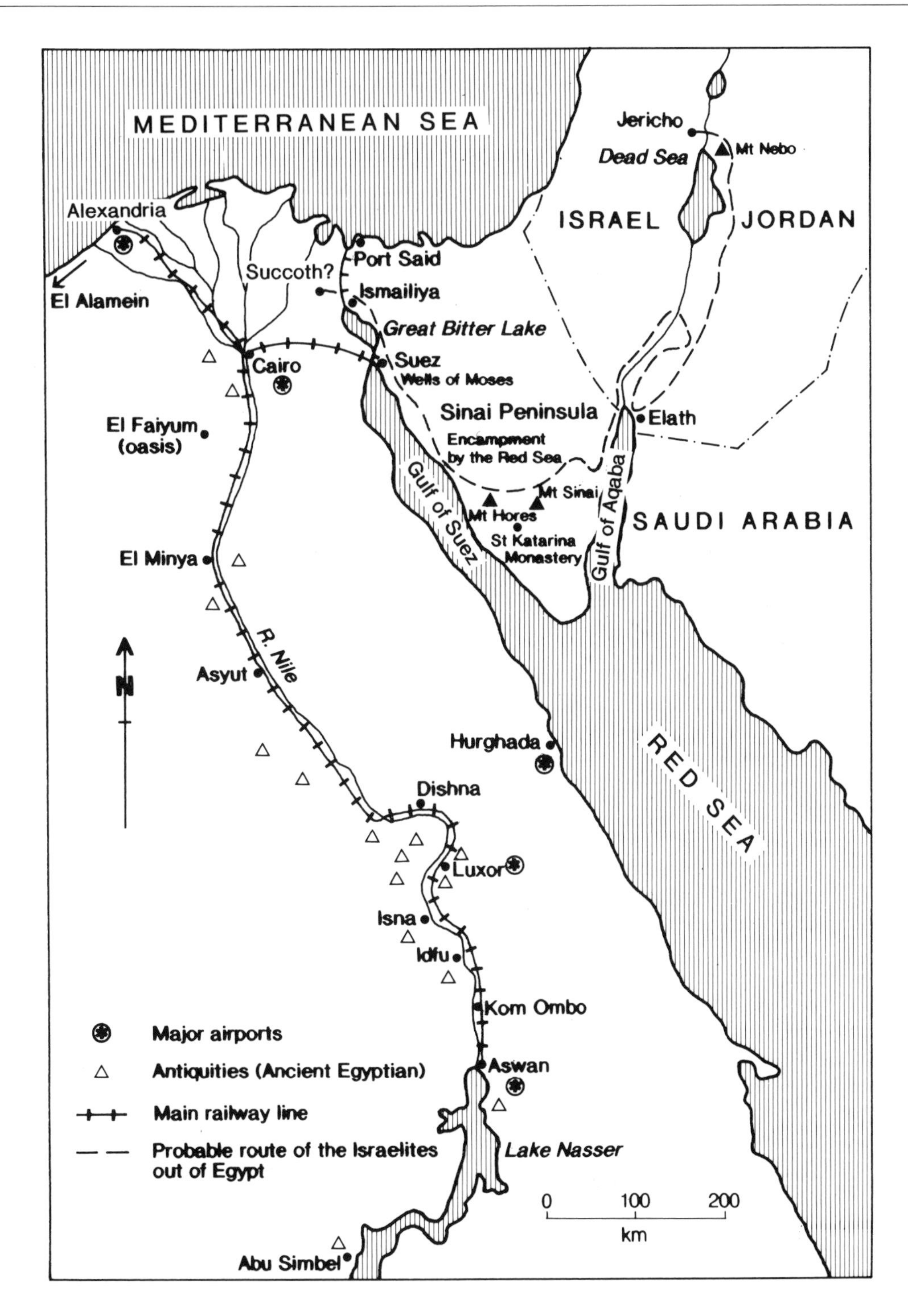

Figure 9.5 The cultural heritage of Egypt.
Based on Gamble, 1989.

The British railway network has lost about 13,600 km. of passenger line through closure. New uses are linear (preserving the form of the routeway) or non-linear in form (Appleton, 1970; White, 1986). Recreational uses for the former include footpaths (such as the Potteries Loopline Greenway; Tissington Trail; Wirral Country Park), bridle and cycleways (HMSO, 1982) or as private railways (Grimshaw, 1976; Turnock, 1982, 214–27; Patmore, 1983, 50–1). Preserved steam railways are a growth area in many countries. British examples include the well-marketed steam railway, electric trams and horse trams on the Isle of Man, and the Welsh group of narrow gauge railways, including the UK's only steam-rack railway on Snowdon, comparable with Swiss examples. In the English Lake District, the Lakeside and Haverthwaite Railway connects with vintage steamships on Lake Windermere, as it did from its completion in 1869. The East Lancashire Railway traverses the Irwell Valley north of Bury, and is marketed as a route linking other recreation sites including such diverse attractions as Helmshore textile museum and the Rawtenstall ski-slope.

Main-line railways are also much used for recreational trips although these are difficult to disaggregate from other passenger traffic. On many networks – including the national systems of Britain, France, Holland, Germany and Canada – trips are stimulated by provision of cheap time/zone based tickets, or national travel passes (Vacances Pass in France). In Britain, commuters can use season tickets for additional trips outside their regular journeys to and from work. British Rail have designated lines in north and west upland Britain, the south-west and Cotswold areas as 'Scenic Railways' to generate recreational trips on peripheral sections of the network.

Similarly, it is difficult to distinguish tourist and leisure traffic from overall bus and coach provision, long-established in the recreational market, except where services are specifically recreational. These include minibus services in Humberside serving the Humber river banks and Flamborough Head supported by local authorities and the Countryside Commission, 'Guide Friday' tours in centres such as Stratford-upon-Avon (their original location), Bath, Oxford, Edinburgh and the 'Fellrunner' community minibus guided tours in the Cumbrian Pennines and Eden Valley. In Glasgow, taxi tours make provision for wheelchair passengers. Many coach firms have a long history of recreational service: in Canada, Brewster Tours originated as Guides to the British Columbian icefields in 1892. By 1916, the firm replaced its 70 horse-drawn carriages by five motor coaches and established chalets in ski areas in the 1930s. The enterprise survives as a component of Greyhound Lines of Canada, and continues to run tourist services on 'Snocoaches' to the Athabasca Glacier.

Coastal and lake ships also survive to offer general and leisure services. Caledonian-MacBrayne services on the west coast of Scotland give vital services with car ferries which much reduce the distances by road (e.g. Dunoon-Gourock, a popular service with competition from Western Ferries). A similar situation is that of the Norwegian coastal services.

Implications of increased car ownership and usage

The flexibility of mobility and access that private cars confer upon their owners and users is reflected in the marked rises in car ownership and greatly expanded patterns of associated recreational activity in affluent Western societies. In the USA, camping (motorized) is a major recreation activity of middle-income households. The car has opened up the countryside in the UK, but created severe, although often localized, problems of over-use. Changing transport modes progressively increase fears for the environment (Figure 9.6). The car's flexibility increases the scale and dispersion of the problem (Chapter 4). By 1980, 76 per cent of recreation trips in the UK were made by car. Car ownership thus creates a

Photograph 9.1 The Settle-Carlisle railway: a former Southern Railway express locomotive, 'City of Wells', approaches the summit of the railway at Ais Gill (356m) beneath the slopes of Wild Boar Fell with the southbound 'Cumbrian Mountain Pullman'. The photograph gives an indication of the scenic grandeur and gradients (at this point 1 in 100) of the S & CR (photograph by John Shuttleworth).

Figure 9.6 'Stop complaining, William. I bet one day somebody will invent something much worse than railways'. Cartoon drawn by 12-year old Chris Shaw with reference to the intrusion of the Kendal-Windermere Railway into the Lake District, strongly opposed by William Wordsworth, 1844 (*Lake District Guardian*, 1991).

Reproduced by permission of Ian Shaw.

'mobility threshold' (Sidaway, 1982, 4–5), in accord with more general rising mobility levels (Hillman et al., 1976). Such trends encourage those who can participate in car-based recreation trips to increase their frequency while the carless become relatively more disadvantaged. With the closure of many uneconomic rural transport services, they become absolutely so, although higher mobility does not necessarily *determine* increased participation in recreational activities, but greatly eases access (Patmore, 1983, 80–1). The car has considerably reduced time constraints giving increasing choice of more distant sites, but more importantly allowing greater temporal and spatial flexibility of opportunities within the destination area (Wall, 1971, 111). Through the considerable increases in caravans and camping usage, both on permanent sites (further growth is now much curtailed by planning authorities in the UK) and especially in touring (Coppock and Duffield, 1975, 171–91; Turton, 1982) this flexibility has been extended to the allied area of accommodation for staying visitors. Growing car ownership and use reflect the car's increasing dominance in passive recreation, especially in day trips, but leads to the over-use, congestion and pollution of recreational sites.

The effects of the increased costs of motoring appear to have little lasting impact

(Shucksmith, 1980; Patmore, 1983, 125). It seems that the frequency of recreational trips does not vary with socioeconomic characteristics of car owners, but those with higher incomes and higher levels of education travel further (Wall, 1972).

Walking and cycling

Patmore (1983, 242) comments that 'the private car may serve most needs, but it does not serve all'. Public transport for those without cars, and alternative forms of travel for those who wish to enjoy those modes for their own sake, remain important. Thus some choose to ride on steam trains, for example, while 'the most popular outdoor recreation by far is simply going for a walk' (Patmore, 1983, 201). Mapping and signing of rights of way and the provision of long-distance footpaths grew from the 1949 National Parks and Access to the Countryside Act in Britain, greatly aided by support from the Countryside Commission which has set up a programme of extension of the national network of paths (Figure 9.7; Countryside Commission, 1988). In rural areas, policies have aimed to generate the use of bicycles and walking to ease car-induced congestion (Banister and Groome, 1984). The Dumfries and Galloway Region Area Tourist Board, together with the Scottish Tourist Board, the Forestry Commission and the Countryside Commission for Scotland have recently initiated a pilot study aimed to encourage cycling using disused railway-track beds in Galloway and Upper Nithsdale, establishing a long-distance cycle route from Loch Doon (Ayrshire) to Rhinus of Kells, Clatteringshaws Loch and Raider's Road in Galloway, and establishing cycling tracks through Galloway Forest Park and Ae and Mabie Forests (*Sunday Times Scotland*, 7 April 1991).

Carrying capacity

In simple terms, carrying capacity is 'reached when the demand for a resource is just matched by the supply of that resource', but in reality is far more complex, through aspects of physical, ecological, perceptual, economic and optimum capacity (Patmore, 1983, 222–33). Since the concept applies to 'the ability of land to support people' (Mitchell, 1979, 177) there are clear implications for the ingress and egress of visitors, in terms of traffic flows and of the carrying capacity of the site itself.

Mitchell (1979) reviews work on the management of the roadless Boundary Waters Canoe Area (BWCA) in northern Minnesota, stressing variations in demand and conflicts between recreational mobility within the area – using canoes or motor boats – and the need to predict travel behaviour between camp-sites to reduce the over-use of particular sites. The problem of wilderness-use is incorporated in a model of BWCA travel behaviour to prevent excessive use. All travel modes, trip purposes and types were incorporated, including circular trips within the area, camp-site to camp-site migration and 'day-use' trips in and out of the area within the day: a system of entry quotas using permits at each entry point redistributed the pressures of use. Elsewhere, Jaakson et al. (1976) further develop model-based policies: their study of the Emma Lakes and Christopher Lakes, Saskatchewan, shows the role of transport initiatives within a range of management policies.

In the UK attempts to manage traffic flows suggest links with the concept of carrying capacity in traffic terms. These are reviewed below.

Urban and rural recreation and problems

Idealized representations of recreational trip-making linking town and country are based upon time and distance criteria. Bryant et al. (1982, 126–40) identify four zones: of daily leisure outside the home extending into the urban fringe, of day trips to accessible surrounding countryside, of the countryside used for longer trips for weekend stays, for

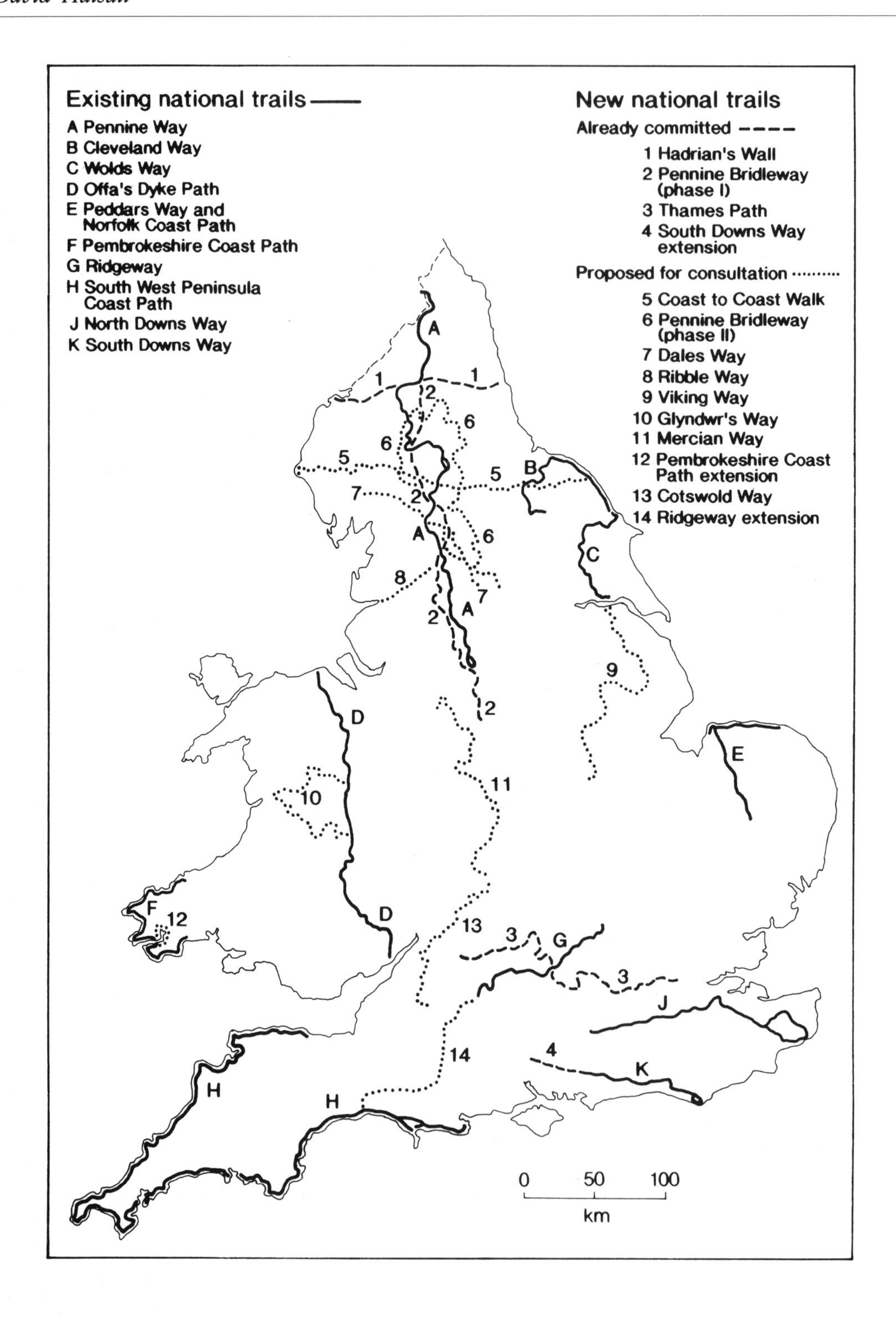

Figure 9.7 National trails in England and Wales, 1988.
Source: Countryside Commission, 1988, 19.

example in second homes, and of longer visits to holiday/wilderness areas. Each zone's boundary reflects a particular level of journey time and effort. Pigram (1983, 30–7) highlights Razotte's idea of concentric zones of urban zone (few hours), intensive recreation (one day), rural holiday resorts (weekend), extensive recreation (long weekend to several days) and national and international resorts (for long vacations and retirement). Such classifications encompass all possible areas! But the extension of recreational activities outwards from the city despite the increasing leisure potential within the home (Davidson and Wibberley, 1977; Patmore, 1983), together with the processes of decentralization and counter-urbanization, has entailed specific policies of providing recreational opportunities (including transport facilities). In both urban and rural areas, schemes to improve access by public transport to recreational opportunities often help to increase or indeed maintain or re-introduce public transport services to given areas (Dobbs, 1984; Groome, 1984; Speakman, 1984). However, many recreational resources have only local catchment areas. Evidence suggests, for example, that the urban fringe draws few inner-city residents; nor does it act as a substitute for more distant areas. Rather, it serves a local population (Harrison, 1983; Sidaway and Duffield, 1984).

In urban areas, expansion of recreational land-use is often an integral part of urban regeneration. In the UK, the former Greater Manchester County Council (GMCC) increased access for non-car owners to reclaimed valleys, desolated by earlier industrialization (Newman, 1982), as part of the Countryside Commission Wayfarer project (Lumsden and Speakman, 1982). Williams and Tanner (1982) plot the availability and encourage the use of public transport for day trips to recreation sites in the West Midlands (Figure 9.8). In Merseyside, access by public transport to a variety of rural-urban fringe sites ranging from valued natural environments on the south-west Lancashire coast to reclaimed quarry sites (Pex Hill, Widnes) and the city farm at Walton, is encouraged by publicity access guides by Merseyside Passenger Transport Executive, a further scheme supported by the Countryside Commission. Such ideas draw much from the philosophy of the integration of transport drawn by the 1968 Transport Act. Despite the more recent policies of privatization and competition through the deregulation of buses (1986) some vestiges of the former more holistic philosophy remains.

The regeneration of redundant and derelict port areas is an international phenomenon caused by changing maritime technology and trading patterns (Hoyle and Pinder, 1980; Hoyle et al., 1988; Hoyle, 1990). Waterfront regeneration has drawn considerably upon recreation and tourism as a major focus – the Albert Dock in Liverpool is now a top tourist destination. Salford Quays is a similar development and specifically aims to provide dockland leisure centres (Law, 1988). Increasing dockland usage often requires improving access: Albert Dock in Liverpool is served by vast car parks and a MPTE subsidized minibus service into the city centre. In Glasgow considerable investment is required in projects including walkways and public transport to release amenities and open space effectively within a comprehensive planning framework (Parke, 1990). Sydney, Australia, remains a very active port, but as elsewhere has been affected by technological change. Strategies for redevelopment combine economic, social, environmental and heritage elements; the actual impact of revitalization is towards recreational, tourist and residential land-uses (Figure 9.9) (Sant, 1990).

Rural policies encompass schemes to reduce car penetration to sensitive areas – the Goyt valley minibus park-and-ride scheme in Derbyshire – to supplement inadequate capacity for cars which cannot be increased without great detriment. Snowdon Sherpa routes around Snowdon were prompted by 'the inability of the car park at Pen-y-Pass . . . to cope with summer demands' (Dobbs, 1984, 30). They aim to provide opportunities

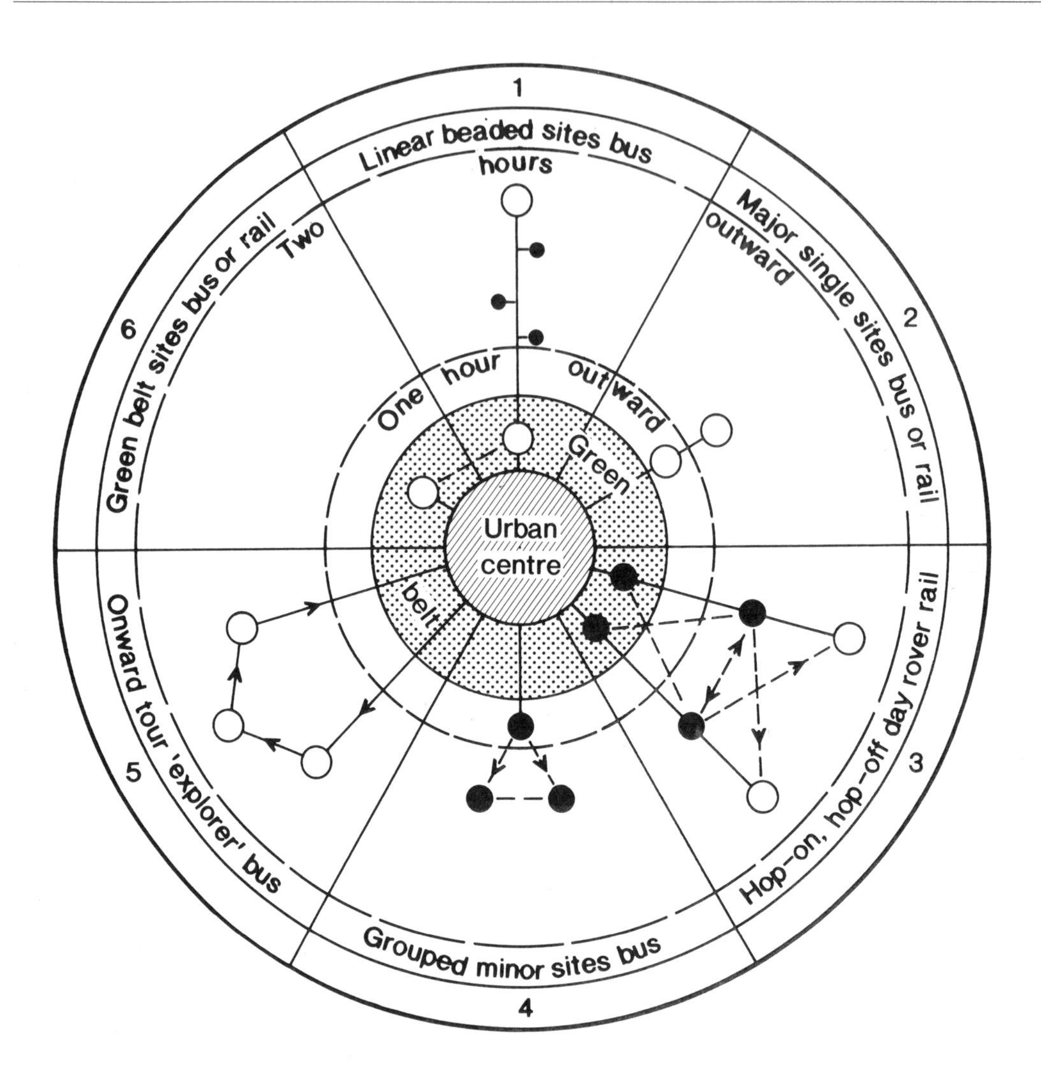

Figure 9.8 'Day out' trip alternatives by public transport: relations between countryside sites, range and access.

Source: Williams and Tanner, 1982, 117.

for car owners to transfer to public transport and non-car owners to gain access to recreational resources – Dalesrail, Parklink and Wayfarer projects (Mulligan, 1979; Speakman, 1982).

Heritage elements

Rising numbers of people enjoy travelling on preserved historic transport modes: data from the Cumbria Tourist Board illustrates this theme with three heritage facilities of transport in first, sixth and seventh order of top ten attractions (Table 9.1). Heritage elements – canals, preserved railways, docklands – are introduced above. Here, the implications of steamboats and steam-hauled excursion trains are studied in more detail. Steam heritage has an increasing fascination demonstrated by the popularity of the three Windermere steamboats, the oldest dating from 1891, and other steamboats in the English Lake District – steam-launch trips from the Steamboat Museum, also on Windermere, and lake cruises on the National

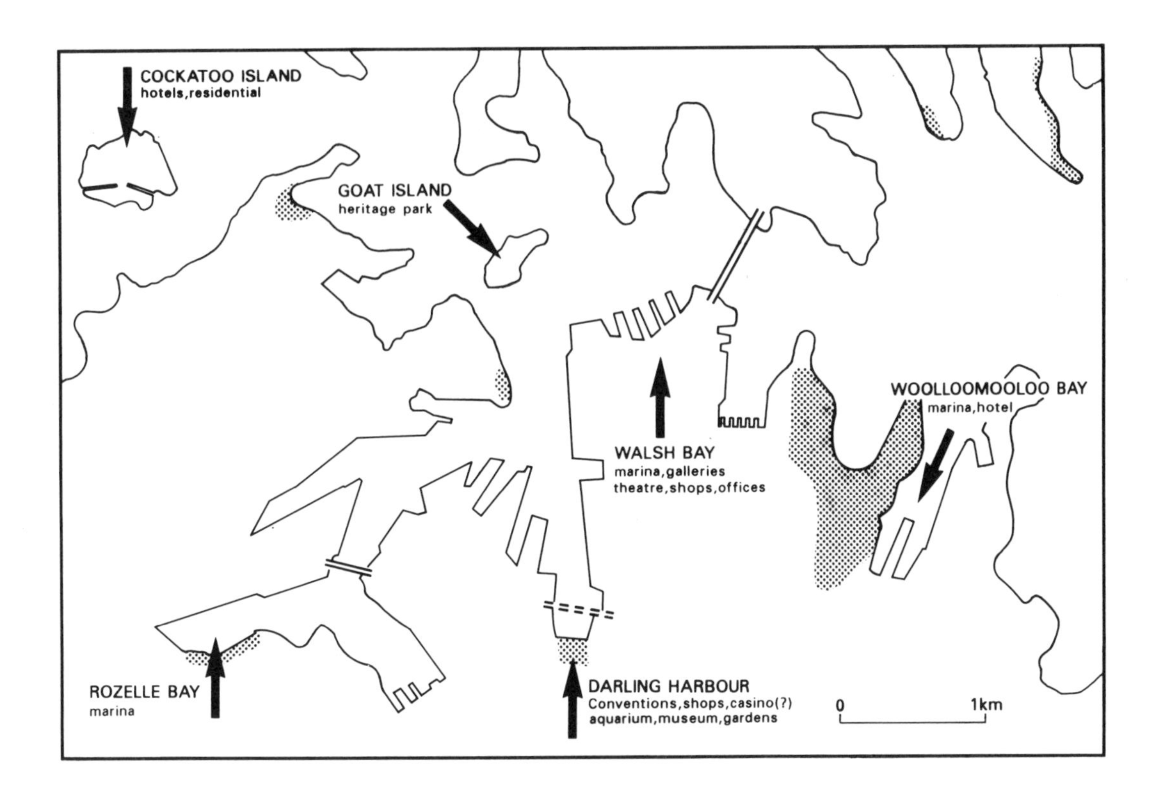

Figure 9.9 Redevelopment of the port of Sydney: projects proposed and implemented in the 1980s.
Source: Sant, 1990, 89.

Table 9.1 Cumbria: Top ten attractions for visitors, 1989

	No. of visitors
*1 Windermere Iron Steamboat Co. Ltd	572,616
2 Talkin Tarn Country Park	200,000+
3 Grizedale Forest Park	160,000+
4 Sellafield (BNFL) Exhibition Centre	159,567
5 Carlisle Cathedral	150,000+
*6 Ravenglass & Eskdale Railway	143,739
*7 Lakeside & Haverthwaite Railway	140,000+
8 Holker Hall and gardens	102,000+
9 Fell Foot Park	100,000+
10 Dove Cottage and Wordsworth Museum	85,900+

* Heritage transport
\+ Estimates

Source: Cumbria Tourist Board, *Regional Tourist Strategy for Cumbria*, 1990.

Trust's steam yacht 'Gondola' (Coniston) and motor conversions of two Victorian steam yachts on Ullswater. In Scotland, the last screw steamer in use on a Scottish loch remains unconverted to diesel because of the dangers of polluting the drinking-water supply in Loch Katrine, and the popular last sea-going paddle steamer 'Waverley' is based at Glasgow from whence cruises around the UK coast are undertaken.

Steam locomotive hauled railtours also feed on nostalgia. In 1991 celebrations of the tenth anniversary of the California State Railroad museum in Sacramento and the Railroad Historical Society's annual convention at Huntington, West Virginia, include steam railtours through spectacular scenery. Equally impressive are the scenic and other attractions incorporated within a 14-day steam tour of New Zealand's railways in 1991. In Britain, an extensive programme of steam-hauled excursion trains on British Rail (BR) scenic routes caters for enthusiasts and day trippers. This is organized by BR in liaison with the private Flying Scotsman Services Ltd (FSSL), with locomotives

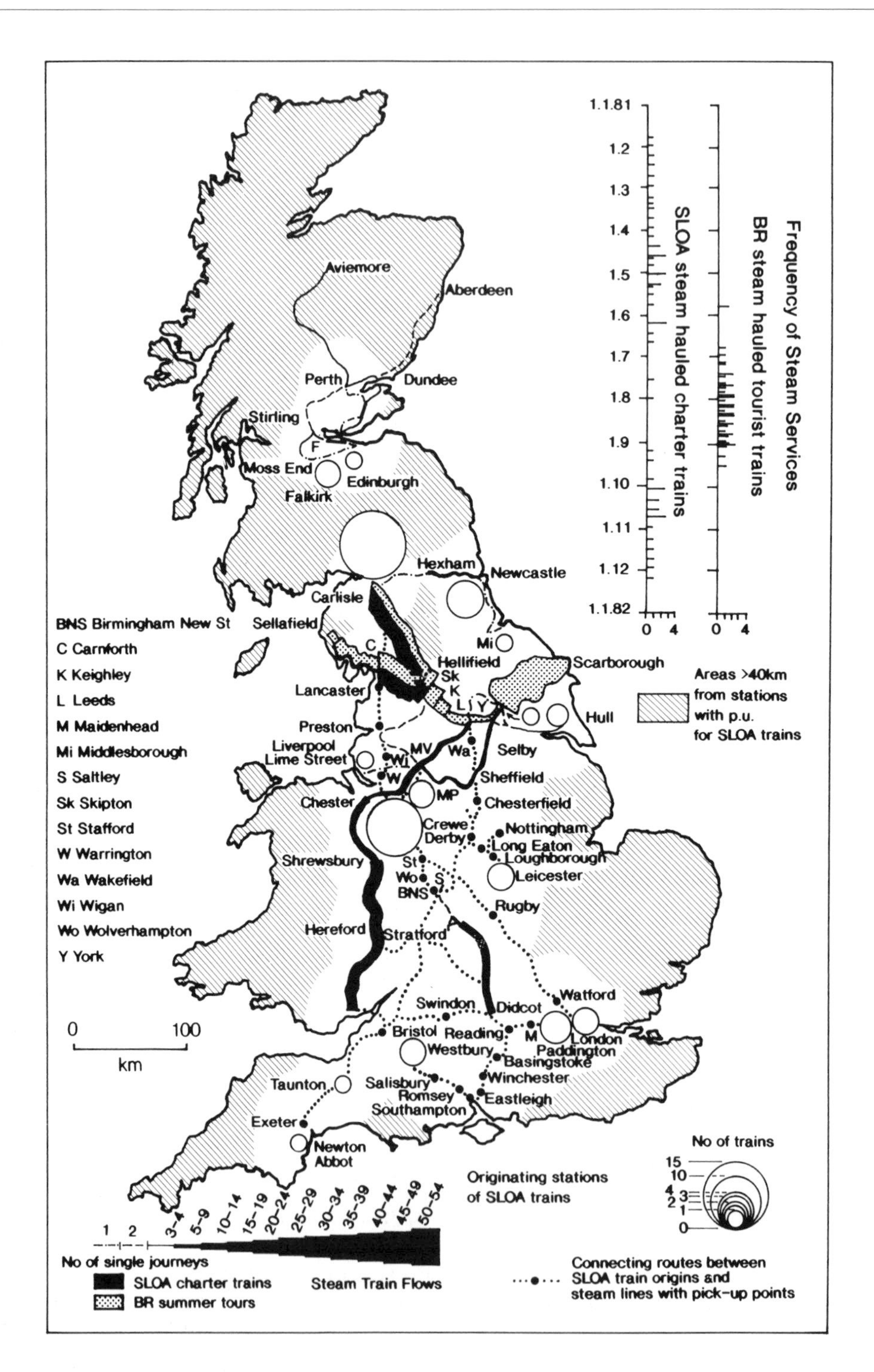

Figure 9.10 Steam-hauled excursion trains on British Rail main lines, organised by public (BR) and private (Steam Locomotive Operators' Association) organisations and operated by BR, 1981.

Source: Halsall, 1982b, 171.

provided by a group of owners, the Steam Locomotive Operators' Association. Figure 9.10 shows the 1981 programme (Halsall, 1982b). By 1991, the lines from Crewe and Chester to Holyhead had replaced the Scarborough steam service and ScotRail runs a frequent steam service from Fort William to Mallaig. Commercial prospects are promising despite the recession. In March 1991 FSSL received 500–900 bookings per week, unprecedented interest after publicity targeted at women, families and retired people in 1990. New activities in the south-east, aimed at the lucrative London, south-east and business market, brought steam trains to Folkestone and Cambridge in 1991, and will possibly go to Windsor and Salisbury-Exeter in 1992. However, further railway modernization and the ever-increasing costs of administration, maintenance, safety tests and the overhaul of ageing locomotives may curtail steam operations in future. It seems likely that the proposals for developing the scenic Settle-Carlisle route (Halsall, 1990) as a corridor of local and thematic tourist attractions (Biggs, 1990) signals a secure future for steam operations on at least one British main line.

Conclusions

The relationships between transport, recreation and tourism are a complex interweaving of interrelated interests. Transport provision is a permissive factor in much tourist/recreation development, itself a product of increasing mobility, leisure time and affluence. The bases of these developments are socio-economic and environmental: recreation and tourism develop through time and in space within social, political, economic and physical contexts. The pressures of extreme overuse threaten not only the physical environment but aspects of society and economy as well. In the transport context, the expression of concern by a former British Secretary of State for Transport, Paul Channon, in 1988 is apt:

> Demand for travel, particularly for leisure travel, is increasing fast. How is this ever-growing demand to be accommodated without doing serious damage to our rural and urban environment?
>
> (Department of Transport, 1988)

It is difficult to provide the definitive answers sought by policy-makers. The geographical contribution is an attempt to combine with a variety of other concerned disciplines in understanding the present situation so that we may attempt to contribute to the well-being of people's future leisure needs and patterns.

References

Appleton, J. H. (1970), *Disused railways in the countryside of England and Wales* (London: HMSO/Countryside Commission).

Banister, C. and Groome, D. (eds) (1984), *Out and about. Promoting access to countryside recreation by bus, train, bicycle and foot* (University of Manchester, Department of Town and Country Planning, Occasional Paper 12).

Baxter, M. and Ewing, G. (1981), 'Models of recreational trip distribution', *Regional Studies* 15, 327–44.

Biggs, W. D. (1990), *Settle-Carlisle Railway. Opportunities for development* (Kendal: Standing Conference for the Settle-Carlisle Railway).

Bruton, M. J. (1985), *Introduction to transportation planning* (London: Hutchinson, 3rd edn.).

Bryant, C. R., Russwurm, L. H. and McLellan, A. G. (1982), *The city's countryside. Land and its management in the rural-urban fringe* (London: Longman).

Coppock, J. T. and Duffield, B. S. (1975), *Recreation in the countryside* (London: Macmillan).

Countryside Commission (1988), *Enjoying the countryside. Paths, routes and trails. A consultation paper* (Cheltenham: Countryside Commission, CCP 253).

Cumbrian Tourist Board (1990), *Regional tourism strategy for Cumbria* (Windermere: CTB).

Daniels, P. W. and Warnes, A. M. (1980), *Movement in cities* (London: Methuen).

Davidson, J. and Wibberley, G. R. (1977), *Planning and the rural environment* (Oxford: Pergamon).

Department of Transport (1988), *Transport and the environment* (London: Department of Transport, 2nd edn.).
Dixey, R. (1988), 'A means to get out of the house: working-class woman, leisure and bingo', in J. Little, L. Peake and P. Richardson (eds), *Women in cities. Gender and the urban environment* (London: Macmillan), 117–32.
Dobbs, B. (1984), 'No tourists – no transport?', in C. Banister and D. Groome (eds), *Out and about. Promoting access to countryside recreation by bus, train, bicycle and foot* (University of Manchester, Department of Town and Country Planning Occasional Paper 12), 25–42.
Duffield, B. S. (1975), 'The nature of recreational travel space', in G. A. C. Searle (ed.), *Recreational economics and analysis* (London: Longman), 15–35.
Elson, M. J. (1979), *State of the art review 12: Countryside trip-making* (London: Sports Council/Social Science Research Council).
Fedler, A. J. (1987), 'Are leisure, recreation and tourism interrelated?', *Annals of Tourism Research* 14, 311–13.
Gamble, W. P. (1989), *Tourism and development in Africa* (London: John Murray).
Grimshaw, P. N. (1976), 'Steam railways: growth points for leisure', *Geography* 61, 83–8.
Groome, D. (1984), 'How leisure can fill those empty seats', *Town and Country Planning*, April 1984, 110–11.
HMSO (1982), *Potential cycle routes and disused railways in England and Wales*, (map) (London: HMSO).
Hall, E. (1969), *Miss Weeton's Journal of a Governess, 2, 1811–1825* (Newton Abbot: David & Charles, revised and reprinted).
Halsall, D. A. (ed.) (1982a), *Transport for recreation* (Lancaster: IBG Transport Geography Study Group).
Halsall, D. A. (1982b), 'Policies and practice of steam train operation on British Rail scenic routes 1981', in D. A. Halsall (ed.), *Transport for recreation* (Lancaster: IBG Transport Geography Study Group), 145–73.
Halsall, D. A. (1990), 'The Settle-Carlisle railway', in C. Park (ed.), *Field excursions in north-west England* (Lancaster: University of Lancaster, Cicerone).
Harrison, C. (1983), 'Countryside recreation and London's urban fringe', *Transactions of the Institute of British Geographers*, New Series 8, 295–313.
Hillman, M., Henderson, I. and Whalley, A. (1976), *Transport realities and planning policy* (London: PEP).
Hoyle, B. S. (ed.) (1990), *Port cities in context: the impact of waterfront regeneration* (Southampton: IBG Transport Geography Study Group).
Hoyle, B. S. and Pinder, D. A. (1980), *Cityport industrialisation and regional development* (Oxford: Pergamon).
Hoyle, B. S., Pinder, D. A. and Husain, M. S. (eds) (1988), *Revitalising the waterfront: international dimensions of dockland redevelopment* (London: Belhaven).
Huff, D. L. (1960), 'A topographical model of consumer space preferences', *Papers & Proceedings of the Regional Science Association* 6, 159–73.
Jaakson, R., Buszinski, M. D. and Botting, D. (1976), 'Carrying capacity and lake recreation planning', *Town Planning Review* 47, 359–73.
Law, C. M. (1988), 'From Manchester Docks to Salford Quays: a progress report on an urban redevelopment project', *Journal of the Manchester Geographical Society*, 2–15.
Lumsden, L. M. and Speakman, C. (1982), 'The Countryside Commission recreational transport projects in Greater Manchester and West Yorkshire: the Wayfarer project', in D. A. Halsall (ed.), *Transport for recreation* (Lancaster: IBG Transport Geography Study Group), 74–83.
Martin, B. and Mason, S. (1987), 'Current trends in leisure', *Leisure Studies* 6, 93–7.
Mercer, D. (1971), 'Perception in outdoor recreation', in P. Lavery (ed.), *Recreational Geography* (Newton Abbot: David & Charles), 51–69.
Mitchell, B. (1979), *Geography and resource analysis* (London: Longman).
Moseley, M. J. (1979), *Accessibility. The rural challenge* (London: Methuen).
Mulligan, C. A. (1979), 'The Snowdon Sherpa: public transport and national park management experiment', in D. A. Halsall and B. J. Turton (eds), *Rural transport problems in Britain, Papers and discussion* (Keele: IBG Transport Geography Study Group), 45–55.
Newman, P. (1982), 'Access to the "urban fringe" for informal countryside recreation: Greater Manchester river valley schemes with particular emphasis on Croal-Irwell', in D. A. Halsall (ed.), *Transport for Recreation* (Lancaster: IBG Transport Geography Study Group), 36–62.
Parke, T. J. (1990), 'Waterfront regeneration in Glasgow: some transportation issues', in B. S. Hoyle (ed.), *Port cities in context: the impact of waterfront regeneration* (Southampton: IBG Transport Geography Study Group), 39–58.
Patmore, J. A. (1970), *Land and Leisure* (Newton Abbot: David & Charles).

Patmore, J. A. (1971), 'Routeways and recreation patterns', in P. Lavery (ed.), *Recreational geography* (Newton Abbot: David & Charles), 70–96.

Patmore, J. A. (1983), *Recreation and resources. Leisure patterns and leisure places* (Oxford: Blackwell).

Peace, S. M. (1982), 'The activity patterns of elderly people in Swansea, South Wales and in South-East England', in A. M. Warnes (ed.), *Geographical perspectives on the elderly* (London: Wiley), 281–301.

Pickup, L. (1988), 'Hard to get around: a study of women's travel mobility', in J. Little, L. Peake and P. Richardson (eds), *Women in cities. Gender and the urban environment* (London: Macmillan), 98–116.

Pigram, J. (1983), *Outdoor recreation and resource management* (London: Croom Helm).

Robson, P. (1982), 'Patterns of activity and mobility among the elderly', in A. M. Warnes (ed.), *Geographical perspectives on the elderly* (London: Wiley), 265–80.

Sant, M. (1990), 'Waterfront revitalisation and the active port: the case of Sydney, Australia', in B. S. Hoyle (ed.), *Port cities in context: the impact of waterfront regeneration* (Southampton: IBG Transport Geography Study Group), 69–94.

Senior, M. L. (1979), 'From gravity modelling to entropy maximising: a pedagogic guide', *Progress in human geography* 3, 175–210.

Shucksmith, D. M. (1980), 'Petrol prices and rural recreation in the 1980s', *National Westminster Quarterly Review*, February 1980, 52–9.

Sidaway, R. M. (1982), 'Mobility and countryside recreation', in D. A. Halsall (ed.), *Transport for recreation* (Lancaster: IBG Transport Geography Study Group), 1–16.

Sidaway, R. M. and Duffield, B. S. (1984), 'A new look at countryside recreation in the urban fringe', *Leisure Studies* 3, 249–71.

Skelton, N. (1982), 'Transport policies and the elderly', in A. M. Warnes (ed.), *Geographical perspectives on the elderly* (London: Wiley), 303–21.

Smith, J. and Gant, R. (1982), 'The elderly's travel in the Cotswolds', in A. M. Warnes (ed.), *Geographical perspectives on the elderly* (London: Wiley), 323-26.

Smith, R. W. (1987), 'Leisure of disabled tourists. Barriers to participation', *Annals of Tourism Research* 14, 376–89.

Smith, S. L. J. (1983), *Recreation geography* (London: Longman).

Smith, S. L. J. and Godbey, G. C. (1991), 'Leisure, recreation and tourism', *Annals of Tourism Research* 18, 85–100.

Speakman, C. (1982), '"Dales Rail" and "Park Link": recreational transport packages in the Yorkshire Dales', in D. A. Halsall (ed.), *Transport for recreation* (Lancaster: IBG Transport Geography Study Group), 63–73.

Speakman, C. (1984), 'Restoring that lost rural mobility', *Town and Country Planning*, April 1984, 110–11.

Thompson, B. (1979), 'Recreational travel: a review and pilot study', in C. S. Van Doren, G. B. Priddle and J. E. Lewis (eds), *Land and leisure. Concepts and methods in outdoor recreation* (London: Methuen, 2nd edn.), 167–79.

Tivers, J. (1988), 'Women with young children: constraints on activities in the urban environment', in J. Little, L. Peake and P. Richardson (eds), *Women in cities. Gender and the urban environment* (London: Macmillan), 84–97.

Turnock, D. (1982), *Railways in the British Isles. Landscape, land use and society* (London: A & C Black).

Turton, B. J. (1982), 'Mobile caravanning in the Peak District: patterns of occupation at Losehill Caravan Club site at Castleton' in D. A. Halsall (ed.), *Transport for recreation* (Lancaster: IBG Transport Geography Study Group), 191–213.

Ullman, E. L. (1954), 'Geography as spatial interaction', reprinted in M. E. Eliot-Hurst (ed.) (1974), *Transportation geography. Comments and readings* (New York: McGraw Hill), 27–40.

Urry, J. (1990), *The tourist gaze. Leisure and travel in contemporary societies* (London: Sage).

Wall, G. (1971), 'Car owners and holiday activities', in P. Lavery (ed.), *Recreational geography* (Newton Abbot: David & Charles), 97–111.

Wall, G. (1972), 'Socioeconomic variations in pleasure trip patterns: the case of Hull car owners', *Transactions of the Institute of British Geographers* 57, 45–58.

White, H. P. (1986), *Forgotten Railways* (Newton Abbot: David & Charles), 195–209.

Williams, A. F. and Tanner, M. F. (1982), 'From conurbation to countryside: a day out by bus or rail in the West Midlands', in D. A. Halsall (ed.), *Transport for recreation* (Lancaster: IBG Transport Geography Study Group), 84–124.

10
Bulk Freight Transport

David Hilling and Michael Browne

Many commodities and products are transported in bulk and sea transport is the dominant mode. Importantly, the economics of bulk transport strongly influence decisions about industrial location and ports are often focal points for economic activity. In northern Europe a small number of ports have emerged as major transhipment hubs for a complicated mix of reasons. Among the most significant are: economies of scale in bulk sea transport, government policies and commercial practices.

Introduction

In maritime transport it has been normal to distinguish between two main types of cargo: bulk and break-bulk (Bird, 1971). The former comprises homogeneous materials without packaging (ores, coal, grain, raw sugar, cement, crude oil and oil products, etc.) usually for a single consignee and destination, while the latter, often known as general cargo, consists of an almost infinite variety of freight, usually in small consignments for numerous consignees and packaged in a variety of bags, bales, boxes, crates and drums of diverse shape and size.

The term 'bulk' cargo is sometimes used (Stopford, 1988) for any commodity that may be carried in ship loads – bananas, refrigerated meat, cars – while the term 'neo-bulk' is used for cargoes such as forest products when carried in large quantities in a standardized form. The question of shipment size is important in determining the method of movements: large consignments of grain will be carried in bulk but small quantities of rice or malting grain may be bagged. Similarly, small quantities of oil products (e.g. lubricants) may be carried in barrels but large volumes in bulk.

Recent developments in cargo-handling technology for both bulk and break-bulk cargo would suggest a refinement of the above classification. Thus, much of the break-bulk cargo is now carried in a unitized form with individual items grouped or 'bulked' into larger, standardized units. This is done by *packaging* (crating, strapping together of plank timber), the use of *pallets* (wooden or metal platforms on which goods can be stacked and strapped), the use of *containers* (internationally standardized boxes) and the lorries, trailers and railway wagons in *roll-on/roll-off* operations. Containers are also used for the bulk rather than bagged movement of commodities such as coffee and cocoa beans; and tank containers are in use for a variety of liquid cargoes such as beer, wine and chemicals. Even lump coal for domestic use is now moved by container between Garston and Northern Ireland to reduce the fragmentation

Table 10.1 Bulk cargoes – handling and movement

Commodity type	*Examples*	*Ship-shore transfer*	*Inland movement*
1 *Liquid*			
a) Normal temperature and pressure	crude oil, most oil products, wine, slurried coal, limestone	pump/pipe	pipeline
b) Other temperature and pressure	liquified gases, heavy oils, latex, bitumen, vegetable oils	pumps, temperature controlled pipelines	temperature controlled pipelines
2 *Dry bulk*			
a) Flowing	grain, sugar, powders (alumina, cement)	pneumatic/ suction, conveyor, grabs	pipes, conveyors, barge, rail wagon, lorry
b) Irregular	coal, iron ores, non-ferrous ores, phosphate rock	grab, conveyor	conveyor, barge, rail wagon, lorry
3 *Neo bulk*	forest products, steel products, baled scrap	lift-on/ lift-off, roll-on/ roll-off	barge, rail wagon, lorry
4 *Wheeled units*	cars, lorries, rail wagons	roll-on/ roll-off	
5 *Refrigerated/Chilled cargo*	meat, fruit, dairy produce	lift-on/ lift-off, conveyor	rail wagon, lorry

Modified from Stopford, 1988.

which is a consequence of bulk handling.

Within the bulk category the physical characteristics of cargoes are very different. Each has specific requirements with respect to stowage in the ship, methods of transfer between ship and shore and the inland transport (Table 10.1). Bird (1971) has suggested that the cargo type is reflected in the associated port activity. For the higher unit value break-bulk cargo (now mainly in unitized form) a port is usually the gateway through which the cargo passes to the hinterland while for the bulk cargo it acts as a terminal – the cargo is stored and often processed before onward movement. This idea of port-related industries associated with the bulk cargoes is one to which we shall return.

While the unit and possibly the total value of bulk cargoes may not compare with those of general cargo, the sheer volumes involved give them a special significance in transport systems. For the railway systems of many countries the volume, regularity and generally simple movement patterns of the bulk traffic provide one of the few areas of profitable operation; most new railway-line construction has to be justified by the availability of such traffic. Recent examples are the Sishen-Saldanha line for iron ore in South Africa, the Cerrejon line for coal in Colombia and the Carejas line for iron ore in Brazil. The survival and improvement of inland-waterway transport in North America, Europe, the Soviet Union and China is explained almost completely by the need to move bulk cargoes.

In 1988 (OECD, 1990) out of a total world seaborne trade of 3,670 million metric tonnes, oil (crude and products) accounted for 1,620 million (44 per cent) and the five principal dry-bulk cargoes (iron ore, coal, grain, bauxite/alumina and phosphate rock) a further 940 million tonnes (26 per cent). This volume of cargo and the impact it has on patterns of trade, port activity and industrial location provides justification for the focus adopted in this chapter. If as is sometimes claimed the shipping industry provides a barometer for world economic conditions, then the bulk trades are the dominant influence. In the context of maritime bulk trades, this chapter examines recent changes in transport technology and operations and assesses their impact on the geography of ports and associated industrial activity.

Transport and industrial location

At a local level for small industries it may be that raw materials are available and can be processed at a location that also provides the market. As industries increase in scale and become more complex and organized on a global scale with diverse inputs and outputs in large volumes, this becomes impossible; transport is the prerequisite for spatial interaction. Models of industrial location have also become more complex and greater reality introduced by the incorporation of behavioural considerations (Greenhut, 1956; Pred, 1967) but it can still be argued that the 'economics of procurement and production distribution combined is the important – even deciding – location consideration' (Hamilton, 1962).

This is not too far removed from the basic thinking behind Weber's (1909) least-cost industrial location model. Weber has rightly been criticized for the unreality of many of his assumptions about freight rate systems and long-haul economics, labour supply and human behaviour, yet the model still has value in the basic distinctions it identifies between 'sporadic', 'ubiquitous', 'gross' and 'pure' raw materials. Thus, fuels and ores, the bulk materials of international trade, are localized (sporadic) while some materials (sands, clays, stone and water) although not ubiquitous are certainly found more widely. It is also the case that some materials, which Weber termed gross, lose weight in processing while others (pure) do not. This was the basis for his 'material index' (the weight of raw material inputs divided by the weight of the end product) and his suggestion that where there is greater weight loss (high material index) the processing is likely to take place nearer the raw material source. He also recognized that this is a significant influence on location where the total weight of inputs is considerable.

For Weber, therefore, the least-cost location was strongly influenced by the relationship between the cost of assembling the raw materials and the cost of distributing the end product – essentially the cost differences in the movement of bulk materials and manufactured goods. Changes in the transport of bulk materials which are considered below have undoubtedly resulted in a modification of the cost differential.

Hoover (1948) provided a more elaborate analysis of the relationship between transport costs and industrial location. In particular he emphasized the difference between the cost to the transporter and the charge to the shipper (freight rates can be and often have been manipulated for political or business reasons), the nature of rate structures (e.g. in relation to distance), and the importance of directional traffic (e.g. low rates on lightly loaded back hauls – the 'Cleveland' effect). He also emphasized the significance of the size of consignments, the total volume of shipments and the relative transferability of the commodities concerned.

Goods of high value per unit weight are able to bear higher transport costs than those of low value. Thus, where the transport cost makes up a high proportion of the delivered price of the good, distance will be reduced to a minimum (e.g. aggregates) and the cheapest possible form of transport will be utilized. Many bulk raw materials have low value in relation to their weight and will be sensitive to transport costs.

In a situation where a production plant has one source of raw materials and one market

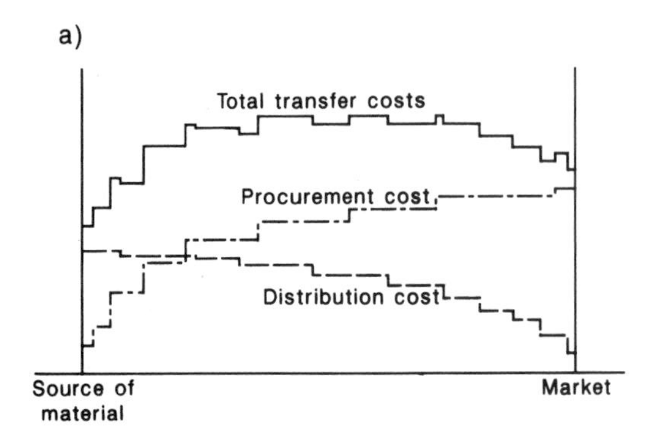

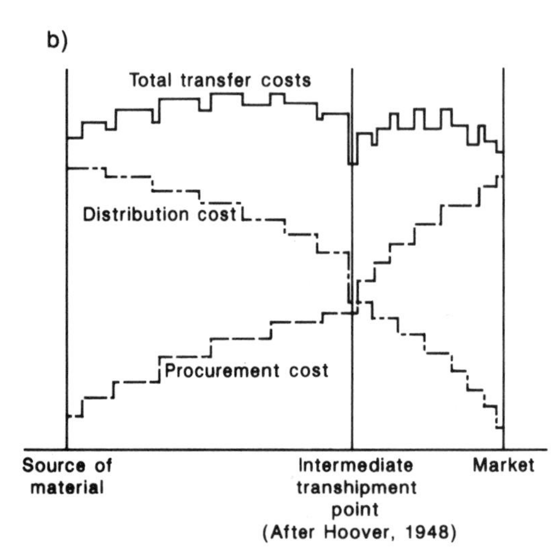

Figure 10.1a and b Transport costs and industrial location (a) with through transport, (b) with transhipment.

(Figure 10.1a), Hoover suggests that generalized procurement and distribution cost curves are likely to produce an ideal location (i.e. least cost in terms of transport) which is at the source of the raw material or the market and not at some intermediate location. If, however, some intermediate point transhipment between transport modes is necessary (Figure 10.1b) it is at this intermediate point that the transport costs are likely to be minimized. The act of transhipment itself greatly increases the cost of transport which is why in recent years so much effort has been put into improving the efficiency and reducing the cost of transhipment for both bulk and break-bulk cargo. For the same reason much industry is located at ports (Bird, 1971; Takel, 1974) especially in the case of heavy industries where the raw materials are in large volumes and transhipment costs are therefore high in relation to the unit value of the commodity.

Transport of bulk cargo

It can be argued that much port industry is a consequence of commodities moving in bulk and that for cargo to be moved in this way certain criteria have to be satisfied (Stopford, 1988). First, the commodity must have physical characteristics that allow it to be handled and moved in bulk. Second, the demand for the commodity must be such that the cost of special ships and handling equipment is justified. The smallest practical consignment size will effectively be that of the smallest bulk carrier available. There has been rapid expansion of the mini-bulker fleet in recent years (Tinsley, 1984; Heinemann and Cheetham, 1990) with vessels of 1,200 to 3,000 tonnes replacing the traditional short-sea vessels of 250 to 700 tonnes capacity. A figure of 1,000 tonnes has been suggested as a minimum threshold for bulk handling (Stopford, 1988).

Third, the bulk-shipping operation must be adapted to the overall transport system. In the case of iron-ore exports from Carejas in Brazil, the railway, port and even the ship (350,000 tonnes) were designed as integral parts of the whole. At the other end of the scale, relief grain has sometimes arrived at African ports with inadequate handling equipment, storage and onward-transport facilities. Fourth, the individual consignment size must be geared to the stocks that can be held at either end of the transport link. This is related to the actual demand at the consuming end, to the storage space available at each end and to the frequency of shipment.

The availability of storage space is an important determinant of the efficiency and productivity of any port since there is an

Photograph 10.1 Opened in 1976, Le Havre-Antifer was a terminal specifically built to accommodate Ultra Large Crude Carriers (ULCCs) carrying in excess of 500,000 tonnes of oil with draughts of 25.5m which excluded them from most other ports. Antifer is now underutilized and used mainly by Very Large Crude Carriers (VLCCs) of 250,000 to 350,000 tonnes which provide greater flexibility but higher unit cost of haul than ULCCs (Port of Le Havre).

almost inevitable mismatch between the rate at which cargo arrives and departs by sea and the rate at which it enters and leaves the port on the landward side (Takel, 1974, 1981). Storage space acts as the essential surge-bin to balance the flows on the sea and land sides. This is important for general cargoes, but it becomes critical for large volumes of bulk cargo. Tankers of 250,000 to 350,000 tonnes are now common. Many dry-bulk carriers are in the 100,000 to 150,000 tonnes range and all can be discharged in several days.

The amount of space for storage is a function of the density of the commodity. A tonne of coal will require 1.4 m^3, bauxite 0.7–0.8 m^3, alumina 0.6 m^3, crude oil 1.2 m^3, latex 1.0 m^3, vegetable oils 1.1 m^3, and cement 0.9–1.5 m^3. Storage space must allow for access and handling equipment such as stackers, cranes, conveyors and reclaimers so that packaged timber will store at 4,375 to 7,500 tonnes per hectare and iron ore at just under 100,000 tonnes. Additional storage space is needed where materials are sorted by grade or type (e.g. coals, ores and crude oils) and possibly to accommodate changes consequent upon conditions (e.g. wet and dry ores or coal).

A regular flow of bulk raw materials is

Table 10.2 Factors influencing ship size

Cargo type	Trade characteristics	Vessel size
General cargo		
a) Conventional	Varied small consignments Numerous consignees Slow handling rates Varied routes Numerous ports	Small
b) Unitized	More uniform cargo Rapid handling Many ports	Small–medium
Dry bulk		
a) Grain	Small–medium consignments Varied handling rates Many restrictive ports	Small–medium
b) Ores/coal	Large consignments Long hauls Moderate–good handling rates Specialized terminals at few ports	Medium–large–very large
Liquid		
a) Crude oil	Very large consignments Long hauls Few main routes Specialized terminals Few ports	Very large–ultra large
b) Oil products	Smaller 'parcels' Numerous consignees Many ports	Small–medium ('handy')

essential for any industrial process. Storage is vital in reducing the effects of flow variations but storage replenishment for a given tonnage can be either by frequent small shipments or by less frequent large shipments. This clearly involves the question of vessel size (Table 10.2) and choice is influenced by an interplay of three main factors: economies of scale; the consignment size in which the cargo is either available at source or acceptable at destination; and the physical constraints provided by the routes, ports and handling equipment (Stopford, 1988).

Shipping is less limited by size constraints than other modes and is able to capitalize on what has been called the 'cube law'. Simply put, this means that for a squaring of a ship's dimensions the carrying capacity is cubed. Also, the design and construction costs and operating costs (crew, fuel) do not increase in proportion with size. A 500,000-tonne tanker is able to operate with a crew no larger than that needed for a vessel one-twentieth the size although there will be variations depending on national flag regulations, the level of automation and company organization (e.g. the amount of emphasis on shipboard maintenance).

The cost savings increase with vessel size and voyage distance and this is why there has been such dramatic growth in the size of bulk carriers and tankers in particular (Table 10.3). These averages conceal the extremes which in the case of oil tankers are in excess of 500,000 tonnes and for dry-bulk carriers around 350,000 tonnes. Many bulk trades are effectively one-way traffic with return voyages in ballast. To reduce this empty running and to increase operational flexibility and revenue there has been an increasing use of combination carriers (e.g. OBO – oil/bulk/ore; CONBULK – container/bulk).

However, while consignments of such size may be available from oil and ore producers and acceptable at the processing plant, this

Table 10.3 Average size of vessel in service

	Tankers over 10,000 dwt	*Dry-bulk carriers*	*Combination carriers*
1963	26,100	-	-
1966	33,000	23,600	-
1971	46,900	32,300	68,900
1976	84,600	33,100	111,100
1979	104,600	36,900	117,100
1985	100,000	38,500	116,200
1986	95,100	39,800	116,700
1989	91,800	42,000	117,300

Source: OECD *Maritime Transport*, annual reports.

would not be the case in many bulk trades. Also, there is no financial advantage in using large vessels if the loading and unloading rate is slow and the vessel is kept unduly long in port. A 250,000-tonne tanker needs to pump at 16,000 tonnes an hour and 22,000 would be appropriate for a 350,000-tonner (Stopford, 1988).

The ultimate constraint on vessel size must be the physical characteristics of the port (channel depths, turning circles, lock gate dimensions and berth lengths) and the routes along which the ships operate. It is normal now to talk of PANAMAX (290 m × 31.7 m × 12.8 m) or the largest vessel that can transit the Panama Canal, a bulk carrier of about 55,000 to 65,000 tonnes; and SUEZMAX where the limit is a draught of 19.5 m which is a loaded tanker of about 150,000 tonnes or a 350,000-tonner in ballast. A tanker of 250,000 tonnes can transit the Malacca Strait on a voyage from the Gulf to Japan but any larger vessel would have to make a much longer voyage by way of the deeper Lombok Strait (Smith, 1973). The port of Rotterdam has had to increase channel depths continuously in order to accommodate ever larger bulk carriers, from 12.5 m in 1960 to 22.6 m in 1987.

Ports as industrial areas

It has been argued (Hoyle and Pinder, 1981) that industrialization at seaports provides the most important sphere in which port functions and urban processes interact. However, this interaction can only be fully understood when seen against the wider relationships that include traffic generation by the hinterland, land-transport systems, foreland links and market factors.

At the outset a port is likely to develop industries relating to the ship itself – shipbuilding and repairs, sail-making, marine-engineering – but as time goes on industries related to the processing of the cargoes is likely to emerge. This may be either the processing of hinterland materials for export or the processing of imports for the hinterland market. The significance of the port as a break-of-bulk point in the transport chain has already been noted. For a variety of economic, technical, administrative and logistical reasons, the act of transhipment provides a basis for the storage and possibly the processing of the commodities concerned. The port may well be as close to the source of the raw materials as an industry can be and any primary processing at the port will provide inputs for secondary processing. The port is ideally located to produce forward and backward linkages and for this reason some planners see ports as potential growth poles (e.g. San Pedro in Côte d'Ivoire and Tema in Ghana) although the spread effects are often slow to develop (Hoyle and Pinder, 1981).

There have been various attempts to classify port industries (Amphoux, 1949; Bird, 1971). A simple distinction can be drawn between *industrial ports* in which industry has dedicated port installations; *general port industry* in which the industry does not have a dedicated terminal but is nevertheless dependent directly on the port as a break-of-bulk point; and *urban industries* which are attracted to the port because it is a centre of population and a market in its own right. In the first category are oil refineries, large metallurgical processing plant, chemical industries, cement works, paper mills and grain mills while in the second group are rubber, tobacco and food processing and timber industries. There is clearly a continuum from those industries for which a dockside location is essential to

Photograph 10.2 Floating equipment at Rotterdam allows flexibility of handling both to shore for storage and sorting before onward movement (coal in background) or direct to barges alongside bulk carriers at anchor 'in stream' (grain in foreground) (Interstevedoring, Rotterdam).

those for which such a location is unnecessary.

The specific location of a particular processing plant in relation to the quayside will also be a function of the physical nature of the cargo and the technology that is required to discharge it from the ship and move it away from the quayside. As a general rule, the more difficult and costly the handling the nearer the processing plant will be (Takel, 1981). However, the handling costs for bulk products are strongly influenced by scale economies and the volumes involved will be a critical factor.

Liquid bulk cargoes can be pumped and piped cheaply and easily over long distances if the volumes are large and this allows for geographical separation of the cargo terminal and the processing plant. Examples include the crude oil pipeline from Loch Long in western Scotland to the refinery at Grangemouth on the east coast and from Rotterdam to the Ruhr. Movement cost by pipeline for small quantities is relatively expensive and the processing of liquid latex, molasses and vegetable oils is often undertaken close to the quayside. Similarly, commodities such as grain or alumina where suction-discharge methods are used are likely to be adjacent to the quayside, although at Tema, Ghana, alumina is moved over two kilometres on an enclosed conveyor belt from the port storage to the smelter.

Photograph 10.3 Completed in 1985, Le Havre's 100 ha multipurpose bulk terminal can accommodate ships of up to 120,000 tonnes and provides open storage for 400,000 tonnes of coal, facilities for screening and pulverizing coal for local industry and export and covered storage for grain substitutes for the animal feed industry. Ships and inland waterway craft can be loaded and unloaded at 1,500 tonnes/hour (Port of Le Havre).

Maritime industrial development areas

Industrial development is therefore a normal characteristic of ports, although to explain all city-port industrialization in terms of the maritime factor is clearly too simplistic (Verlaque, 1970; Winkelmans, 1980). However, in the late 1960s the idea of a Maritime Industrial Development Area (MIDA) was introduced (ZIP – Zone Industrielle Portuaire in French) for planned industrial development in coastal areas where the dominant influence was maritime. In particular this reflected a vast increase in the scale of processing in the oil, iron and steel industries, the need for much larger quantities of the raw materials and the requirement that the sea transport had to be by the largest possible vessels in order to maximize scale economies. The new MIDAs were therefore associated with revolutionary changes in maritime transport and the adoption of Very Large and Ultra Large Crude Carriers (VLCCs and ULCCs) and ever larger dry-bulk carriers to reduce maritime-transport costs (Vigarié, 1981).

The essential requirements of any such MIDA are deep water, large areas of land suitable for storage and industrial plant and a well-developed land-transport system. In a study undertaken for the National Port Council in Britain (Peston and Rees, 1970) the specific criteria adopted were 15.2 m water depth and 2,000 ha of land. Four stages of MIDA development have been identified (Vigarié, 1981).

The first generation of MIDAs were based on what Vigarié has termed the Rhine model. In Rotterdam, starting in 1958 with the Botlek scheme, there has been planned industrial and port development extending seawards through Europoort and Maasvlakte and embracing an area of 10,000 ha with heavy emphasis on oil refining and petrochemicals. A 1955–65 port-development plan for Antwerp was associated with new oil refineries, a wide range of chemical and petrochemical industries (Bayer, Monsanto, Essochem, BASF) and vehicle assembly. In France the 1965–70 port plan resulted in large-scale port-industrial development of iron and steel, oil refining and petrochemicals at Dunkerque (Tuppen, 1981), of oil refining, chemicals, metal refining, wood processing and vehicle assembly at Le Havre (Gay, 1981) and iron and steel, oil and chemicals at Fos (Tuppen, 1975).

In Japan, the need to reclaim land from the sea for large-scale industrial development combined with the need to import nearly all basic raw materials and fuels resulted in port-development plans for 1965–9 and 1971–5 which created large MIDA areas (*kombinato*) such as Mizushima and Kashima (Rimmer, 1984).

There was a reaction to the large-scale industrial development involved in the MIDA on both social and environmental grounds and the recession of the later 1970s slowed their growth. The second-generation developments, as a consequence, have placed less emphasis on heavy industries and more on lighter secondary and tertiary activities including warehousing and distribution. In Rotterdam the development of a series of Distriparks typifies this trend and whereas some 84 per cent of land in early Japanese MIDA developments was devoted to industry the percentage has now dropped to as little as 24 (Vigarié, 1981). MIDA development is also found on a smaller scale in some Third World countries: Vigarié's third-generation MIDA. The varied industrial development at Tema, Ghana (Hilling, 1966) provides an example, as do the port-related phosphate industry at Gabès, Tunisia and plans for MIDA development in Thailand (Robinson, 1981).

If in the cases cited, port-related industrial development has been a result of design and planning, in Britain such areas may be seen as a product of historical process and default. On the Tees, Humber, Thames, Southampton Water and Severn there are elements of port-related industry but at no place has the development been planned in an integrated manner or on a very large scale. At Port Talbot and Redcar existing iron and steel works with inadequate ore-importing terminals were given new deeper-water berths. Spatially separated oil refineries were established on Humberside and on the Thames a planned port-airport-industrial development at Maplin did not find favour with the government. The Peston and Rees report (1970) was quietly buried.

Vigarié's fourth hypothetical stage of MIDA development is one in which some of the traditional heavy industries are maintained but in which there are also a larger number of industries based on imported, semi-finished materials and products, possibly of a high-technology nature and geared to export. There are already in reality some signs of this type of development.

Traditionally, when goods are off-loaded from the ship they become subject to customs duties. Some ports, Hamburg for example, have always claimed to be 'free' ports at which goods can be landed and processed before customs duties are payable. Many ports are now creating such 'duty-free' areas. They take various forms, of which the simplest is a warehouse operating under customs bond – duties are paid on the wine or tobacco, for example, only when they are taken from the *bonded warehouse*. The free zone might consist of no more than one waterside factory at which goods can be landed, processed and re-exported, all

Photograph 10.4 Crescent Shipping's *Stridence* has a length of 85m, beam of 11.5m, draught of 3.45m and a bridge which can be lowered hydraulically to give a maximum air draught of 4.5m. These dimensions give a carrying capacity of 1,800 tonnes but allow the penetration of inland waterways and access to smaller coastal ports which is necessary in short-sea transhipment trades (Cochrane Shipbuilders).

without customs duties being paid. A number of factories might constitute an *export processing zone* while additional functions such as storage and distribution would produce a *special economic zone*. The extreme case is a full free port or comprehensive free-trade zone. Such free-port or foreign-trade zones form a hierarchy based on areal extent and range of functions. The object of such free zones is to attract industry and create employment but they are clearly based on the port's break-of-bulk function and an extension of the MIDA concept.

As a consequence of the developments in sea transport and associated port land-use described above, there is increasing differentiation between ports. New criteria for port selection are leading to the emergence of new port hierarchies based on the ability to accommodate the largest vessels. As in air transport and distribution in general, new hub-and-spoke patterns of activity are emerging.

Examples of bulk trades

Having considered the transport of bulk cargo, its influence on industrial location and the significance of ports as industrial areas, it is appropriate to bring these themes together by looking at some of the principal bulk

Photograph 10.5 Le Havre's Quai Hermann du Pasquier grain terminal has silo storage for 30,000 tonnes and shed storage for 47,000 tonnes and can be used as an import or export facility for bulk carriers up to 30,000 tonnes. The terminal is equipped for transhipment to and from road vehicles, trains and barges (Port of Le Havre).

trades in greater detail. The two chosen for further analysis are the coal and grain trades (in the latter case non-grain animal feeds are also discussed). In all these trades, transhipment plays a role in the bulk-transport system but its significance varies considerably – from a UK perspective, transhipment is more important in the coal trade than in the grain trade.

Importantly the transhipment act itself also varies. For example, transhipment may involve the transfer of cargo direct from a large mother ship to smaller feeder vessels or, more typically, it involves transfer to the quayside or storage area for some degree of processing and then for subsequent shipment in a smaller feeder vessel. The degree of storage and processing has important implications for the role of the transhipment port as an industrial area; the more bulk cargo that is transhipped and processed the greater the opportunity for the port area to develop a wider industrial base as related industries decide to locate their activities nearby.

The coal trade

World trade in coal amounts to over 300 million tonnes a year split almost equally between coking coal which is used in the

Table 10.4 Economies of scale in bulk shipping

Ship size dwt	*Operating cost $'000*	*Bunker costs* $'000*	*Total cost $'000*	*Cost per dwt $ p.a.*
40,000	1,315	1,890	3,205	80
65,000	1,540	2,295	3,385	59
120,000	1,780	3,051	4,831	40
170,000	2,120	3,780	5,900	35

* Assuming 270 days at sea per annum at 14 knots

Source: Stopford, 1988, based on Drewry (Shipping Consultants) Ltd. 1985.

production of steel and steam coal used primarily for generating electricity in thermal power stations (OECD, 1989). However, this represents only 10 per cent of world production, since most coal is not traded internationally but is used to satisfy domestic needs.

The major coal exporting countries are the USA, Canada, Australia and South Africa. Import requirements are greatest in Western Europe and Japan which between them account for 77 per cent of all import tonnage (Fearnleys, 1989). Clearly this pattern of supply and demand results in long-distance transport by sea. On these deep-sea routes the large bulk carrier plays a key role: vessels of over 80,000 dwt carried 51 per cent of coal transported by sea in 1986. Indeed, the general trend in the long-distance coal trades is to use ships in excess of 100,000 dwt.

Large ships dominate the world coal trade reflecting the importance of economies of scale in vessel size. As Table 10.4 indicates, the potential economies of scale are significant: almost doubling the size of the ship from 65,000 to 120,000 dwt only increases total costs by 26 per cent. In addition, both the long ocean voyages and the relatively predictable nature of the coal trade promote the use of very large ships.

Turning from coal trade at the world level to trade concerning the UK highlights the impact on bulk-freight transport of factors such as national policy and commercial practices. Until the 1970s UK imports of steam coal were insignificant – domestic production was perfectly adequate to satisfy the needs of the Central Electricity Generating Board (CEGB) and since both coal and electricity were nationalized industries there was little incentive to change their historically close relationship. The CEGB was for many years by far British Coal's most important customer, accounting for 77 per cent of purchases (MacKerron, 1987). At the end of the 1980s the CEGB continued to have a binding agreement to take 95 per cent of its coal requirements from British Coal, although much of this was at a high price compared with the prevailing price of coal on the international market.

As a result of the agreement, purchases of steam coal from abroad have not risen above a low level; even in 1985 following the miners' strike only 4.4 million tonnes were imported. Most imported coal is transported over long distances and much comes under contract from the USA and Australia, although in the late 1980s spot purchases from Colombia and China increased. Poland is the only truly short-sea source although much of the coal imported from, say, the USA arrives in the UK on board a small vessel from a mainland European port as a result of transhipment. For example, in 1986 3.7 million tonnes of steam coal was imported and an estimated 2 million tonnes of this was transhipped. Transhipment via a mainland European port takes place mainly because there has been little reason to build UK coal-import terminals capable of handling the very large bulk carriers which are now used in long-distance coal transport.

Much of the transhipped coal moves through three major terminals in The Netherlands: Frans Swarttouw and EMO in Rotterdam and OBA in Amsterdam. These terminals are owned by firms experienced in transhipment operations to various European destinations including Scandinavia, the Baltic and Germany. They are also large-scale operations taking Cape-size bulk carriers (120,000–140,000 dwt) and since they handle very high volumes of cargo, they benefit from scale economies. Frans Swarttouw, for example, handles 160 million tonnes of bulks annually, of which coal comprises 45 million tonnes.

However, it has been argued (Browne et al., 1989) that rather than transhipment being seen as a second-best alternative, in the absence of suitable UK facilities, the development of Rotterdam as a European hub for bulk coal movements may provide coal-using industries with a number of significant advantages:

A stock of different grades — industrial customers increasingly require various consignments of different grades of coal. These may come from different mining areas. The major terminals, such as those in Rotterdam, can hold stocks of these different grades only a few hundred miles from the customer and in a strategic location, relatively safe from industrial disruption.

A total quality approach from the terminal — Added value in the forms of screening, blending and washing is provided for half of the current throughput at the main Rotterdam terminals. These activities require specialist skills and equipment which may only be justified at very high levels of use – so focusing on a hub terminal is a logical strategy.

Reduced storage space required — Small consignments can be moved from the transhipment point to the destination but these could not be moved economically directly from the source.

Minimizing the overland haul — Because small vessels are employed for bulk transport after transhipment, it is usually possible to get close to the final destination because there will be a much wider range of ports and terminals able to accept relatively small vessels.

While some or all of these factors may influence decision-makers it is likely that total transport costs will be the most important consideration in comparing transhipment with direct shipment, but the comparison is difficult. Fluctuating costs, particularly freight rates, can radically alter the balance of advantage between transhipment and direct shipping. Lower freight costs achieved by using very large bulk carriers may be largely nullified by extra costs for handling during transhipment.

The hypothetical model shown in Table 10.5 illustrates how transhipment could be a preferred option for a UK coal importer and shows how changes in handling and transport costs can strongly influence the decision. The model compares rates for coal from Australia and since no small vessels would be used on such a lengthy voyage the options are:

a) Large bulk carriers transhipping on the Continent with transfer to final destination on small bulkers. (The model considers three typical sizes for large bulk carriers.)
b) Panamax vessels sailing directly to the UK with land or sea transfer to the final destination. (The model considers two typical Panamax-size vessels.)

Using the costs illustrated, transhipment appears to be the cheaper option. The price-ranges per tonne are: a) for transhipment, $13.75–$20.00; and b) for direct shipment, $20.00–$23.00. But it is evident that there is some susceptibility to minor fluctuations, since the top of the range for (a) is the same as the bottom of the range for (b).

With this narrow margin there is clearly sensitivity to cost changes in all parts of the model. Continental handling costs rose from $2.50 per tonne in 1984 to $3.00 in 1988 and further rises would make transhipment less competitive. The cost of the small bulker varies with market fluctuations and there is also a wide difference between rates for geared and gearless vessels. Most significant of all are fluctuations in the charter market for deep-sea bulk carriers. While upward or downward freight-rate trends will be approximately followed by the various sizes of vessel, the key to the model as presented here may be the differential between rates for Panamax and large bulk carriers. The narrower the gap, the weaker the case for

Table 10.5 Transport and handling cost model – coal from Australia to UK

Option (a) Transhipment

Vessel	Freight rate – port of origin to transhipment port	Transhipment port handling cost	Freight rate – transhipment port to UK power station	Total transport cost per tonne
110,000 dwt	12.00	3.00	3.00–5.00	18.00–20.00
140,000 dwt	8.15–10.00	3.00	3.00–5.00	14.15–18.00
160,000 dwt	7.75–8.00	3.00	3.00–5.00	13.75–16.00

Option (b) Direct shipment

Vessel	Freight rate – port of origin to UK port	Freight rate UK port to power station	Total transport cost per tonne
60,000 dwt	20.00	2.00–3.00	22.00–23.00
70,000 dwt	18.00	2.00–3.00	20.00–21.00

Notes
All costs US$ per tonne of cargo.

UK stevedoring costs are assumed to be the same in each case and are therefore omitted from the calculations. The ocean freight rates are based on actual fixtures recorded by Lloyd's Maritime Information Services in 1988.

Source: Browne et al., 1989.

transhipment from a large vessel. Indeed, fluctuations in rates in this way may be partly responsible for changes in the volume of bulk transhipment from year to year.

Yet it is not so much changes in freight rates and ship- or cargo-handling technology that will affect the volume of imported coal. Of far greater importance are changes of an institutional type: in this case the ownership of one of the main buyers, the CEGB. Privatization of the CEGB and the inevitable loosening of the coal-supply agreements between the generating companies and British Coal could lead to a dramatic increase in coal imports in the second half of the 1990s (*The Economist*, 1989).

Steam coal has been available on the open market at prices much lower than the average paid by the CEGB to British Coal. The future level of steam-coal imports will depend on how competitive domestically produced coal is in comparison with overseas supplies. While the price has been low, it is important to note that the market for internationally traded steam coal is quite small. The total world volume of traded steam coal is less than twice the UK demand (MacKerron, 1988) and any large-scale buying by the UK electricity industry could cause significant price increases. On the other hand, new, low-cost sources including Sumatra, Borneo and China could increase supply. British coal prices could also change significantly as high-cost pits close and low-cost areas increase production.

Within an uncertain picture for the coal trade as a whole, the volume of transhipment is difficult to forecast. The level of utilization of larger-than-Panamax bulkers for coal is increasing. If such vessels cannot berth at UK ports, transhipment may well increase. However, if common-user terminals taking large vessels were built, such as those proposed at Immingham or the Isle of Grain, the case for transhipment would be weakened. Alternatively, large vessels could be accommodated at British Steel terminals (normally used solely for imports of coking coal and iron ore for their own use) and steam coal could be transferred from, say, Hunterston or Redcar to power stations.

The grain trade

The grain trade provides a contrast to coal in

Table 10.6 Major trade routes for grain shipments

Route	*Volume (billion-tonne-miles)*
United States to Far East (including 203b-t-m to Japan)	345
United States to Europe	205
Argentina to Europe	74
United States to other Americas	64
Canada to Europe	45
Australia to Far East	32
Canada to Far East	32

Source: Lloyd's Maritime Atlas, 1990.

the size of ships used, the key trading routes and the scope for transhipment. When used in the context of bulk shipping the term 'grain' comprises a variety of produce ranging from wheat to soya beans. Wheat and maize dominate seaborne trade in grains, accounting for almost 90 per cent of tonnages shipped (Lloyd's of London, 1990). The major exporting regions are North America, Argentina and Australia and the main trade routes are shown in Table 10.6. World trade in grain ran at about 200 million tonnes a year during much of the 1980s. However, trade volumes are more erratic than in the case of coal, since variations in weather conditions and harvests can have a dramatic impact on supply and demand.

Almost half (48 per cent) of world trade in grain is transported on ships of less than 40,000 dwt and only six per cent is moved on those over 80,000 dwt; a very different pattern from that found in the coal trade and one which clearly has implications for the need to tranship cargo. Although there are economies of scale in the bulk shipment of grain, just as there are for coal shipments, a number of factors limit the use of very large vessels. Among the most important are:

1 Draught limitations in many of the export loading ports. For example, the depth of the Mississippi is a critical limiting factor on the optimum ship size for use in trades between the USA and Europe.
2 The grain trade has been described as opportunistic, reflecting the vagaries of supply and demand together with its seasonality. This degree of uncertainty makes it difficult to schedule the use of very large vessels. Often it may be better to have the opportunity to use, say, five 30,000-dwt ships rather than one 150,000-dwt vessel. The use of several smaller vessels means grain shippers and buyers can respond in a more flexible way to changing market conditions.
3 The complex market structure. Shipments are often made up of consignments from a number of different sellers which leads to coordination difficulties for very large shipments.

European trade in grain has been heavily influenced by the agricultural policies of the European Community (EC) and the impact of these policies on the UK has been marked. Since the early 1970s the UK grain trade has experienced substantial changes in both volume and direction resulting from British entry to the EC and the increase in grain production which has turned the UK from a net grain importer to a net exporter. Between 1970 and 1975 the UK imported up to 9 million tonnes of grain annually, principally from distant countries such as the USA and Canada. But the expansion of UK production led to a dramatic fall in imports and by 1980 they were down to 5.2 million tonnes with a further fall to 3.3 million tonnes in 1985. Of the latter figure only 1.3 million tonnes was shipped from distant sources of supply (Department of Transport, 1988).

UK exports rose to over 2.5 million tonnes by 1980 and continued to increase so that by 1985 they reached almost 4.0 million tonnes. While much of this goes to other EC countries, markets have also been established elsewhere in Europe, North Africa and the Middle East (Department of Transport, 1988). Most of the European cargo is shipped in small vessels from small ports near the production areas. Larger vessels had been required in earlier years to serve more distant markets economically and the lack of grain-exporting facilities for larger vessels in the

UK had led to transhipment via the Continent. More recently, however, facilities for direct shipment of larger volumes have been available, so vessels of 15,000 to 50,000 tonnes have been used. Nevertheless, transhipment of grain exports to long-distance overseas markets continues and since 1982 the annual volume has averaged 300,000 tonnes.

The requirement of some grain buyers for the grain to be bagged can also encourage transhipment. Saudi Arabian buyers of barley for animal feed form a significant part of the non-European export market and they buy bagged grain for ease of distribution locally. This is shipped in bulk to Antwerp and Ghent, which have a good reputation for bagging, then continues break-bulk to Saudi Arabia. This system also enables small shipments from small UK ports to be grouped on the Continent for sale as larger consignments.

The case for transhipment is more marginal for grain than for coal as very large bulk carriers are not normally used. A large freight-rate differential between vessels able to call in the UK directly and those used in a transhipment operation would be necessary for transhipment to be cheaper. The following example shows that this may only rarely be the case. In 1990 the charter rates per tonne for grain shipments from the US Gulf to Europe varied considerably (Lloyd's List, 1991). The range for two different sizes of vessel were quoted as: (a) for a 50,000 to 60,000 dwt vessel, $11.25 to $16.00; and (b) for a 70,000 to 85,000 dwt vessel, $11.00 to $14.50. While the largest differential, $5.00, would cover the cost of transhipment and onward carriage from the Continent to the UK, in most cases direct shipment to the UK would have been the cheaper option.

Non-grain animal feed

One of the consequences of the higher levels of UK grain exports has been a shortage of domestically produced grain for animal feed. Exchange rates coupled with EC agricultural policy have made it attractive for UK compound-feed manufacturers to import non-grain substitutes for UK grain. This has become a large part of bulk import transhipment since 1981 (Browne et al., 1989).

Compound-feed manufacturers produce several products such as oil cake, cattle feeds, pig feeds and poultry feeds. Each product has perhaps seven or eight main ingredients together with small quantities of additives. The main ingredients can be grains such as barley and maize but increasingly a variety of non-grains is used. This includes soya, sunflower, copra, corn gluten and tapioca. Non-grain animal feeds are imported from distant sources to Europe in approximately the following proportions (Browne et al., 1989): North America, 40 per cent; Far East, especially Thailand, 30 per cent; South America, especially Brazil and Argentina, 30 per cent. Large vessels are often used, particularly from Thailand where single loads of up to 140,000 tonnes may be exported.

The compound-feed business is intensely competitive. There are a large number of buyers showing little brand loyalty and several hundred manufacturers in the UK. The prices of the main raw materials fluctuate wildly but to some extent they are substitutes for one another. This causes manufacturers to buy late and use linear-programming techniques to calculate the right mix at lowest prices. Rotterdam is the dominant market place for the compound-animal-feed business. All price quotations are based on 'cif Rotterdam'. The size of the market allows buyers and sellers to trade with anonymity and enables last-minute buying because of the large volumes available. It is also a traditional main grain port able to receive the largest vessels. Various stevedoring companies handle the cargoes but the largest is Graan Elevator Maatschappij (GEM) which handles 17 million tonnes annually. Only about one-third of Rotterdam's agribusiness is for Dutch consumption: it is the major transhipment centre for non-grain animal feed throughout Europe. GEM alone annually tranships over three million tonnes of compound animal feed by sea and the UK market accounts for up to 70 per cent of this.

There is a strong case for transhipment in the compound feed industry since each shipload is split and sold in lots of 10,000 tonnes or less to different compound-feed companies and each of these buyers may need only 800 to 2,000 tonnes delivered at a time. Each consignment may represent just one of, say, eight ingredients needed to make one of a range of animal-feed products. It would clearly not be practicable to ship such small amounts direct from the distant country of origin. The need to divide and then combine various bulk commodities in this way provides a powerful rationale for the use of a central hub such as Rotterdam. Furthermore, compound-feed manufacturers prefer to buy in Europe. A purchase made in South America may have become relatively expensive by the time it reaches Europe because of price fluctuations. This could make the firm's product uncompetitive since 80 per cent of compound-feed manufacturers' total costs are raw materials.

An additional factor which favours the use of a major hub port is concerned with the behaviour of the sellers. Feedstuff shippers in, say, Chicago may prefer to use the major ports and terminals for large consignments of compound feed: they are worried that if only part of the shipload can be sold during the voyage then at a small destination port they will not find the buying infrastructure necessary to deal with the unsold part of the cargo.

As a result of this complex mix of commercial, locational and technical advantages Rotterdam is unlikely to lose its position as the focal point for compound-animal-feed business to European rivals. Hamburg and Antwerp specialize in grain but do not have the expertise in compound animal feed. Bremen and Ghent cannot receive the larger vessels now used. However, the future for grain and compound-animal-feed transhipment depends to a considerable extent on EC policy. Attempts to eliminate grain surplus would be likely to reduce exports and therefore cut export transhipment. As larger vessels can now be handled in the UK, grain export transhipment is likely to fall anyway. Furthermore, countries currently demanding bagged grain are likely to be able to handle bulk cargo in future. This will cut any transhipment caused by the use of continental bagging plants.

Grain exports will also be hit by EC limitations imposed by agreement with the USA. If grain cannot be exported because of these agreements on subsidies, then the EC surplus will have to be dealt with within the Community and it could be used for animal feed. As a consequence there could be a drop in the demand for imported compound animal feed and a reduced role for transhipment.

Conclusion

The development of bulk transport by sea has been strongly influenced by technology. Economies of scale have been achieved through the construction of larger ships and this has affected the optimum parcel size, vessel routeings and port-selection criteria. Port selection is especially relevant because of the strong link between ports and industrial activity. In some cases this activity has been the result of deliberate government policy, as in the case of MIDAs, while in others it merely reflects the commercial and economic advantages of locating activities at the transhipment point. However, technology and vessel design are by no means the only factors at work to influence the patterns of the world bulk trades; government policy, commercial buying practices and physical constraints such as water depth in ports also play a key role.

The transhipment point for bulk cargo may become the focal point for further processing of the bulk commodity. The processing of a cargo at the intermediate transhipment point is of considerable importance. When coal is transhipped it may be screened, washed, blended and graded at the transhipment port, while animal feeds may be tested or inspected during transhipment. Indeed, the scope to add value to the product or cargo at the transhipment port is important to the terminal operators for it enables them to earn revenue, to differentiate their services from those of competitors and to reduce unit prices by spreading fixed costs over greater terminal throughput.

Although there are important similarities between the bulk shipment and transhipment of grain and coal, there are also important differences. These differences arise in large part from the different pattern of supply and demand and the different type and size of vessel used to service these major world bulk trades. Since coal is shipped over long distances in very large vessels, the use of a north-European hub at Rotterdam is often the lowest-cost choice. Transhipment by feeder vessel can then be made to smaller ports in the UK and Scandinavia located near the ultimate consuming destination, frequently a power station. A coal-producing country such as the UK has had little need for imported coal, but changes in the ownership of utilities – the privatization of the CEGB – means that the desire to purchase coal on the low-cost open world market may override the traditional arrangements for using almost entirely British coal.

By contrast with the coal trade, grain is shipped in much smaller vessels and transhipment is, as a consequence, far less important. Interestingly in the case of non-grain animal feed, some of the same arguments which lead to the hub concept in coal movement can be applied; again, a logical outcome is increased transhipment. This is first because there are physical limitations to the ports that can accommodate these large ships and second because there are economic advantages in having a central stockholding point and then distributing smaller parcels of cargo to the buyers. This enables buyers to minimize their inventory without sacrificing to too great an extent the safety or security of buffer stocks held relatively close at hand.

Until recently the individual elements of the transport chain while functionally related were operated in a largely disparate way. In the bulk trades, as in maritime transport in general, there is now a realization that the effective operation of through-transport from the producer via the processing industry to the consumer requires a high level of organizational interdependence. It is realistic now, as it has not been before, to view bulk transport as a 'system' embracing not only the movement but also the intermediate handling, storage and processing of the commodities involved. It follows from the variety of decision-makers and external influences involved that this system is highly dynamic.

In bulk as in general cargo trades the reduction of inventory and storage costs by Just-in-Time (JIT) shipments seems set to increase in significance. It seems likely that closer economic integration and the removal of customs barriers in the European Community will stimulate the reorganization of industrial production and distribution and strengthen hub-and-spoke patterns. This is clearly linked with the economies derived from the combined use of the largest possible ocean carriers and efficient mini-bulkers on feeder services. However, as this chapter has demonstrated, within this broad framework, the responses in different trades will vary in detail depending on the nature of the commodities involved, the organization of the processing industries and the geography of the market for their products.

References

Amphoux, M. (1949), 'Les fonctions portuaires' *Revue de la porte océane* (Le Havre), no. 54, 19–22.

Bird, J. (1971), *Seaports and seaport terminals* (London: Hutchinson).

Browne, M., Doganis, R. and Bergstrand, S. (1989), *Transhipment of UK trade* (London: British Ports Federation and Department of Transport).

Department of Transport (1988), *Short sea bulk shipping: an analysis of UK performance* (London: Department of Transport).

Drewry (Shipping Consultants) Ltd. (1985), *Dry bulk operating costs, past, present and future* (London).

Economist, The (1989), 'Keep the home fires burning' (London, 26 August 1989).

Fearnleys (1989), *Annual review and World bulk trades* (Oslo).

Gay, F. J. (1981), 'Urban decision makers and the development of an industrial port: the example of Le Havre' in B. S. Hoyle and D. A. Pinder (eds), *Cityport industrialisation and regional development* (Oxford: Pergamon), 201–22.

Greenhut, M. L. (1956), *Plant location in theory and practice* (Chapel Hill: University of North Carolina).

Hamilton, F. I. (1962), 'Models of industrial location', in R. Chorley and P. Haggett (eds), *Socio-economic models in geography* (London: Methuen), 362–424.

Hilling, D. (1966), 'Tema – the geography of a new port' *Geography*, 52(2), 111–12.

Heinemann, M. and Cheetham, C. (1987, 1990), *Modern Rhine-sea ships* (London: Fairplay).

Hoover, E. M. (1948), *The location of economic activity* (New York: McGraw Hill).

Hoyle, B. S. and Pinder, D. A. (eds) (1981), *Cityport industrialisation and regional development* (Oxford: Pergamon).

Lloyd's List (1991), 'Weekly freight table' (London: Lloyd's List).

Lloyd's of London (1990), *Lloyd's maritime atlas* (London: Lloyd's).

MacKerron, G. (1988), 'The international steam coal market and UK coal' in *Privatising electricity: impact on the UK energy market* (London: Institute for Fiscal Studies).

Organisation for Economic Cooperation and Development (1989, 1990), *Maritime transport* (Paris: OECD).

Peston, M. H. and Rees, R. (1970), *Maritime industrial development areas* (London: National Ports Council).

Pred, A. (1967), *Behaviour and location* (Lund: University of Lund).

Rimmer, P. J. (1984), 'Japanese seaports: economic development and state intervention' in B. S. Hoyle and D. Hilling (eds), *Seaport systems and spatial change* (Chichester: Wiley), 99–133.

Robinson, R. (1981), 'Industrialisation, new ports and spatial development strategies in less-developed countries: the case of Thailand' in B. S. Hoyle and D. A. Pinder, op. cit., 305–21.

Smith, S. R. (1973), 'Ocean-borne shipments of petroleum and the impact of straits', *Maritime Studies and Management* 1(2), 119–30.

Stopford, M. (1988), *Maritime economics* (London: Unwin Hyman).

Takel, R. E. (1974), *Industrial port development* (Bristol: Scientechnica).

Takel, R. E. (1981), 'The spatial demands of ports and related industry and their relationship with the community' in B. S. Hoyle and D. A. Pinder, op. cit., 47–68.

Tinsley, D. (1984), *Short-sea bulk traders* (London: Fairplay).

Tuppen, J. N. (1975), 'Fos – Europort of the south' *Geography*, 60(3), 213–17.

Tuppen, J. N. (1981), 'The role of Dunkerque in the industrial economy of Nord-Pas de Calais' in B. S. Hoyle and D. A. Pinder, op. cit., 365–79.

Verlaque, C. (1970), *L'industrialisation des ports de la Méditerranée occidentale* (Montpellier: L'Université).

Vigarié, A. (1981), 'Maritime industrial development areas: structural evolution and implications for regional development' in B. S. Hoyle and D. A. Pinder, op. cit., 23–6.

Weber, A. (1909), *Uber den standort der industrien* translated by C. J. Friedrich (1929), *Alfred Weber's theory of the location of industries* (Chicago).

Winkelmans, W. (1980), 'Transport and location: an inquiry into principal evolutions' in J. B. Polak and J. B. Kamp (eds) *Changes in the field of transport studies* (The Hague: Niihoff), 202–11.

11

Multimodal Freight Transport

Yehuda Hayuth

Freight transportation is gradually shifting toward an integrated transport system in which individual transport modes contribute to making the entire transport journey more time- and cost-efficient. Organization and coordination comprise the focus of the new multimodal companies, landbridge operations and changes in many conventional transport concepts.

Introduction

Transportation has traditionally been viewed and analysed as a segmented system, comprised of individual transport modes. It has also been looked at as a component of the spatial organization of economic activities (Ullman, 1956). The dichotomy of passenger and freight transportation is a well-established practice in the literature, and the separate attention given to sub-modes of the freight-transport system, such as bulk transportation and general cargo-liner services, is quite common and not without its rationale.

The underlying approach to most transportation studies has been the separate treatment of individual modes, such as sea transportation (Couper, 1972), seaports (Bird, 1971), railroads (O'Dell and Richards, 1971), barge transportation (Mellor, 1983) and trucking or air transportation (Sealy, 1968). Most transportation textbooks adopted this concept, as have, needless to say, indices and library catalogues. Certainly, each one of these transport modes has its own characteristics, rationale and competitive environment. Justifying such an approach, therefore, is not an uphill task.

To what extent is this view of the transport industry as a segmented system an artificial if not a misleading division? This question has come into sharper focus with the global development of transportation, particularly in the second half of the 1980s. One of the fundamental objectives of transportation is to overcome the spatial gap between points of demand and centres of supply. Transport systems are there to move passengers or freight from their origins to their final destinations. We might need to change transport modes along the way, but the rationale behind the entire operation and the principal objective is to overcome the total route in the most cost- and time-effective manner. With the development of containerization and especially with the advance of intermodality, a reconsideration of the traditional unimodal approach to transportation becomes inevitable.

As its title suggests, this chapter concentrates on freight transportation and primarily on general cargo trade. Although some local and regional issues are raised, the emphasis is placed on the international arena and the adoption of a global approach to the subject.

Unitization and containerization

For generations, liner shipping and conventional breakbulk ships, with multiple ports-of-call, dominated the seaborne general cargo trade. The rapidly increasing rate of international trade after the Second World War, however, and the growing volume of commodities demanding to be transferred overseas could not have found a proper solution with the shipping-transport system of the 1950s. Small vessels, inefficient cargo-handling methods and old port facilities could not cope with the ever-growing volumes of ocean-going trade. The inevitable result was a slowdown in the turn-around time of ships and heavy port congestion.

Unitization of small consignments into larger loading units, utilizing pallets, and more significantly into standard containers, marked the revolution in cargo-handling methods for breakbulk commodities in international trade. Sea-Land's *Ideal X*, plying the route from New York to Puerto Rico in 1956 and Matson's *Hawaiian Merchant*, sailing from San Francisco to Honolulu in 1958, are commonly regarded as the starting points of containerization (Kendall, 1986). The movement of goods in a portable reusable container by more than one mode of transportation is by no means a new or modern idea. In fact, it might have been a common practice in transportation as early as the age of waterborne trade itself. The old humble barrel and the modern drums, the iron crates for hauling coal in England in the 1790s, and the use of containers by the US Post Office in the 1920s provide a few of the many examples of the early uses of containers in one form or another (McKenzie et al., 1989).

The diffusion of the present form of containerization was rapid and extensive. In the late 1950s and early 1960s, seaborne container service became established between continental North American ports and Hawaii, Puerto Rico and Alaska. In the mid-1960s containerized trade between the United States East Coast and Northern Europe began to take hold and the major shipping consortiums, such as OCL and ACT, were formed. In the second half of the 1960s, the trades between North America and the Far East and between Europe and Australia entered the containerization era. Since the early 1970s, containerization has become the dominant method of transporting general cargo commodities in the international trade arena. In 1971, over seven million TEU (20-foot equivalent unit) container movements were recorded in ports around the world. This figure leaped to over 36 million TEUs in 1980 and nearly doubled again to 67 million TEUs by 1987 (*Containerisation International Yearbook*, 1990).

Containerization greatly facilitated the operation, management and logistics of conventional oceanborne, general cargo, liner trading. Moreover, the impact of the container revolution went far beyond the shipping segments of international trade. A list of some of the many direct impacts of containerization on domestic and international trade includes newly designed cellular vessels, much faster ship turn-around times in port, greatly improved cargo-handling productivity at ports and an expanded relationship between water and land transportation (Johnson and Garnett, 1971; Mayer, 1973; Kendall, 1986).

In addition to the contribution to a much greater efficiency of the transport system, containerization brought about a major change in the conceptual level of transport organization as well as in the competitive structure of transportation modes. The container unit became the common denominator for ocean, land and even air transportation. The freight transport industry, consequently, has gradually been shifting from the traditional port-to-port or station-to-station transport practice to a more total-system approach. One effect was greatly to enhance the level and intensity of both cooperation and competition among transport modes.

Throughout the 1960s and part of the

Photograph 11.1 On-dock intermodal terminal at the Port of Seattle. The double-stack train, with some 500 TEU containers, is ready to depart from Terminal 18 at the Port of Seattle for a journey of several thousand kilometres to destinations in the US Midwest and the Atlantic Coast. These trains have greatly enhanced the economy of scale of rail transportation.

1970s, the container revolution was primarily confined to sea transportation and port operations: there was very little interaction between ocean and land transportation. Containerization had been introduced as a direct result of obsolete cargo-handling methods and an intolerably slow loading and unloading process in ports. Indeed, during the first decade of containerization, the pattern of port-hinterland delineation and inland cargo distribution had followed to a great extent the traditional patterns, long established by the conventional break-of-bulk trade. Relatively few containers penetrated deep into the interior of continents; the majority of containers were basically confined to the immediate hinterland of each port. It was only later in the 1970s, with the emergence of new cargo-distribution schemes incorporating a variety of land-bridge concepts, that the conventional notion of a hinterland was challenged.

Multimodality

By the early 1990s, containerization was definitely considered a mature transport system. A great majority of the liner services between industrial countries were containerized, and the level of penetration of containerization into the liner trade between developed and developing countries had surpassed the half-way point. Moreover, no

Photograph 11.2 Transtainer loading containers on a double-stack train. In the new On-dock intermodal terminal at the Port of Seattle, transtainers are loading containers into double-stack trains directly from the container terminal. Such an operation is essential for efficient intermodal transportation of large volumes of containers penetrating deep into the hinterland.

radical changes in the design of container vessels, gantry cranes or methods of port operation had been introduced for over a decade.

The domestic and international freight-transport systems, however, were by then confronting a new phase of the container revolution – the concept of intermodal transport. Intermodality envisions the uninhibited movement of individual containers by at least two different transport modes (Kendall, 1986). Unlike containerization, which was characterized primarily by technological features and design, intermodality focuses on the organization and synchronization of the transport system. In the conceptual metamorphosis of the transport system, from a unimodal basis to a multimodal approach, cargo movements came to be viewed and analysed in the light of the total distribution system.

Included in such a total system are producer, shipper, ocean, land and air carriers, ports, freight-forwarding organizations and warehousing services. The physical distribution of cargo, then, involves an integrated logistical system, in which cooperation and coordination among all the participating elements provide the central focus of the concept (Hayuth, 1987). In such a 'through-transport' system that this 'total' concept implies, each mode serves as a link in an entire transport chain. Individual transport modes do not lose their identities and importance; rather, the role of each is determined by the objectives of the total system. As a system, intermodality attempts to bridge the gap between the points of origin of supply and the final destination points of demand by more rational means of conveyance in terms of cost, time and efficiency.

Cooperation between transport modes is not a new phenomenon. Rail and steamship cooperation existed in Europe and in the USA in the late nineteenth century. Most modal-split freight analyses undertaken in transportation research, however, refer mainly to the share of the total trade held by each mode. Rarely was there any attempt to evaluate the degree of cooperation among various modes. With intermodal transportation, in which the strength and efficiency of the transport chain is in effect determined by the weakest link, cooperation and coordination are essential; it is the system as a whole that must be assessed.

As the volume of intermodal movements of containers grew, especially in the later half of the 1980s, shipping lines were faced with the growing need to expand operations beyond their traditional responsibilities as ocean carriers. For years, container lines had been involved in the overland movement of some of their cargo; but this was mainly in order to control and manage their inventories – the container units – so as to ensure that their boxes would return to the ports of loading. Intermodality intensified and accelerated the engagement of the ocean carrier in land transportation.

The advance of intermodal transportation throughout the 1980s has encouraged shipping lines to break away functionally and geographically from their conventional task of serving merely the ocean-going trade. Gradually, shipping lines have assumed greater responsibility and control through vertical integration of larger portions of the total distribution of cargo (Casson, 1986). Such a move was motivated by a variety of reasons: through their involvement in inland transportation, shipping companies wanted to ensure traffic volumes for themselves on the ocean leg; to capture a fair share of the expanded inland transport market that was opened up for them; and to establish a long-term role in the intermodal transport chain.

Multimodal companies

One of the most significant manifestations of the vertical integration introduced into the freight-transport system as a result of intermodality is the evolution of large multimodal transport companies (McKenzie et al., 1989). These total transport companies, so-called, provide shippers with a controlled and coordinated international movement of

Photograph 11.3 Burlington Northern railways intermodal yard. At Burlington Northern's Seattle International gateway, seaborne containers are being loaded and discharged in a quick interface operation between land and sea transportation.

containerized cargo through a system that links together all three surface transportation modes – ocean, rail and truck – and possibly air transportation as well.

Shipping lines that were involved in intermodal transportation, as well as railroads, started to buy trucking companies, inland terminal operations and freight-forwarding agencies. The rationale for this move lies in the economics and the marketing of their operations. A multimodal company can channel the flow of cargo through the various modes under its control with better coordination and improved efficiency than can the different segmented transport-mode companies. It can eliminate duplicate administration expenses and combine marketing efforts. With the consolidation of various transport modes under one managerial umbrella, a multimodal company can potentially benefit from the relative strength that each mode brings from its own region, area of specialty and traditional clientele (Hayuth, 1987).

The changing perspective of shipping lines, from being ocean carriers to acting as 'total' freight-transport companies, performing a diverse range of activities in transportation, as well as in related services and operating over an extended geographical arena, represents a structural change in the freight-transport industry. Not only did it have significant impact on the organization, operation and marketing strategy of transport companies, it also had a widespread effect on many supporting services, such as insurance, legal coverage and documentation (UNCTAD, 1986).

American President Lines was among the pioneering transportation companies to adopt the new perspective and become an intermodal company. APL started to take steps toward this end as early as 1979. Its first involvement with inland transportation was to contract for dedicated trains to carry its own maritime containers. In consequence of a 1983 merger agreement, APL became part of a publicly traded corporation known as APC (American President Companies). In 1985 the company added to its organization its own freight-forwarding agency and formed API (American President Intermodal) to operate the innovative double-stack train system. In 1987 APD (American President Domestic) was established to manage all the domestic activities of the company: its distribution services, trucking, as well as API. Currently, APC offers the gamut of surface transport services – ships, railcars, trucks, freight-forwarding, port and inland terminals (McKenzie et al., 1989).

Another prominent illustration of a multimodal transport company is provided by the acquisition in 1987 of Sea-Land, one of the first and largest container shipping lines, by CSX, now a giant conglomerate but once a railroad company, and the approval given by the US Interstate Commerce Commission to enable the two different surface carriers to merge into CSL Intermodal. The CSX/Sea-Land and APC cases represent a growing trend in international freight transportation to create a 'supermarket-type' company, in which the shipper can find a full range of transportation and distribution services under one corporate roof. By using such a 'total' transport company, a shipper may obtain a single bill of lading and also a single rate for the entire journey of his cargo, thus avoiding the need for multiple negotiations with various transport carriers and services in order to transfer his goods from their origin to their final destination.

The multimodal company phenomenon still must prove its long-term viability and vitality. It is already clear, however, that the concept has a great impact on the competitive structure of the transport industry. The traditional competition among shipping lines, railroad companies and trucking firms in any given trade, whether within or between the various transport modes, is being replaced by competition between multimodal companies on entire transport systems. Moreover, the long-standing rivalry between shipping lines and railroads on some trade routes and even the more common competition between railroads and trucking firms are gradually disappearing in the intermodal era. Instead, there is increasing complementarity between these conventional rivals, as each transport mode adds its own relative advantages on some segments of the route, thereby contributing to a more efficient, total transport chain.

Deregulation

The deregulation of the United States transport industry in the late 1970s and early 1980s proved a most significant catalyst for intermodal transportation. The US railroads had been subject to government regulation as early as the mid-nineteenth century; this was followed by the regulation of steamship lines in the early 1900s and of motor carriers in the mid-1930s (Muller, 1989). The legislative authority and the regulating responsibility had represented a segregated approach to the transport industry: the ICC regulated trucklines, railroads, inland waterways and pipelines; the Federal Maritime Commission (FMC), the shipping lines; and the Civil Aeronautics Board (CAB), air-transport services.

The US transport legislation allowed very little room for competition, whether within or between transport modes. Any intermodal activity was prohibited. Thus, when in the early 1970s volumes of containers started to be moved under the control of single shipping lines from the Far East to the US Mid-West via US West Coast ports, documents still needed to be broken up to satisfy separate regulatory authorities.

Then came the series of deregulation acts. The Motor Carrier Act of 1980 eliminated many of the restrictions that had been placed on truck routes, areas served, rate structures and entry into the trucking business. The Staggers Rail Act of 1980 reinstated the rate competition among railroads and provided more flexibility in the operation of rail services. The US Shipping Act of 1984 allowed, *inter alia*, greater rationalization of combined transport operation (Kagan, 1990).

Deregulation seemed to inject a new sense of dramatic activity into the transport industry and revive a competitive environment. The shippers, from their perspective, now had the choice of a much greater option than ever before of rates routes, mode combinations and ports (Talley, 1988). The carriers now tried to exploit the greater flexibility allowed their operations, including the competitive pricing of container movements. Since deregulation, railroad container freight rates have dropped some 30 per cent (Fleming, 1989). The incentive had been provided for efficiency of operation and more innovative marketing strategies.

Deregulation was instrumental in the development of intermodality. It helped carriers in one mode to own and operate carriers in other modes; it initiated mergers, but at the same time enhanced cooperation and coordination among railroads, trucking firms and ocean carriers. The liberalization set in motion by the deregulation acts greatly altered the structure of competition in the transport industry (Muller, 1989).

Inland intermodal systems

The advance of the intermodal concept did not change the basic characteristics of the individual transport modes. Water and rail transportation, which have to endure a high proportion of fixed costs, had and still have an advantage on the long-distance hauls. Trucks, on the other hand, offering more flexible operation, maintain an edge on the short-distance trips. Intermodality did not alter this basic notion; however, it did modify the relative advantages of some transport modes *vis-à-vis* the total transport chain and changed the competitive status of other transport modes on significant sections of the transport journey.

Accepted wisdom among both transport researchers and operators has traditionally been that, per ton of cargo, water transportation supplied the cheapest mode of transportation; thus, every effort was made to extend the ocean voyage as deep into the interior as possible in order to reduce land-transportation costs. A traditional criterion for port location was the deepest point in the interior to which an ocean-going vessel could navigate. Although the assumption of the cost effectiveness of water transportation still prevails in most cases, its validity has been made conditional.

Perhaps one of the most significant manifestations of the intermodal concept – and a true realization of the changing competitive relations among transport modes – was the development of so-called 'landbridges'. Figure 11.1 illustrates the evolution of the North American landbridge system. It may be argued that long-haul overland trade routes competing with all-water transport movements was not started with either containerization or intermodality. Some may take the view that the idea of a continental United States as a landbridge goes back to the first transcontinental railroad in the second half of the nineteenth century. For others, the ancient caravan trails between Europe and the Far East, known as the 'silk trail', may be considered a much earlier operation of a 'landbridge'. Nevertheless, it was the intensity and volume of traffic that began to move through a variety of landbridges in the 1980s, in direct competition to ocean transportation, that proved unprecedented.

The various landbridge schemes not only presented a challenge to the long-standing cost advantages of waterborne freight over formerly alternative surface transport modes, but also greatly modified the concept of port hinterland. The deep penetration of inland distribution concepts, such as landbridges, extended the traditional port tributary areas, which now stretch over entire countries and

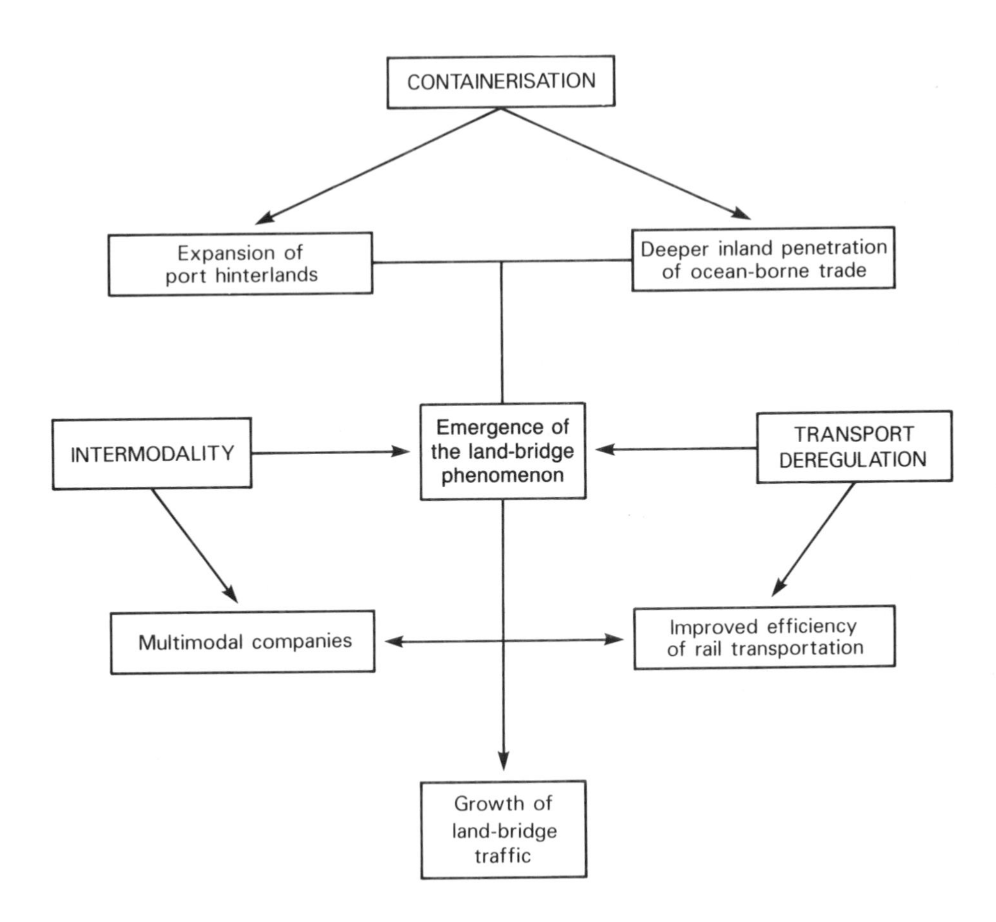

Figure 11.1 The evolution of the North American landbridge system.
Based on Hayuth, 1982.

even continents (Hayuth, 1982). In the intermodal era, the whole of North America forms a common, single hinterland, and does so along both coasts for more ports than ever before. Moreover, with the advance of the integrated transport system, the separation of the foreland and hinterland relationships of a port into two labelled packages represents, to a great extent, a false dichotomy (Robinson, 1970).

The growth of intermodal movements and of landbridge operations is not restricted to only one region of the world or to specific trade routes. Long-haul overland movements of containers, as an alternative transport system competing with or complementing other transport modes, have become an integral feature of trade patterns in many parts of the world. The Canadian landbridges, the US minilandbridge operation, the Trans-Siberian landbridge, as well as landbridges in the Middle East, South East Asia and Central America, present dynamic evidence of the intermodal era.

North American intermodal

Of all the intermodal overland services in the world, perhaps none is more extensive than those offered in the USA since the mid-1980s (Figure 11.2). This outcome could not have been achieved without a major change in attitude by the railroad systems. Ironically,

Photograph 11.4 Container terminal at the Port of Seattle. Terminal 5 of the American President Lines illustrates the vast back-up area needed for load centre container operations. The On-dock rail terminals can greatly improve the turn-around time of containers in the port and thus reduce the need for back-up space.

American railways were slow to adapt themselves to containerization and intermodality. It was not until the early 1980s that the US railroad industry finally made the move, and the railway map of America has since changed radically. Deregulation, mergers, new rolling stock, dedicated cross-country trains and most significantly increased cooperation with seaports and shipping lines now characterize the US rail-transport system.

The rapid development of double-stack container trains in the USA demonstrates most visibly – together with its important economic features – the new era of rail intermodality. Since the early 1980s, when the first double-stack trains began rolling across the United States, infrastructure clearance restrictions on moving such trains have been largely eliminated. By the end of 1989, 14 intermodal train operators were offering some 284 double-stack trains per week in the United States, Canada and Mexico, each train with a capacity of 270–280 40-foot equivalent unit (FEU) containers (UNCTAD, 1990).

The double-stack trains afforded large savings on long-haul runs. The reduced cost stemmed from a saving in crew labour, as the number of containers per train was doubled, and from lower capital costs per payload, as fewer platforms and engines are needed per container. Operational savings can also be gained as a result of the use of much lighter cars. A standard railroad flatcar has a tare weight of about 32.5 tons; some double-

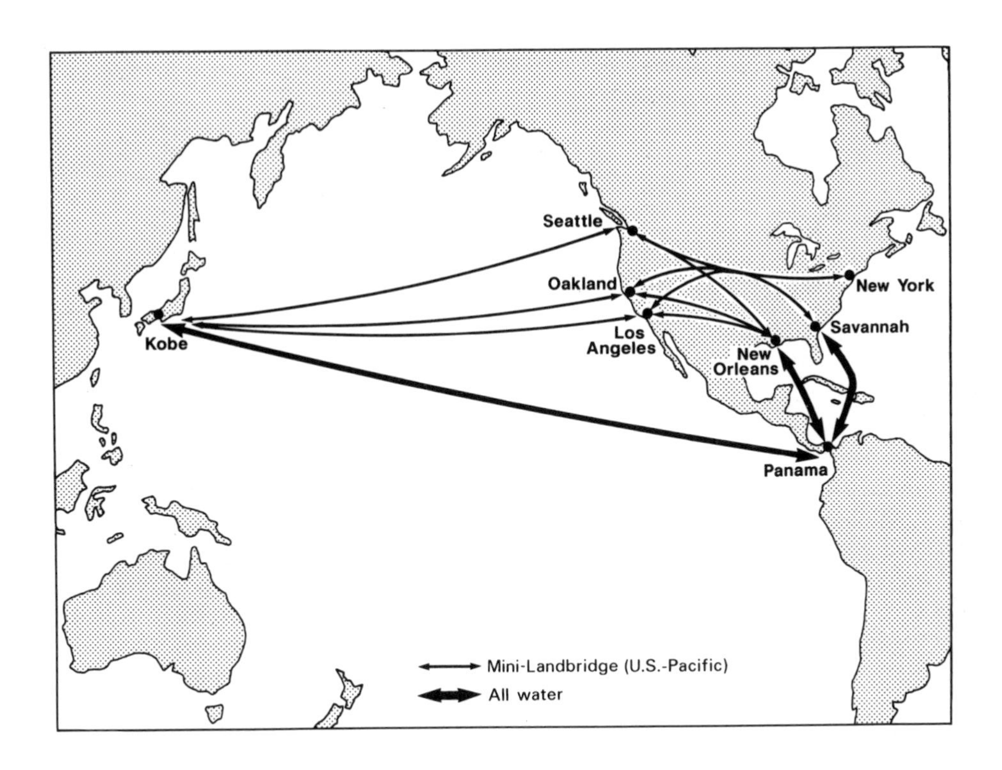

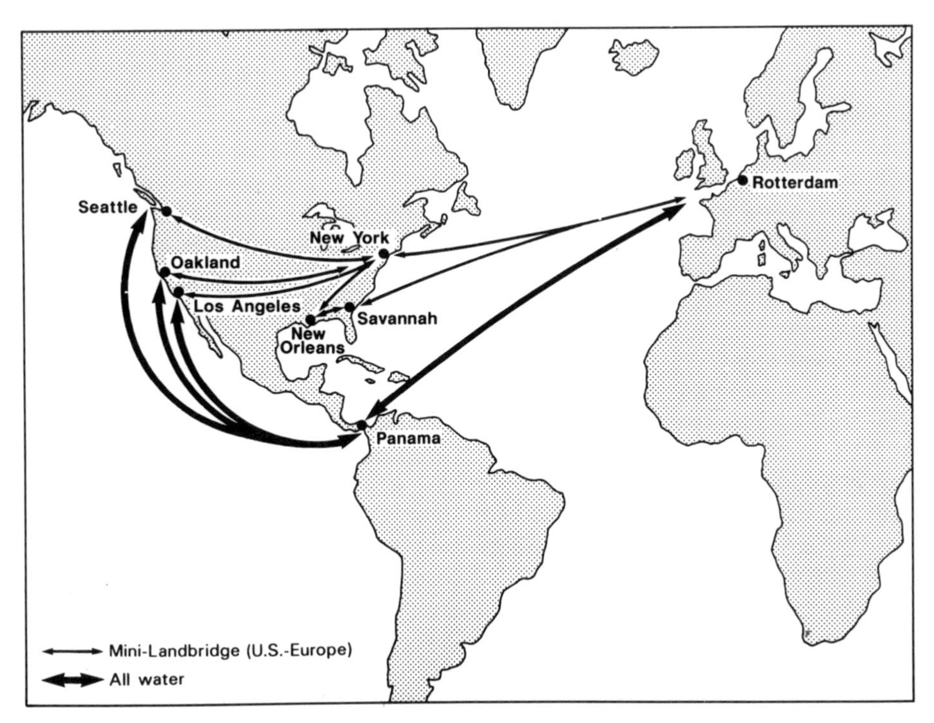

Figure 11.2 US landbridge transport routes.
Based on Hayuth, 1982.

Table 11.1 Asian minibridge imports to the United States, 1978–1988 (thousand short tons)

	West coast custom district-unloading					
Destination	1978	1980	1982	1984	1986	1988
North Atlantic	597.5	656.1	778.7	1142.7	1680.9	1612.5
South Atlantic	87.9	113.4	138.5	194.3	244.7	287.2
Gulf Coast	477.2	480.8	592.4	797.8	817.5	898.5
Total	1162.6	1250.3	1509.6	2134.8	2743.1	2798.2

Source: Port of Oakland, 1989.

stack train platforms weigh less than half that, about 14.5 tons. The reduction in weight translates into a 41 per cent saving in fuel costs (Muller, 1989). The economies of operation of the double-stack trains have greatly improved the competitive position of this land-transport operation *vis-à-vis* the all-water transport service through the Panama Canal. The cost saving attributed to double-stack trains is estimated at between 25 and 40 per cent. The total land-transport cost from using such a train may run, on average, to three cents per ton-mile, about the same as for transporting bulk cargo (Eyre, 1989).

Landbridges first emerged in North America in the 1960s, as Canadian National, Canadian Pacific and US railroads were searching for a more efficient transportation service between the Far East and Europe. Successfully implemented were minibridge services for shipments between overseas and US points. The movement of cargo along these landbridges is covered by a single bill of lading. The significance of their operation may be illustrated by the fact that in 1988 minibridge traffic from the US West Coast captured 37.4 per cent of liner imports that previously had gone to Atlantic and Gulf ports via the Panama Canal (Port of Oakland, 1989). Asian minibridge imports to the USA are illustrated in Table 11.1.

European intermodal

Unlike containerization, in which the technology was adapted in standardized form throughout the world, the diffusion of intermodality, which emphasizes organization and logistics, has been highly influenced by the geographical setting and the spatial organization of the region or continent. In Europe, distances between the largest cities or between the major production centres and the seaports are much shorter than in the United States. It is not surprising, then, that the implementation of intermodality takes a different direction in Europe than it does in North America. Thus, flexibility and frequency of service receive more attention from European shippers and consignees than do economies of a long-distance haul. Until recently the politically segmented European map and the complexity of rules and regulations in each country, as well as border restrictions and customs inspection, seemed almost to contradict one of the basic notions of intermodality – the continuous flow of goods throughout the journey. In the context of 1992, however, considerable effort is being invested to reduce many of these obstacles within the European Community.

Despite the constraints, intermodality has not been ignored in Europe. Double-stack trains may not present a valid option in most parts of Europe, either because of the large number of restrictive clearance points in passes, tunnels and mountainous terrains, or because of the electrification of many railway systems. On the other hand, intermodal movements involving barge transportation are found at a more advanced stage than in North America. Almost all of the national railway systems in Europe operate their own intermodal units. British Rail's Freightliner system and the German Federal Railways

Transfracht operation are two examples. These intermodal systems, however, handle only domestic containers within their own railway systems. Indeed, the distinction between domestic and international traffic flows is much more evident in Europe than in North America. Not only are the two trades functionally separated, they are also visibly distinguished. In Europe, domestic and across-the-border trade is moved in special containers, called 'swap bodies'. These lightweight, non-standardized containers are too fragile to be stackable; therefore, they cannot be shipped aboard the standard cellular vessels serving international seaborne trade.

In the international trade market, containerized cargo moves in and out of Europe in ISO-standard marine containers, mostly 20 and 40-feet long. European railways in 1967 established Intercontainer, a company that operates in both cross-border traffic in Europe and in international trade. A Basel-based intermodal operator, Intercontainer, generates an estimated yearly traffic flow of about one million TEUs; however, most of this is in the inter-European market.

Trans-Siberian container services

An old transport route connecting the Far East and Europe overland has been revived in the intermodal era. This came about with the introduction in 1971 of container movement along the Trans-Siberian Landbridge. In this intermodal system, containers originating in the Far East are shipped to the East Coast Soviet ports of Nakhodka and Vostochny. From there they travel by rail to a variety of destinations along the Baltic Sea, the Black Sea or Moscow to transfer points and then to final destinations in Western and Central Europe or the Middle East (Figure 11.3).

The Trans-Siberian Landbridge faces problems, though, that are not at all marginal. The low-level, unreliable service offered by the Soviets, the unbalanced trade, the harsh Siberian winters and the inadequate handling of containers at transfer points raise serious obstacles for the further development of this intermodal service. In 1983, the Trans-Siberian Landbridge handled about 110,000 TEUs; over 80,000 of these were west-bound traffic (UNCTAD, 1990). Since then, total traffic over the route has gradually dropped, until it was nearly halved by 1989 when it stood at about 60,000 TEUs. This reduction stemmed mainly from the fierce competition offered by shipping lines in the Far East-Europe trade (Table 11.2).

The constraints on this long, complex landbridge have prevented new customers from diverting their cargo from the traditional shipping lines to this overland route. In the light of the growing trade between the Far East and Europe, however, Soviet authorities secured the cooperation of Western experts to help improve the effectiveness of the operation, particularly its time-efficiency. An attempt was made, late in 1990, to achieve a total transport time between Asia and Europe of between 20 and 23 days. At present, the fastest all-water service between the two continents is 25 to 30 days (Berniker, 1990).

Minor landbridges

Several landbridges operate on a smaller scale, or are being planned, in other parts of the world. The government of China has plans to create a new 'Euro-Asia landbridge' by connecting 2,000 km. of railway from the Port of Lianyungang, on the Pacific Coast, to the Trans-Siberian landbridge, for a service to Rotterdam. A Turkish landbridge was set up in the mid-1970s in order to overcome major congestion at Middle East Gulf ports. Containers shipped to the Gulf region were unloaded at Turkey's Mediterranean port of Mersin and then hauled by truck across the country to major centres in the Middle East such as Baghdad, Tehran, Kuwait and Riyadh. The validity of the bridge was successfully tested a little later when the Suez Canal was fully reopened for navigation. The

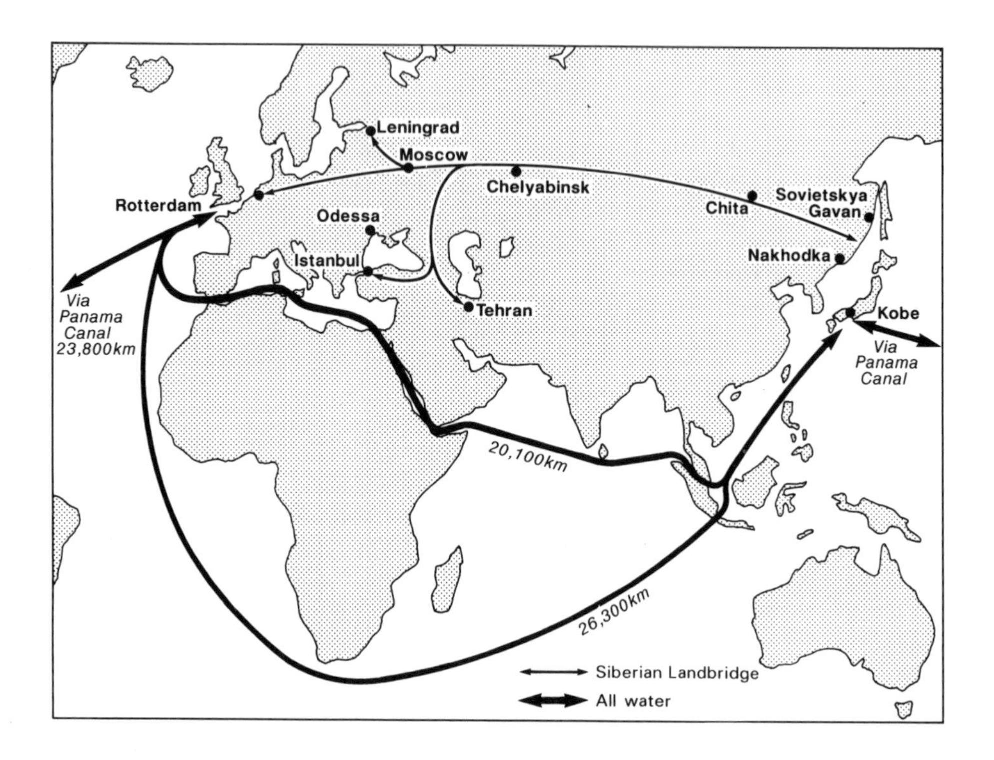

Figure 11.3 The trans-Siberian landbridge transport route.
Based on Hayuth, 1982.

Table 11.2 Trans-Siberian landbridge traffic ('000 TEUs)

	1984	1985	1986	1987	1988	1989
Eastbound	22	17	20	21	28	20
Westbound	62	60	58	57	55	40
Total	84	87	78	78	83	60

Source: Shipping and Trade News, Tokyo, 12 December 1989.

movement of goods on this route was disrupted only by the war in the Gulf in 1991.

A smaller landbridge, both in scale and distance, has been operating for more than two decades in Israel. A combined truck and rail operation bridges the land gap of 350 km. between the Red Sea at the Port of Eilat and the Mediterranean at the Port of Ashdod. This intermodal landbridge continues to function, despite the operation of the Suez Canal. An even shorter landbridge operates in Mexico, where two ports, one on the Atlantic Coast and one on the Pacific, and a railway system are trying to provide some competition to the Panama Canal. In Thailand, the government approved the creation of a new 180 km. landbridge that will shorten the distance between Europe and the Far East by at least 1,000 km.

Inland container terminals

The deep penetration of maritime containers into the interior of countries and continents, the intensification of this trend through the growing volumes of intermodal movements and the extended size of port hinterlands have had far-reaching implications for the organization of traffic flow, logistics and physical distribution. The containerization revolution primarily focused on the maritime segment. With intermodality, the centre of gravity of the transport system has been shifting towards the land-transport modes. In order to facilitate the long-distance overland routes and to rationalize the rapidly developing inland transport schemes, a network of inland transfer points has sprung up.

Inland container freight stations or depots emerged in many parts of the world during the 1970s, for the most part as a direct response to the lack of back-up space for container stripping, stuffing and clearance within port boundaries. Inland Container Depots (ICDs) in the UK had an earlier form, as customs-clearance points for railway ferry wagons, since they offered a more convenient and less costly way of clearing the wagons than did the ports. In the USA, container freight stations were set up at first in close vicinity to large ports.

As the volume of intermodal containers grew, increasing need was felt to facilitate the major inland routes by means of inland depots. Although a few such facilities operated at a considerable distance from any port almost from the early days of containerization, such as the Clearfield Freeport Center in Utah, it was during the 1980s that many new inland centres began to be established at the end of and along major rail corridors. These inland terminals also took on additional functions, such as cargo consolidation, container repairs, repacking and other customer services; they also acted as customs-clearance centres despite being hundreds of miles from a seaport.

As seaports have changed through intermodality from their basic role as break-of-bulk points to become links in an entire transport chain, so, too, inland container terminals have gradually changed their status from being mere transfer points to become true logistical centres. In the process, they have carved for themselves an important niche in the intermodal transport chain (Hayuth, 1980). In North America, the multiple transfer-point system that supported the railroads and the trucking industry in the early days of containerization has undergone major streamlining. The concentration of containerized cargo in a few large ports – the load-centre phenomenon – and the channelling of inland container movements along a limited number of major corridors required a process of rationalization, which brought about the inland depots network. Indeed, consolidation of the inland centre system in both Canada and the USA has taken place in the last few years. The resultant network of much fewer, but considerably larger, inland load centres has gradually been assuming a predominant role in the inland transport system of North America (Slack, 1990). But inland container terminals, with their regional transfer, logistics and marketing functions, are fast becoming a common feature of intermodal systems in other parts of the world. Examples of the trend include the Nei Li inland centre situated in the heart of Taiwan's northern industrial area; the Kano inland terminal in northern Nigeria; and the inland container terminals in Warsaw and Krakov in Poland.

Conclusion

Multimodal transportation represents an advanced stage of containerization. The new phase is characterized not so much by its technological innovations as by an emphasis on the organization, control and management of the transport system. For the freight-transport industry, intermodality had several major consequences. It brought about a shift in thinking from the conventional, unimodal approach to transport to a systems concept, in which each one of the various transport modes, while not losing its own identity, becomes a link in a transport chain. The efficiency and reliability of the entire system are determined by the weakest link in this chain. Thus, cooperation and synchronization of modes comprise an integral part of the concept.

The emergence of multimodal companies represents a trend toward the increasing vertical integration of the transport industry, a development that can lead to better control and coordination of the entire system, including transport-related services. The 'market-to-market' viewpoint replaces the 'port-to-port' approach as the essential rule of the game. If containerization focused primarily on the ocean segment of the journey, then land transportation is the recipient of considerably more weight and importance in the intermodal era. The development of a variety of landbridges around the world manifests this trend. Intermodality involves, to a great extent, both a functional and a spatial organization of the transport system. It is not surprising, therefore, that the geography and the geopolitical structure of regions have had a great effect on the implementation of the new concept. Consequently, intermodality takes different directions in North America, Europe and the Far East. Recent and current developments provide ample examples to support this hypothesis.

References

Berniker, M. D. (1990), 'Trans-Siberian landbridge links Europe and Asia' *Journal of Commerce*, 10 December 1990.

Bird, J. (1971), *Seaports and seaport terminals* (London: Hutchinson).

Casson, M. (1986), 'The role of vertical integration in the shipping industries', *Journal of Transport Economics and Policy* 20, 7–29.

Containerisation International Yearbook (1990) (London: National Magazine).

Couper, A. D. (1972), *The geography of sea transport* (London: Hutchinson).

Eyre, J. L. (1989), 'The containerships of 1999', *Maritime Policy and Management* 16, 133–45.

Fleming, D. K. (1989), 'On the beaten track: a view of US West Coast container port competition', *Maritime Policy and Management* 16, 93–107.

Hayuth, Y. (1980), 'Inland container terminal: function and rationale', *Maritime Policy and Management* 7, 283–9.

Hayuth, Y. (1982), 'Intermodal transportation and the hinterland concept', *Tijdschrift Voor Economische en Sociale Geografie* 73, 13–21.

Hayuth, Y. (1987), *Intermodality: concept and practice. Structural changes in the ocean freight transport industry* (London: Lloyds of London Press).

Johnson, K. M. and Garnett, H. C. (1971), *The economics of containerisation* (London: Allen & Unwin).

Kagan, R. A. (1990), *Patterns of port development* (Berkeley: Institute of Transport Studies).

Kendall, L. C. (1986), *The business of shipping* (Centreville, MD: Cornell).

McKenzie, D. R., North, M. C. and Smith, D. S. (1989), *Intermodal transportation: the whole story* (Omaha: Simmons).

Mayer, H. M. (1973), 'Some geographic aspects of technological changes in maritime transportation', *Economic Geography* 49, 145–55.

Mellor, R. E. H. (1983), *The Rhine, a study in the geography of water transport*, University of Aberdeen, Department of Geography, *O'Dell Memorial Monograph* 16.

Muller, G. (1989), *Intermodal freight transportation* (Westport, CN: ENO).

O'Dell, A. D. and Richards, P. S. (1971), *Railways and geography* (London: Hutchinson).

Port of Oakland (1989), *Liner trade analysis: Pacific basin countries* (Oakland: Maritime Division).

Robinson, R. (1970), 'The hinterland-foreland continuum: concept and methodology', *Professional Geographer* 12, 307–10.

Sealy, K. R. (1968), *The geography of air transport* (Chicago: Aldine).

Slack, B. (1990), 'Intermodal transportation in North America and the development of inland load centers', *Professional Geographer* 42, 72–83.

Talley, W. K. (1988), 'The role of US ocean ports in promoting an efficient ocean transportation system', *Maritime Policy and Management* 15, 147–55.

Ullman, E. L. (1956), 'The role of transportation and the bases for interaction', in W. L. Thomas (ed.), *Man's role in changing the face of the earth* (Chicago: University of Chicago Press).

United Nations Conference on Trade and Development (UNCTAD) (1986), *Multimodal transport and containerization* (Geneva: United Nations).

United Nations Conference on Trade and Development (UNCTAD) (1990), *Review of maritime transport, 1989* (New York: United Nations).

12

International Surface Passenger Transport: Prospects and Potential

Richard Gibb and Clive Charlton

This chapter examines the nature and evolution of international surface passenger transport. A conceptual framework is developed around the two basic concepts associated with interaction: complementarity and distance. This highlights the structural similarities in international surface passenger transport while at the same time recognizing their great diversity. The constraints and opportunities for the development of this form of transport are then examined using examples from throughout the world and the European Community in particular. Both for economic and political reasons, international surface passenger transport is likely to experience substantial further growth.

Introduction

It is neither necessary nor possible to provide a comprehensive world review of the nature and form of International Surface Passenger Transport (ISPT). The aim of this chapter is to provide an introduction to the character, problems and associated solutions of ISPT movements. In order to achieve this goal, various scale-levels of analysis are used to illustrate key issues. To begin with, the overall character of ISPT and the constraints on its development are examined using examples from throughout the world. This highlights the structural similarities in ISPT markets, while recognizing their great diversity. Focus is then moved from the global to the supranational, with an evaluation of the European Community's (EC) attempts to remove the structural impediments to the development of ISPT movements. An examination of the Common Transport Policy (CPT) reveals the serious problems facing the development of ISPT within a region of growing economic and political unity. Finally, the chapter concentrates upon perhaps the most ambitious recent development in ISPT, the development of high-speed rail passenger services in Europe.

To examine the factors that control ISPT, it is possible to develop two basic concepts associated with interaction: complementarity and distance (Ullman, 1973). Complementarity represents the demand for travel across borders, while distance is interpreted very broadly as the ability to meet that demand. Both contain conditions that either facilitate and encourage travel or constrain it.

Complementarity – the demand for ISPT

The complementarity between states that generates passenger flows is extremely complex. Levels of economic development, economic structure and trade patterns are important. Higher levels of income will sustain a stronger propensity to travel, although increasing wealth may divert traffic from surface public transport. This effect is very evident across the US-Canada border, where bus and train are of minor importance, and to a growing degree in Western Europe. As national economies mature, the demand for business and professional travel may expand faster than aggregate growth and trade, following a shift towards advanced consumer and service-based structures. The prominence in the global economy of supranational organizations such as transnational companies and international agencies also encourages work-related travel via the interchange of personnel. This is accompanied by trade liberalization and the internationalization of industries and markets, as is evident with the Single European Market, for which the EC sees enhanced passenger-transport connections as a prerequisite.

International complementarity is also expressed through tourism, although ISPT has played a diminishing role, despite the importance of coach travel and to a lesser extent rail in certain sectors in Western Europe, such as the 55 + and 18-26 year age groups.

Distance – the constraints on ISPT

'Distance', as the force that determines how far the demand for travel is met, can be translated as a series of constraints on ISPT. To varying degrees, these may be modified (normally reduced). A fundamental factor is the spatial organization of states relative to each other, and especially the locations of principal concentrations of population and economic activity. Europe has numerous relatively small but developed and populous political units in close proximity, so generating complex patterns of interaction, whereas for isolated states such as Madagascar or New Zealand, surface passenger travel is insignificant. The capital cities of Congo and Zaire, Brazzaville and Kinshasa, are a 20-minute ferry ride apart every 30 minutes across the Congo River, but African centres like Mogadishu and Addis Ababa are isolated by hundreds of kms. from complementary centres beyond their national frontiers.

National territories are often partly defined and divided by physical barriers, which accentuate the basic spatial constraint of distance. Although major rivers and stretches of sea are readily crossed by ferry services, these are relatively slow, inconvenient and costly, so that water is normally seen as an additional constraint to ISPT. Where technically feasible, international fixed-link water crossings have been developed. The most notable example is the Channel Tunnel, with major projects in prospect between Denmark and Sweden, and Denmark and Germany. Elsewhere, new construction, including such developments as the recently opened international bridge across the Parana River between Encarnacion in Paraguay and Posadas in Argentina, and crossings of the River Uruguay, have eased ISPT in the Plate Basin of South America. Mountain ranges operate effectively to inhibit ISPT directly and via their influence on political, demographic and economic patterns. Major massifs such as the Andes, Alps and Pyrenees have a profound impact on surface travel; significant improvements are extremely expensive, as would be the case for the major 'base tunnels' necessary to upgrade trans-Alpine connections.

Political constraints are of central importance in assessing ISPT flows and can take various forms. Many states exercise a direct influence by controlling the freedom of their citizens to travel abroad, and, to a lesser extent, the entry of foreign visitors: there are restrictions on tourism in many countries, including Oman, Burma and Bhutan. The fractured pattern of ISPT in many parts of the world is a stark testament to the extent of political antipathy between states. Despite strong potential complementarity and physical proximity, travel between the USA and Cuba is normally prohibited: indeed, the island has no ferry services to any of its neighbours. Similarly, international travel has been closely controlled within and to and from the Soviet bloc. Recent political transformations are likely to produce a marked rise in travel demand in Eastern Europe, even though economic and infrastructural constraints will persist. The virtual absence of public transport into Albania and the remarkable paucity of links between Greece and Turkey are further European examples. The pattern is repeated in the Middle East and parts of Asia. Indeed, as regards railways, post-colonial schisms have severely curtailed services that once operated (Din, 1990). Even before the 1990/1 confrontation in the region, through-rail services between Turkey and Iraq had been abandoned.

Although spatial, physical and economic constraints also apply, political factors have ensured that the majority of Asian States have very weak through-international surface passenger services. Exceptions are few but they include Thailand-Malaysia, as well as China-Hong Kong, where the now intensive rail and water services reflect the profound political change under way.

The influence of past political and economic patterns is visible in the underdeveloped status of international transport infrastructure of some parts of the Third World. This is the case in West Africa, where main transport axes have tended to focus on major ports, serving the needs of export-oriented economies, rather than being oriented to connecting adjacent states (Omiunu, 1987). The proposal for a railway linking ECOWAS countries (Economic Community of West African States) from Nigeria round to Senegal and Mauritania is likely to remain a distant prospect for many years.

Many transborder journeys are made more difficult through politically-motivated legal or administrative controls. This extends beyond lengthy border formalities and visa restrictions. A striking example applies at the United States-Mexico frontier, where all public-transport travel still requires a change of vehicle – there are no through bus or train services from, for example, major Texas cities to Monterrey or Mexico City, despite the considerable demand for travel. A similar situation holds at the Mexico-Guatemala frontier. Concern about migration control and differences in technical standards may underlie what would seem to be an inefficient and inconvenient constraint.

Other modes and functions of transport will affect the status of ISPT. Wealthy societies or those where international travel involves long distances and difficult terrain will depend to a high degree on air transport. In some specialized cases, railways may be built to serve almost exclusively mineral transport and passenger travel given very low priority. Alternatively, many important infrastructural investments are justified principally for their role in improving export and import flows, but will bring important subsidiary benefits for passenger services.

Removing the barriers: the European Community's common transport policy

ISPT within the European Community (EC) is still influenced by the existence of political boundaries. The boundaries of Europe, with their political, economic, cultural and religious symbolism, fragment the Community economy, enabling a clear distinction to be made between international and national surface passenger transport. However, the boundary in itself does not create the problem: it is the divergent character of the states it divides. The task of removing a boundary's impact therefore involves much more than technical legislation designed to eliminate frontier formalities; it involves the approximation, if not the harmonization, of political economies.

The 12 Member States of the EC (Figure 12.1) have developed transport systems tailored to suit their own national interests. Nevertheless, the Community, and Western Europe more generally, is one of the most promising markets for the development of ISPT. With disposable incomes growing in a

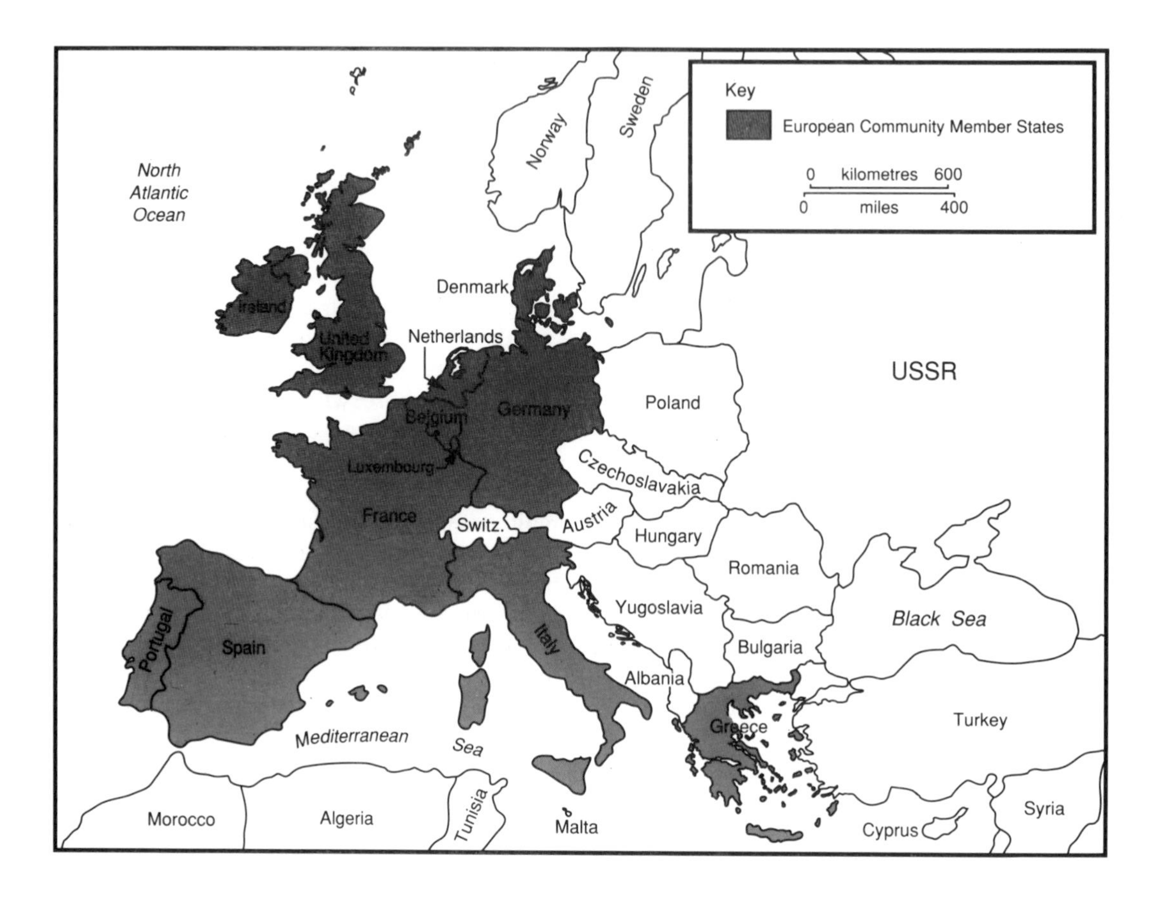

Figure 12.1 The twelve member states of the European Community, 1991.

region with a highly mobile but concentrated population, ISPT has in the past experienced immense growth. In the 1980s, international traffic as a whole increased twice as fast as national traffic in the Community (Budd, 1987).

Despite this growth in the development of ISPT, EC Member States continue to base their transport policies on national criteria. The different approaches to transport policy are also a product of the unique population and geographical characteristics of member states. The end result is less ISPT than would otherwise be the case. The 1977 OECD study of international passenger transport concluded:

> If Europe were one nation, sharing a common culture, there would in general be five to ten times as much traffic crossing those lines which mark today's frontiers. (OECD, 1977, 264)

This chapter now goes on to examine the EC's Common Transport Policy (CTP) and how it proposes to remove the obstacles facing ISPT in order to exploit Europe's transport infrastructure on a supranational basis.

The Treaty of Rome

The founding fathers of the EC recognized the importance of transport to the establishment of an integrated European economy. Transport is an important industry in its own right, employing over 15 per cent of the Community workforce and representing more of the EC's GDP than agriculture. However, it is more vital to the functioning of a truly integrated and free market than these figures suggest. The foundation to the Treaty of Rome is the provision allowing for the free movement of goods and people. The organization and control of the transport sector is therefore critical to the Treaty's objective. It is that sector upon which most other aspects of the common market depend.

The Schaus memorandum: 1961–73

Given the Treaty of Rome's lack of clarity as to the substance of the proposed CTP, it was left to the Commission to establish the principles of operation. The Commission's thoughts were set down in the Schaus memorandum, named after the first EC Commissioner for Transport, M. Lambert Schaus, of 10 April 1961. At this initial stage of policy formulation it was the Commission's intention to eliminate the influence of different transport policies in order to create a Community-wide transport system. As the memorandum states:

> The problem is now to create a Community economy . . . A transport policy applying to the transport of the Community as a whole must, therefore, gradually replace national transport policies. In the very process of economic integration, the difference between national and international transport within the Community will disappear. (As cited in Despicht, 1969, 33)

The Commission's memorandum was an ambitious proposal with the aim of eliminating all obstacles preventing the development of a free market in transport. ISPT would therefore be indistinguishable from national-transport movements. This was to be achieved by three policy initiatives. First, by the removal of national discriminatory policies and rates. Second, by the elimination of policies distorting transport movements and third, by the creation of 'healthy competition' in its widest scope. While the first two methods of ensuring a free market in transport were practical and politically realistic, the idea of creating free market competition was altogether too ambitious. Given the diversity in the structure, regulatory controls and objectives of the different transport policies adopted by member states, it was unrealistic to assume immediate unification. Consequently the Council of Ministers, that EC institution responsible for ratifying legislation, never reached a consensus on the memorandum.

1973–83: new policy initiatives following the enlargement of the Community

Following the 1973 accession of Ireland, the UK and Denmark, the objectives of the CTP were reviewed producing a subtle change of emphasis. Although the measures proposed were similar to those identified by the Schaus memorandum, there were a number of important differences. Most notable was the move away from harmonization as a necessary prerequisite to a common policy. More emphasis was placed on infrastructure planning and investment as a means of integrating the diverse transport systems of member states. The idea was to promote Community intervention in the planning and financing of an integrated network so that resources could be put to their best possible use. As a result, the Community emphasized the role of dovetailing separate national transport policies in order to create an infrastructure and transport market suited to the needs of the Community. The objective was to align the national transport systems for the carriage of passengers at Community level.

Unfortunately, the 1973 proposals met with the same fate as the 1961 memorandum. At the beginning of the 1980s, few of the objectives established in 1973 had been achieved and the EC has progressed no further in its attempts to create either a common market in transport or a CTP. The failure of the EC to develop a CTP and liberate ISPT, treating it as an equal alongside national transport, re-emphasizes many of the obstacles to the development of

ISPT outlined at the beginning of this chapter. Perhaps the single most important difficulty is the long-term persistence of substantial divergencies in the national-transport policies of member states. These divergencies are often the product of sound, justifiable national policies arising from unique economic, geographical and historical reasons. As Bayliss (1979) points out, each member state considers its own transport policy to be more suitable than any common policy which would inevitably involve compromise and costs for them all.

Formulating the CTP has also suffered from the conflicting philosophies of 'social intervention' and 'liberal free trade'. The Treaty of Rome was negotiated during the great moves towards liberal trade policies. However, there was no corresponding move in the transport sector which was highly regulated and managed by member states. Consequently, to accept the original Community proposals, based on free market principles, would involve major changes in attitudes and, more seriously, economic dislocation. The polarization between the need to liberalize transport and harmonize the conditions of competition has been the major source of conflict in the development of a CTP. The end result has been over 30 years of inaction: the most characteristic feature of the CTP has been its failure to affect the transport policies and markets of member states.

1992 and its impact on ISPT

In the early 1980s, European Community Governments recognized that if the Community was to regain its competitive position in the world economy, all restrictions to the free movement of goods and people had to be abolished. Furthermore, as a result of the uncoordinated nature of national policies segmenting the market, it was agreed that common policies had to replace the divergent actions of individual member states. Following the European Council meeting at Fontainebleau in 1984, Lord Cockfield presented his White Paper, entitled 'Completing the internal market', to the Council of Ministers. The White Paper outlined a programme of legislative measures deemed necessary to achieve a single integrated European market. The fragmented nature of the transport market and the need to promote a CTP, based upon free market thinking and competition, featured highly in both the White Paper and the Single European Act (SEA) passed in June 1987. The future of the CTP is now dependent on the success of the 1992 programme being implemented in full. While there are many aspects of the CTP not covered in the SEA, such as railways, financing and the coordination of infrastructure, the establishment of an integrated market should provide the basic foundation to a common market in transport.

The SEA, like the 1961 memorandum and the 1973 Communication, considers the establishment of a common market in transport to be essential to the development of a free internal market. It also considers free market policies, driven by ever-increasing competition, to be the most appropriate way to achieve a CTP. The Commission's aim is to allow for the freedom to provide cross-border services in road, inland waterways and air transportation. Again, the problem of applying and then enforcing free market principles to a CTP is reflected in the SEA's coverage and wording. The large discrepancies in the levels of liberalization, government support and public service obligations connected to the national railway systems precluded the establishment of a free market for this mode of transport. As a consequence of the special status of rail transport, there will be no free market access in international rail transport services. Notwithstanding this exception, the SEA focused on the need to deregulate and privatize as a means of cost cutting and implementing a common market in transport.

As far as ISPT by road is concerned, the SEA established common rules for the international carriage of passengers by coach and bus. Any carrier will have the right to provide transport services freely throughout the Community as long as a number of conditions are met. The most important of these are:

1 national and international road-safety requirements;
2 to be established in a member state in conformity with the national regulations of that state;
3 to be authorized by that state to carry out the international carriage of passengers;
4 the carrying company must belong, or have the majority of its shares, with nationals of member states. (European Commission, 1985)

For the operation of regular scheduled services and shuttle services, the member states of destination would have to be consulted whereas member states whose territories were crossed in transit would only have to be informed. However, the member state of destination would need to provide the necessary authorization for the service to proceed. In the event of this authorization being refused, the matter would be referred to the Commission for arbitration. For operators providing occasional services, no authorization would be required. Furthermore, there is a proposal to allow any carrier to provide regular, shuttle or occasional services in the domestic markets of other member states, subject to a list of conditions similar to those governing international movements. Carriers would therefore be allowed to undertake operations in any member state as long as they respected the regulations enforced in the country of operation. These regulations are in compliance with the Treaty of Rome which outlaws discrimination on the grounds of nationality.

The SEA therefore continues a well-established tradition of trying to implement a CTP via the free market. In both its objectives and underlying philosophy, the SEA replicates the 1961 Schaus memorandum and the 1973 communication. Furthermore, it has returned to the idea of creating a CTP by grand design as opposed to incremental *ad hoc* decision-making. Given the SEA's similarity with past initiatives to create a CTP, will the SEA be met with a similar obstructive response in the Council of Ministers? The answer to this question is by no means straightforward. In favour of the White Paper proposals succeeding is the fact that it has already been passed, in principle, by the Council of Ministers.

The proposals forwarded by Lord Cockfield are also of a highly technical and detailed nature; they do not leave much ground for debate or legal interpretation. Most importantly, however, the White Paper and the Cecchini Report (1988) predict that the Single European Market will produce a new competitive environment creating opportunities for growth, job creation, economies of scale and improved productivity. As these benefits will be available only when the European economy becomes a truly integrated market, based on free market principles, it is in the interests of member states to enact the SEA legislation in full. On the other hand, member states will continue to be the key providers of most public transport and infrastructure. For good or bad, states will continue to adopt policies, sometimes unconsciously, that hinder the creation of a competitive market in transport. The negative consequences arising from a move towards unfettered market forces in the transport sector may also prove to be politically unacceptable. Both Cutler (1989) and Lodge (1989) predict the dislocation of economic activities arising from the SEM programme. Such a prospect has in the past persuaded member states to adopt interventionist policies designed to obstruct competitive forces moulding the shape of the transport sector. Whatever the precise outcome of the SEA's impact on the development of a CTP, it seems likely that the SEM will have a beneficial impact on the ISPT market within the EC. The SEM proposes to eliminate all physical barriers within the 12 member states, including immigration controls, customs checks and other administrative procedures. The removal of these barriers will constitute one of the most direct and visible benefits of the Community to its citizens. If the free movement of people is accompanied by increased living standards, higher levels of disposable income, further European integration and enhanced working conditions, as proposed by the 'social Europe' legislation, the ISPT market within the EC should experience a considerable growth in demand.

The failure of the EC's CTP over the past 30 years to create a common market in the transport sector or integrate European infrastructure is testament to the very real obstacles that exist in the development of ISPT systems. As the European Parliament noted in 1982: 'Unfortunately it is a well known and distressing fact that, a quarter of a century after the founding of the EC, most bottlenecks still occur on the internal frontiers of the Community' (European Parliament, 1982).

In 1986/7 the Commission prioritized the development of high-speed rail links between the major cities of member states. This policy to enable rail to realize its full potential was strengthened when member states began to plan high-speed links on a continental scale in order to exploit the opportunities created by the Channel Tunnel.

The development of high-speed rail passenger services in Europe

The striking progress with high-speed rail services in Western Europe has arisen through a series of autonomous initiatives with national, rather than international objectives. In contrast, European international rail passenger services have, until recently, seen only modest improvement, or have even stagnated (Savelberg and Vogelaar, 1987). While the general advance to higher speeds is a continuous process, a number of benchmarks stand out. However, these can only be understood when examined in a national context.

France

The highly successful French TGV (Train à Grande Vitesse) programme, although primarily an internal venture, seems certain to be a cornerstone of any wider European high-speed system. Springing from an urgent need to expand capacity in the Paris-Dijon-Lyon corridor, the TGV is seen both as a potent tool in French regional-development strategy and a prestigious symbol of national technical prowess. Three years after the first TGV services began running from Paris to Lyon in 1981, rail traffic between the two cities had risen by 45 per cent, much of it diverted from air transport.

The TGV-Sud Est set several fundamental markers in the design and operation of high-speed rail services. The TGV concept is based on the construction of new railway lines, exclusive to high-speed trains, although these are not confined to the new tracks. The absence of slower, heavier traffic and the use of high-powered electric traction allows high speeds to be maintained on steeper gradients than normal, thus reducing line-construction costs. Line speeds on the TGV-Sud Est are up to 270 km/hr, rising to 300 km/hr on the more recent TGV-Atlantique. The TGV operates as a high-capacity, high-quality service, with obligatory reservations and supplements payable, between a limited number of access points on the high-speed line proper (Le Creusot is the only stop on the Paris-Lyon services).

Despite being designed for operation on a specialized infrastructure, TGV trains also run, at lower speeds, on other electrified lines. This critical flexibility allows a much wider range of destinations to be linked directly into the high-speed network than would be the case with entirely specific trains and track (Figure 12.2). The TGV-Sud Est carries through-services to points in the south of France such as Nice and Montpellier and to Switzerland, while TGV-Atlantique trains run as far south as Hendaye on the Spanish frontier. Although the dynamic behind the growth of the system is essentially internal, serving French domestic needs, the connections to Switzerland and Spain suggest how high-speed rail development can enhance international services even in the absence of large-scale, specifically trans-border investment.

The extremely ambitious and costly (FFr 188 bn) master plan revealed by the French government in June 1990 (*Railway Gazette*, 1990) provides for a number of international services to operate as extensions of additional TGV lines radiating from Paris to provincial cities (Figure 12.3). These include routes to Germany and Luxembourg on the TGV-Est to Strasbourg; to Italy via Modane and via Nice and two routes into Spain. Work is already under way on the most overtly 'international' TGV line, the TGV-Nord to Lille, with its future connections to the UK via the Channel Tunnel, as well as to Belgium, The Netherlands and Germany. Completion of the master plan is likely to be many years distant, but it clearly has the potential to accelerate many European international journeys by the early twenty-first century.

Germany

Several strategic new links are under construction or planned in Germany with the primary aim of improving network structure, especially on the north-south axes. With line speeds up to 250 km/hr, the new links (*Neubaustrecken*) will cater both for passenger and freight services, highlighting the fact that in the majority of cases, planning and operation of high-speed services cannot proceed in isolation from other forms of rail transport. To permit both fast and heavy haulage, the new German lines are very costly. The Hannover-Wurzburg *Neubaustrecke* is due for completion in 1991, to be followed by that from Mannheim to Stuttgart, with a new all-passenger connection between Cologne and Frankfurt in prospect for 1998. These last two projects should offer opportunities to accelerate international services in the future, such as Paris-Munich and Amsterdam-Zurich.

Italy

High-speed train development in Italy is based on improving internal intercity connections, although some of the technology chosen to achieve this has aroused interest elsewhere in Europe. Although new line construction is basic to Italian plans, notably the Rome-Florence-Bologna *Direttissima* (and later, a Bologna-Milan high-speed route), difficult terrain has made this extremely costly. A partial solution has been the development of tilting-body trains that can negotiate curves at higher speeds. These *Pendolino* trains are already in service; several other European railways have shown interest in tilt-body stock, including Germany, Sweden, Austria, Yugoslavia and Switzerland (Ford, 1990). Although tilting technology is not without problems, it could permit higher-speed international connections through mountainous areas, notably the Alps, where the costs of constructing conventional high-speed lines would be immense.

Spain

The first modern high-speed railway in Spain (*Alta Velocidad Español* (AVE)) is now under construction, with the purpose of dramatically cutting travel times between Madrid and Seville. This is an essentially 'internal' objective, but Spanish railway aspirations also reflect a keen awareness of the need for more efficient connections with partners in the EC. This is shown by the decision to build the Madrid-Seville line to the standard 1435 mm gauge, rather than the Spanish 1668 mm gauge. Spain evidently regards preparation for through-running to other European states as worth the shorter-term inconvenience of isolation of the new line from the rest of its national network. The move will only be valid if further standard-gauge lines to the French border are completed. The first step is likely to be a TGV-standard line from Barcelona to Perpignan to join the expanding French network, which appears to meet the

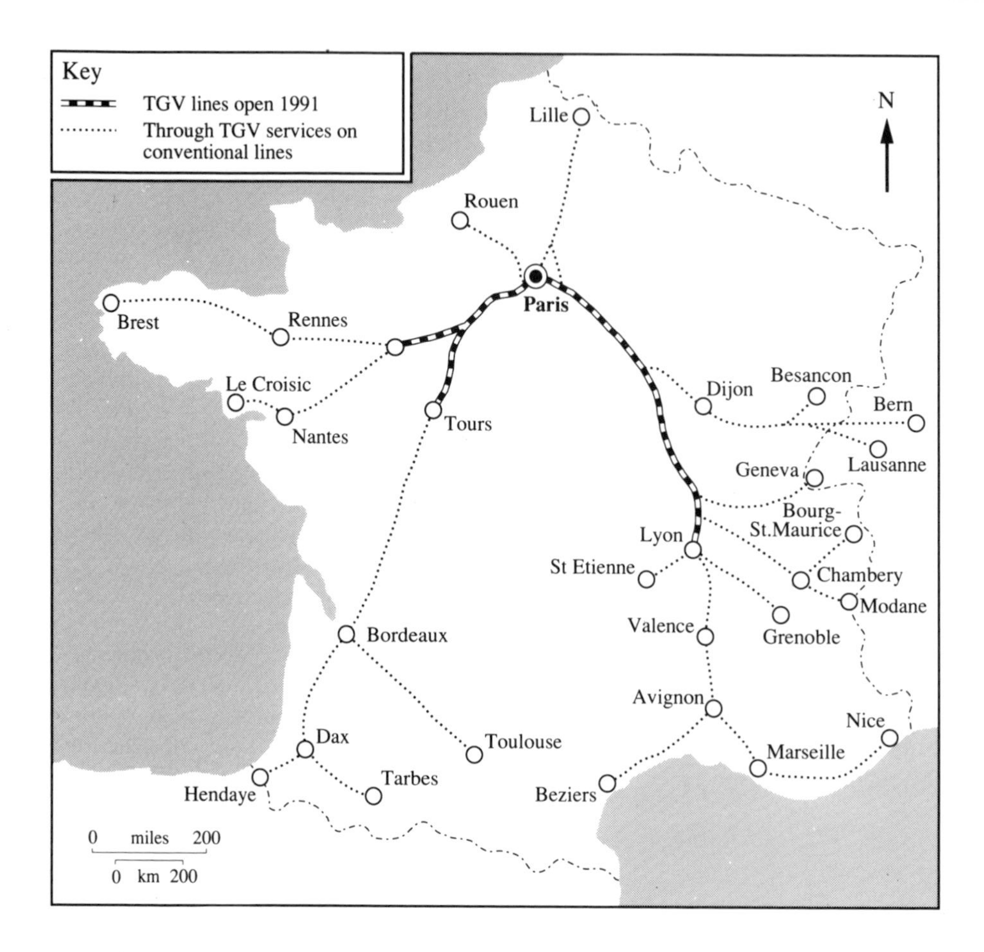

Figure 12.2 The French TGV network, 1991.

aspirations of Catalonia as an autonomous Spanish region within the European Community even more than would a new connection to Madrid (Lopez Pita and Turro Calvet, 1987).

European international high-speed projects

The growing maturity of high-speed rail passenger services has coincided with the advance of economic, social and political integration in Europe. An international high-speed passenger network is expected to play a central role in the transport systems necessary to foster the achievement of a functional single European market. Much of this high-speed network is in place or planned as part of national rail strategies, but its completion will require a more overt commitment to specifically international action. This should include the coordination of investment plans, technical standards and operational and marketing procedures (European Study Service, 1988). Specific matters for agreement include optimum operating speeds, the voltage of electric traction, axleloads, loading gauge, common signalling standards, whether new lines should be exclusively for high-speed passenger services or should be shared with other traffic and the roles of public and private sectors in funding and operation. Further collaboration

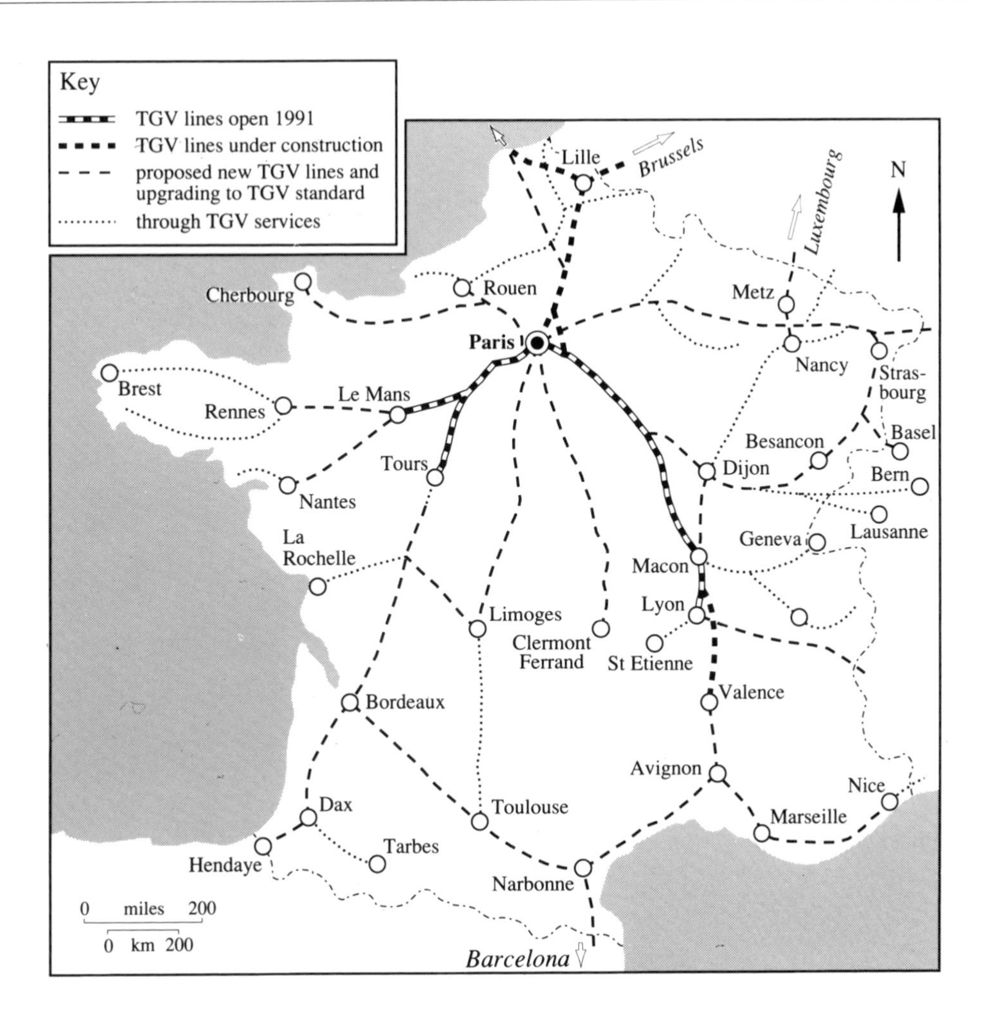

Figure 12.3 The French masterplan for the future TGV network.

between governments, railway organizations and the EC will be essential. Several organizations already operate to coordinate and promote international rail services, including the Union Internationale des Chemins de Fer (UIC), through which considerable technical standardization has already been achieved (O'Brien, 1986).

Among several critical steps in progress towards an international high-speed system is the Channel Tunnel, which will transform the opportunities for surface travel from the UK and become a basic element in the emerging international network in north-west Europe. New high-speed rail services between the Tunnel and Paris/Brussels will enable passengers to reach Paris from London in just over three hours; and Brussels, headquarters of the EC, in two hours 45 minutes. This compares favourably with the equivalent air and ferry journey times (Figure 12.4). At the same time, the connecting rail links illustrate rather too effectively the discontinuities that remain across international frontiers. The full potential for high-speed rail development is certain to be inhibited for many years by inadequate connections beyond the Tunnel mouth on the UK side. There is a sharp contrast between the policies of the British and French

Photograph 12.1 The *Train à Grande Vitesse*.

governments on state investment in public transport. On the British side, the reluctance to proceed with a new, dedicated line to London means that the trainsets operating from Paris and Brussels will use existing tracks, which are shared with commuter services and use the third-rail 750 volt dc electrification system. At this voltage, the operating speeds of the 18-coach Tunnel trains will be much reduced, not only adding to overall journey times but possibly impeding the progress of suburban services (Semmens, 1990b). Slow service speeds will also reduce the overall utilization of expensive rolling stock and thereby add to costs. In contrast, French investment in the costly new TGV line, electrified at 25,000 volts ac, will allow the trainsets to operate up to 300 km/hr from 1993. If a high-speed line with this voltage were built between Folkestone and London, the London-to-Paris journey could be completed in a matchless two hours 30 minutes.

A central concept in the emergence of a network for north-west Europe is the proposed high-speed line from Paris via the TGV-Nord to Lille and Brussels, with onward branches to Köln and the Ruhr, and via Antwerp and Rotterdam to Amsterdam (Baumgartner, 1986; Ruhl, 1987) (Figure 12.5). Although the route and its financing have yet to be agreed, such a network would clearly be of fundamental importance and be profitable, running through much of the economic and demographic heartland of the EC. It would also connect the French and German domestic high-speed systems, as well as having a mutually reinforcing relationship with the Channel Tunnel via the line from Lille. A further addition to the network

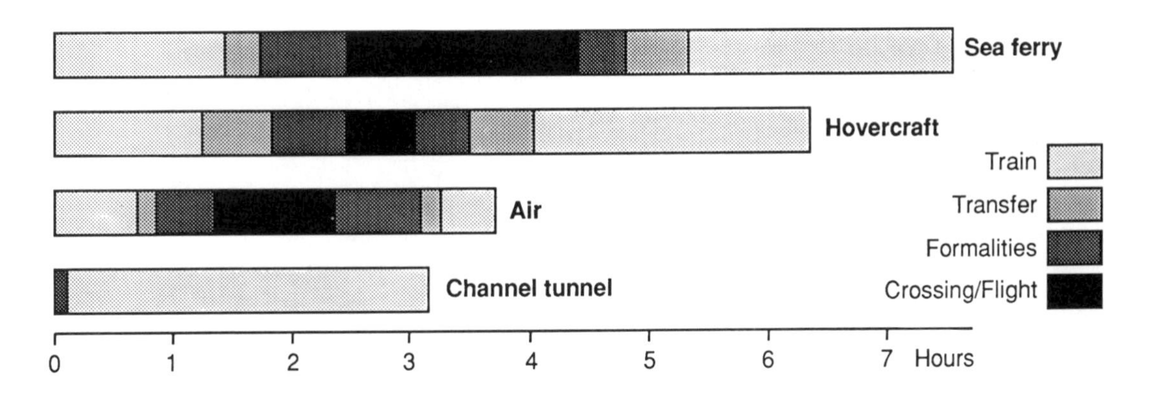

Figure 12.4 Journey times between London and Paris for ferry, air, hovercraft and Channel Tunnel trains.

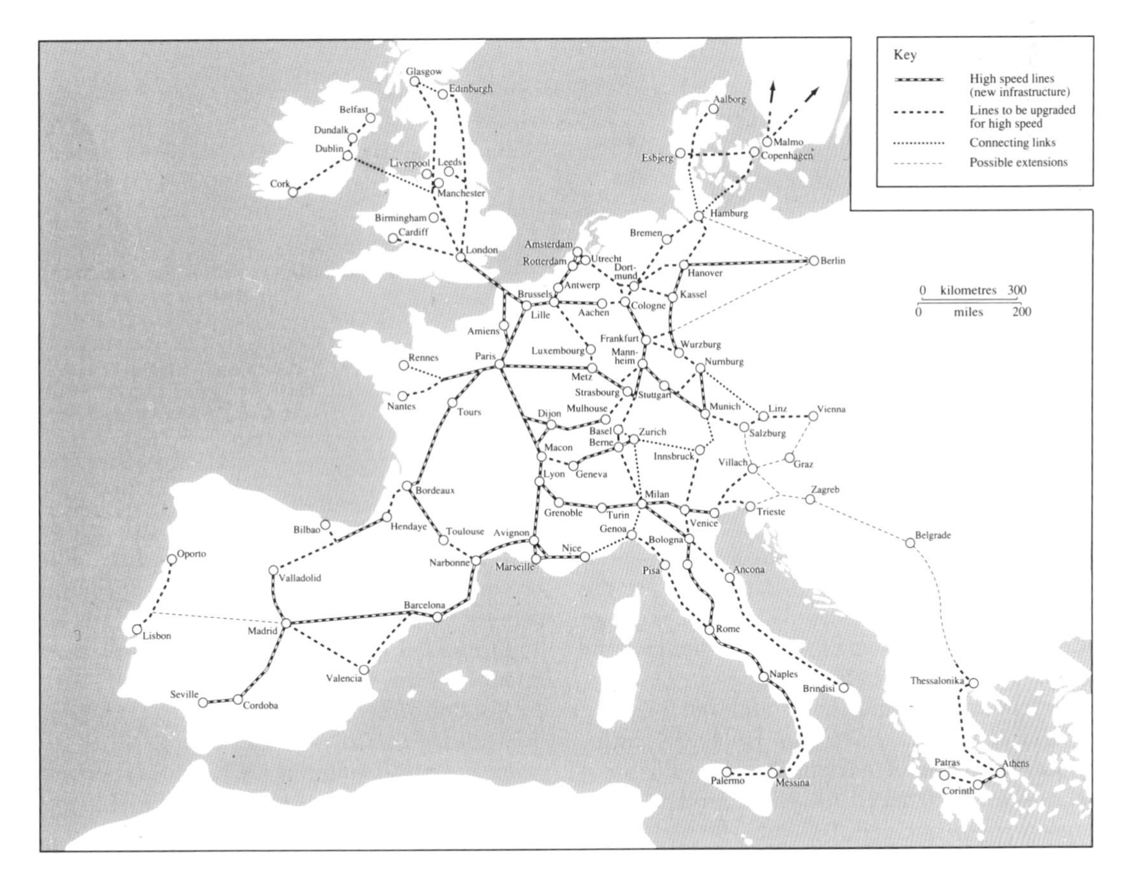

Figure 12.5 The proposed European high-speed rail network.

Photograph 12.2 The Channel Tunnel Project, under construction in 1991.

might provide direct high-speed Netherlands-Germany services.

A comprehensive international high-speed network is now firmly established on the European transport agenda. The European Commission of the EC has promoted the need for a thoroughly integrated development of trains, marketing and tariff-setting policies (ECMT, 1986). In 1990, the European Commission adopted a proposal for a European high-speed network, an ambitious vision of some 18,500 km. of new and upgraded high-speed links (Figure 12.5) (Allen, 1991). It followed a similar vision produced by the Community of European Railways (CER), which represents the EC 'Twelve' and Austria and Switzerland (Kormoss, 1989; *Modern Railways*, 1989). The emerging strategy involves not only the more viable routes between core centres, but also incorporates more peripheral Western Europe, including connections to Lisbon and Sicily, and the Belfast-Dublin-Cork line. The network is clearly tentative and does not entirely coincide with national strategies (Figures 12.3 and 12.5). Some dramatic time savings by the year 2005 are suggested, such as London-Basel in five hours five minutes (13 hours in 1989) and London-Milan in seven hours 50 minutes. A fourfold increase in rail traffic is forecast over the 30-year development period envisaged.

Some possible problems and constraints

There is a formidable gulf between the ideals embodied in the EC and CER proposals and what is realistically attainable. A range of factors can be identified that might impede the radical expansion of the high-speed network or at least challenge its validity. Most daunting are the immense investment costs required, even in the presence of positive backing by European governments. Transport will be competing with heavy demands for capital in energy readjustment, industrial restructuring, environmental management and urban redevelopment. Within the transport sector, there will be parallel claims on private and public capital sources from road and air transport (which both foresee an urgent need to expand their capacities in the coming decade), as well as from other rail sub-sectors, notably freight and commuter services. Domestic priorities may override plans for international transport investment, as in Germany, which has strong incentives to improve transport to the east of the reunified country.

The viability of international rail passenger services will depend in part on the competition from other modes – air transport, international coach services and not least the private car. The steady extension of the European motorway network will reinforce the competitive strength of international private and public road transport. While an enduring rise in energy prices might give rail an advantage for many intra-European journeys, responses can be expected by its competitors. Also, without the infrastructural adjustment and better integration of public transport, the steady decentralization of population and industry will tend to favour the flexibility of road transport.

New high-speed railway construction in Europe is likely to encounter increasingly trenchant opposition on environmental grounds, despite its longer-term capacity to reduce energy consumption and road congestion. The first two French TGV lines were built with relatively low perceived environmental impact, but more recent projects are more problematic. There have been strong protests over the potential effects of the proposed line from London to the Channel Tunnel, the Spanish AVE from Madrid to Seville (*Guardian*, 19 March 1990), and the TGV-Mediterranée extension into Provence (Allen, 1991). The growing need to respond to such opposition will push up construction costs.

Further questions focus on the justification for concentrating resources on *high-speed* intercity services and involve issues of opportunity cost, and spatial and social equity. Whitelegg (1990) challenges assumptions on the implications of faster rail services and points out the disproportionate advantages accruing to places and groups in European society that are already well favoured. Although some high-speed services reach the periphery (for example Paris-Brest), the great majority will be concentrated in the European 'centre'. Here they converge with air transport in terms of service standards and cost, and elevated price levels are necessary for acceptable returns on the great capital investments involved. Newly constructed lines such as the French TGV have few points of access. Accessibility differentials will grow, so that well-connected cities like Lille look forward to buoyant economic development (Gabb, 1990), while others, such as Amiens, protest their disadvantage at being excluded from the network. There is also a possibility that the demands on capital for high-speed projects will be such that investment in other elements of European railway systems will be excessively curtailed.

Alternatives to high-speed trains

The very high investment costs necessary for new high-speed lines could render them suspect on economic as well as social and environmental grounds. Greater benefits could flow from a widespread upgrading of the existing system. British Rail's reduction of journey times from London to Edinburgh to around four hours from July 1991, following electrification but without costly new construction, is evidence of what can be achieved at lower cost (Semmens, 1990a). Worthwhile service improvements on 'second-rank' services are possible, as shown by British Rail Regional Railways' 'Express' network and the German 'InterRegio' services (Allen, 1989). This concept could be applied to a variety of international city pairs: hypothetical examples are Liège-Luxembourg, Stuttgart-Zürich and Toulouse-Barcelona. This 'incremental' approach could be especially applicable in transforming international rail services with and within Eastern Europe. Here, there are hopes of high-speed connections but more modest relative improvements are much more feasible.

There are a range of other possible steps towards improved European international rail passenger services in advance (or in lieu) of full-scale high-speed lines. Close cooperation on timetabling via the European Passenger Train Timetable Conference (CEH) has facilitated the 'EuroCity' network of international trains. Over 90 daily services connect major European cities, all with a minimum commercial speed of 90 km/hr, including stops. An emerging complement to these existing links would be a fully integrated ticketing and reservations system and properly coordinated marketing.

Conclusion

A striking feature of ISPT is the lack of both academic and professional attention devoted to it. Generally, studies of the subject depend on an exceptionalist and empiricist methodological framework, a stricture that extends to the material presented here. While the individuality of the different transport modes and networks must be recognized, this should not preclude generalizations. Similarities in the forces behind what have been expressed here as complementarity and distance suggest that progress towards a more rigorous organizing framework for the study of ISPT should be possible.

Despite the various physical, technical, political and sociocultural obstacles to the development of ISPT, the market has been expanding to meet a growing demand, even though this has been a somewhat uncoordinated and uneven process. For various reasons, the strength of complementarity across the frontiers is growing relative to the constraints that maintain distance, generating ever-increasing international surface-passenger flows on public transport. ISPT is likely to be a key beneficiary of an era characterized by a new world order in which the importance of international boundaries is devalued. This is particularly the case in the EC where densely populated countries are experiencing increased living standards, higher levels of disposable income and enhanced working conditions, as well as the liberalization of economic and labour markets. A similar effect is possible in the longer term in other areas, not least Eastern Europe. ISPT could also gain relative to air transport and the private motor car as these encounter increasing problems from congestion and higher fuel prices. The overstretched capacity of major airports and airspace in Europe and North America is already causing serious concern, while the prospect of a secular rise in fuel costs could allow more energy-efficient forms of international travel, particularly train and ferry, to regain a competitive advantage.

Both for economic and political reasons, ISPT is likely to experience further growth and rising relative as well as absolute importance in the international travel market. Nevertheless, there will continue to be serious impediments to the full realization of the potential of ISPT. As the EC case-study proves, the coordination of infrastructure and services among separate sovereign states is fraught with difficulties. It is a task not to be underestimated, even in an era characterized by enhanced political and economic cooperation. Equally, the experience building up with ISPT planning, development and operation in Europe could provide valuable lessons for other parts of the world.

References

Allen, G. F. (1989), 'D. B. InterRegio cars', *Modern Railways* 46 (491), 412–15.

Allen, G. F. (1991), 'EEC High-speed rail plan', *Modern Railways* 48 (511), 188–97.

Baumgartner, J-P. (1986), 'National infrastructure projects as elements of a European high-speed rail network', European Conference of Ministers of Transport, *High-speed traffic on the railway network of Europe* (Paris), 22–39.

Bayliss, B. T. (1979), 'Transport in the ECs', *Journal of Transport Economics and Policy* 13 (1), 28–43.

Budd, S. (1987), *The EEC* (Kogan Page, London).

Cecchini, P. (1988), *1992: The European Challenge* (Wildwood House, Aldershot).

Cutler, T., Haslan, C., Williams, J. and Williams, K. (1989), *1992: The Struggle for Europe* (Berg, Oxford).

Despicht, N. (1969), *The transport policy of the ECs* (Chatham House, London).

Din, M. A. E. (1990), 'The transport importance of the Arabian (Persian) Gulf', *Transport Reviews* 10 (2), 127–48.

ECMT (1986), *High-speed traffic on the railway network of Europe*, (Paris: European Conference of Ministers of Transport).

European Commission (1979), *A transport network for Europe; outline of a policy*, Bulletin 8/79, EC Office for Official Publications, Luxembourg.

European Commission (1985), *Completing the internal market*, White Paper from the Commission to the European Council (Luxembourg, EC Office for Official Publications).

European Parliament (1982), 'Report drawn up on behalf of the Committee on Transport', *EP Working Documents*, DOC. 1-214/82, Luxembourg.

European Study Service (1988), *EEC transport policy within the context of the Single Market* (Rixensart, Belgium).

Flight International (1990), 8 August.

Ford, R. (1990), 'Tilting train technology matures', *Modern Railways* 47 (502), 368–71.

Gabb, A. (1990), 'Plight at the end of the Tunnel', *Management Today*, September, 48–53.

Guardian (1990a), *The Guardian*, 9 March 1990.

Guardian (1990b), *The Guardian*, 19 March 1990.

Gwilliam, K. M. (1984), 'The Future of the Common Transport Policy', in *Britain within the EC*, ed. El-Agraa, A. M. (Macmillan, London).

HMSO (1967), *The Treaty of Rome*, HMSO, London.

Kormoss, I. B. F. (1989), 'Future developments in North West European tourism: impact of transport trends', *Tourism Management* 10 (4), 310–19.

Lodge, J. (1989), *The EC and the challenge of the future* (Pinter, London).

Lopez Pita, A. and Turro Calvet, M. (1987), 'Faisabilité d'une ligne de chemin de fer à grande vitesse de Barcelone jusqu'à la frontière française', PTRC, *Proceedings of Seminar A, Summer Annual Meeting, 7–11 September 1987* (London: Planning and Transport Research and Computation), Volume P288, 79–92.

Modern Railways (1989), 46 (489), 314.

O'Brien, J. J. (1986), 'Rail passenger transport', Institute of Civil Engineers, *European transport – challenges and responses* (London: Thomas Telford), 7–19.

Omiunu, F. (1987), 'Towards a transport policy for the ECOWAS sub-region', *Transport Reviews* 7 (4), 327–40.

Railway Gazette (1990), 146 (7), 503.

Ruhl, A. (1987), 'The Paris-Brussels-Köln/Amsterdam high-speed link', PTRC, *Proceedings of Seminar A, Summer Annual Meeting, 7–11 September 1987* (London: Planning and Transport Research and Computation), Volume P288, 93–8.

Savelberg, F. and Vogelaar, H. (1987), 'Determinants of a high-speed railway', *Transportation* 14, 97–111.

Semmens, P. W. B. (1990a), 'Five miles a minute', *Railway Magazine*, 136 (1066), 84–8.

Semmens, P. W. B. (1990b), 'Government "no" to high-speed link', *Railway Magazine*, 136 (1072).

Ullman, E. L. (1973), *Geography as spatial interaction: studies in regional development, cities and transportation*, (University of Washington Press).

Whitelegg, J. (1990), 'The society now departing . . . ', *Guardian*, 15 June 1990, 25.

13
International Air Transport

Kenneth Sealy

In this chapter, the controversy over the regulation of airlines forms the basis for this discussion of the world's airlines and airports. Particular emphasis is given to Europe and the USA, where the problems have been most acute.

Introduction

The first major fact about modern air transport is the rapidity of its growth since the late 1960s, even though there have been short-term fluctuations. Thus, following the advance in the 1960s, traffic remained remarkably buoyant over the recession years of the 1970s, whereas the Gulf war of 1991 has seen a dramatic fall in traffic. This has affected tourist traffic particularly, a section of the market more affected by political than economic events. Most analysts expect the long-term upward trend to assert itself eventually, even if the timing is difficult to forecast.

The aircraft's freedom of movement enabling it to overfly both the physical boundaries of land and sea as well as political boundaries makes international operations a dominant part of its operating spectrum. Since it is also a high cost mode, chiefly because in making use of the low resistance of the atmosphere for movement a lot of expensive work must be done to sustain flight, freedom of operation is essential. Restrictions on movement usually mean a need for subsidies and favourable operating agreements. Recent changes towards greater freedom from regulation have not come easily and there is certainly as yet no universal agreement; the battle of the 1990s will continue where theories concerning control and freedom mix with the legacies of history. This, then, is the second major element in the modern air-transport scene.

Increasing traffic makes severe demands upon airports and their associated air-traffic control (ATC) systems, not to mention the problems of surface access to the airports. The battle here revolves around the need for expansion with problems of site evaluation, bringing with it conflicts with the local environment. Not only are there difficulties in managing aircraft movements from the airport, but ATC problems have also increased in proportion. The spectre of the 'near miss' has become a reality. Surface access at major airports has also been affected and the need for high-speed rail systems has assumed greater importance with the increasing difficulties posed by road access in many countries.

Geography influences all these aspects of the air transport system from the easily perceived basic orientation of the world's regions being the origins of the major markets for transport to more subtle impacts on route and system structure or airport systems and the problems in meshing the various ground and air networks to form a workable whole. In this chapter, the basic elements will be taken as given. There remains the question of the order in which the detailed discussion should proceed. Since the regulatory atmosphere is crucial to the economic operations, we shall begin the discussion by looking at this aspect before proceeding to airline and airport matters.

Airline regulation

Airline regulation is a long and tortuous story which cannot be detailed here, but a brief statement is necessary as a prelude to an economic and geographical discussion of airlines. For details see Pryke (1987) and Barrett (1987). Internationally there are two options: either one may claim 'open skies', comparable to the freedom of the seas, whereby multilateral agreements between states permit the freedom to carry virtually without restriction, or a system based on the 'sovereignty of airspace', which means that agreements between countries are negotiated bilaterally. The possibility of regional rather than national bilateralism is a relatively recent variation being put forward, for example, by the European Community (EC), but owing to wartime fears of American dominance in 1944, the formative Chicago Conference in that year opted for a bilateral system. This has persisted with only a few cracks until very recently. Most states also instituted their own form of regulation, none more so than the USA.

International agreements

Internationally, the first bilateral agreement was between the USA and the UK, signed in Bermuda in 1946; it was important because it set the pattern for most later agreements worldwide. Operationally, agreements revolve around the 'five freedoms'. The first two permit aircraft to fly over a state's territory, either without landing, or to land for technical reasons only, so most of the 52 states who attended the Chicago convention agreed this 'Two Freedoms' agreement without comment. Freedoms Three and Four allow airlines to take on and discharge passengers and cargo within the negotiating states, while the elusive Fifth Freedom bestows traffic rights between two states foreign to the operator's country. The last named has always been very restricted. Even in 1982 the European Civil Aviation Conference (ECAC) estimated that Fifth Freedom traffic within Europe accounted for only 3.5 per cent of the total (ECAC, 1982).

Under the bilateral system, routes into the two countries are agreed and a review of capacity on the routes is monitored to avoid 'excessive capacity'. Fares and rates are not controlled, but were to be negotiated by the airlines' own organization, the International Air Transport Association (IATA), subject eventually to governmental approval. Geographically this usually meant only one operator on each route from each country plying between named 'gateway' airports. An exception to the rule was the two-operator agreement between the USA and the UK. Fifth Freedom traffic was only permitted if it was related to the remaining traffic potential on the route. When similar national controls on route entry and price control are added, airlines had very little room to 'compete', usually through control of the frequency of operations and the quality of service. Hence arose a familiar pattern: a linear, or grid, route system of direct links between major cities, usually capitals, together with struggles to provide the best equipment, serve the best meals and provide a frequency of timetable to suit the customer. Pooling agreements on capacity and revenue also grew up, while odd and anomalous fare systems baffled customers. Thus, for example, the basic rule was that a fare from A to C via B must not be cheaper than the direct route A to C. When this is extended to cover more than two sectors, not only route but also regional differences in fare levels can be considerable over comparable distances. For more details, studies by Rosenberg (1970) and the Association of European Airlines (AEA, 1977) written before the upsets of the 1980s, cover most of the ground for Western Europe and Scandinavia.

This system worked because the main source of traffic in the period up to the early 1960s was from the business community, who were time-conscious and demanded high-frequency operations and a high quality of service. Fares were usually paid by the company and were less vital than time costs. Finally, services were almost exclusively operated by scheduled airlines.

The deregulation of airline operations

Charter services were hardly covered by regulations and they provided the impetus for the massive rise in tourist services after 1963, spearheaded by the inclusive tour. Until the mid-1970s the scheduled carriers were hamstrung in their attempts to compete, most notably on the North Atlantic, while the rapidly growing tourist networks of the charter carriers provided an obvious economic and geographical base for expansion into scheduled services long dominated by scheduled carriers. The advent of Laker Airways 'Skytrain' services over the North Atlantic, following an absence of agreement on fares, forced the established scheduled carriers to fight to re-establish themselves, something they successfully accomplished in the 1980s. Restlessness internationally spurred national changes; the most important was the deregulation of the US domestic air services following the implementation of the Airlines Deregulation Act passed in October 1978. These developments received a mixed reception elsewhere. The EC has haltingly made some advances, notably so far through the Regional Services Agreement, but against this must be set recent mergers between European airlines effectively 'sealing off' certain areas. Beyond the Western scene, growth by Pacific rim countries, spearheaded by Japan and including Singapore, Hong Kong, South Korea and Australia, have made the Pacific the fastest growing sector in the airline world of 1990.

Other countries, fearful of the effects of deregulation upon their national carriers, have felt cause to oppose rapid change. Bilateral agreements here usually ensure a 50 per cent access to traffic and a similar proportion of revenue and restricted entry for foreign carriers. The fear that under deregulation their carriers would soon lose out to the large 'megacarrier', depriving them of foreign exchange as well as control of their air services, is a problem we shall return to in the next section. Within this group must be included most of the African and many Latin American countries. The old stager, the Bermuda agreement between the USA and the UK, went through a meagre liberalization in 1979; more recently, further revisions were agreed in March 1991. Finally, there is the recent opening-up of the Eastern European market, together with a more liberal attitude by the USSR. Apart from an unholy scramble by Western airlines to buy-up or join-up with Eastern airlines, little has yet been finalized politically.

The political geography of the airline world is, then, in a state of flux. The initial euphoria over the success story of deregulation in the USA is being increasingly disturbed by more recent moves towards a monopolistic situation with the growth of the giant 'megacarrier'. In the words of Ed Beauvais of the US airline, America West, 'Deregulation means free enterprise, but not *laissez-faire*.' Maybe we need regulation to ensure freedom of competition.

Airline operations

Theoretical basis

The previous discussion underlined the regulatory background to airline operations. Here we shall consider the economic models involved and their consequences before considering the characteristics and performance of the world's airlines.

The regulatory model is based on the fact that entry into the airline business under competition is relatively easy. In fact, it only requires the buying or leasing of an aircraft; and since all operations of passenger vehicles carry empty seats, the short-run marginal costs (SRMC) will be low. Regulation is advisable because a new operator can often undercut existing fare levels and still make a profit. Since SRMC is low, the incumbent operator will be tempted to reduce his fares in reply to the competition and we have the beginnings of a 'fare war', which may mean that the airlines concerned do not cover their total costs in the long run. There may also be an incentive to lower safety standards to cut down on cost levels, and to increase capacity beyond peak requirements in order to increase the market share. Taken to a final conclusion, bankruptcies, or more likely, mergers and takeovers, will occur leading eventually to a monopoly, or at best an oligopoly. The 'megacarriers' that emerge may exploit their geographical advantage by dominating not only routes, but also airport (hub) capacity, making monopoly profits that may be used to cross-subsidize competitive operations elsewhere. Unless there is some control, less-populated areas will lose services, whether or not there is a social case for their retention. Since such routes cannot support multi-operator services there is nothing to foster competition or prevent eventual monopoly conditions.

For the Third World competition by powerful megacarriers would lead either to the eclipse of the national carrier, or possibly, worse still, control would pass to the big carrier through a merger with the national airline. The result is a serious loss in foreign-exchange earnings and a loss of control and national stature in the world. Finally, for deregulation to succeed internationally, it must extend to all countries. Otherwise countries that retain regulatory powers could 'invade' the free airspace of other countries while protecting their own airspace to the advantage of their own airlines and the disadvantage of other carriers.

Following these arguments it is clear that deregulation would mean that airlines would have to defend themselves against opposition, which may best be done by consolidating their operations around a sound base where they could dominate a section of the market, whether business or tourist oriented. This means domination of a hub airport and lucrative routes from it to other major centres, fed by feeder routes into the hub. The result is a 'hub-spoke' network, which may be expanded to take in other hubs and to buy out feeder airlines where appropriate (Figure 13.3). The result in the long-run is an oligopoly of large airlines dominating most of the major airports. While this model does not see civil aviation as a true public utility – it does not have any natural monopoly like water or gas – it is still, as Wheatcroft put it in the heyday of regulation, 'an industry of real public importance' and one which forms part of a nation's social overhead capital and should not be left to unregulated market forces (Wheatcroft, 1964).

The case against regulation rests on the assumption that air services are no different from other forms of trade or services and as such should be treated in the same way. This does not mean that there are no rules. Thus, for example, air transport would come under the anti-dumping laws as applied by the General Agreement on Tariffs and Trade (GATT), but otherwise normal economic theory should be applicable. This would mean free entry to the market, whether on routes or into airports, and the freedom to set prices to cover costs within the markets available. It is sometimes claimed that routes are contestable and where any operator tries to charge excessive fares this will attract other operators on to the route and restore a competitive fare. Price competition would reduce most fares in the short term, since airlines under regulation competed through quality of service and had therefore built up large fleets of aircraft many of which would not be suitable, or even needed, under a competitive situation. Promotional and sales, including labour costs, had also grown out of proportion to need, while the regulatory cushion enabled cabin staff to negotiate highly favourable incomes. Once this transition period was complete and adjustments were made to suit the new conditions, fares should reflect cost and market propensities. In high-density tourist markets, reductions in fares would be the greatest, while for some thinner, or more demanding, business sectors, fares would rise. Even so, this model recognizes the fact that some socially desirable routes would not pay and would need support. But the method of achieving this support, once the case had been proved, would be by public tender and the service operated by the airline offering the required level of service at the lowest price.

The position with respect to Third World airlines is more complicated. Many of these are government controlled, or at least government supported, and enjoy very favourable terms under the existing bilateral agreements. As noted earlier, the fear is that competition would mean the demise of the airline. There appears to be a case, then, for some protection of these 'infant industries'. The truth is that many of these airlines are rather aged infants and have been protected for decades with little incentive to provide a cost-effective service. As long as the country provides a potential market to support airline service, then the national airline ought to have the basis for a competitive service, particularly as some costs, for example labour costs, would be lower than those facing overseas competitors. Geographically this means not only a resource base, but the opportunity to build a strong hub from which to base a radial network that maximizes the use of that hub. Once this sector of the airline's network is established, international services would be expanded as the carrier's ability to compete increased. Many current services operated at very low-load factors for prestige rather than economic reasons would seemingly be abandoned. If there still remained social reasons for operating certain routes, then government subsidy, governed by the establishment of need and efficiency of service, would be in order. After all, many airlines have become established with practically no national hinterland of any size at all: for example, the phenomenal growth of Cathay Pacific Airlines, Hong Kong's flag carrier, or Singapore Airlines covering 58 cities in 38 countries.

The geographical implications of this model are important and can be decisive since there is no regulatory cushion to protect faltering routes. The most important issue must be the relation of the airline's structure to the geographical realities of its network. Indeed, an airline's competitive position depends far more on this relationship than the size of the airline or any technical superiority it may possess. Density of traffic is the key which depends ultimately on the economic geography of the catchment area and its potential for business, tourist and other non-business traffic, passenger and freight. The principle should now be familiar: feed from small beginnings into larger centres and thence to a wider network of hubs, each with a profitable hinterland. The trouble with the regulated systems that controlled route entry and reduced the number of operators on any route to one, or at the most two airlines, is that it favoured direct links between city pairs, regardless of market structure. This led to a grid pattern based on the geographical model network that connects all points in a network by the shortest distance between points: accessible, but expensive! As noted in the previous section, this favoured the time-conscious business market. The fact that route allocations were also done piecemeal reinforces the pattern. On many long-haul, intercontinental routes, there is little difference in pattern, since market opportunities are widely spaced and long-sector, linear patterns are normal.

Control of the hubs of this system is, then, of vital interest to all airlines; hence the ascendancy of strategic international hubs like London, New York or Hong Kong. For the traveller, total journey time is a relevant factor, tempered by cost; direct services favour the grid, but cost may be more important. It follows, then, that a multisector network is cheaper, but may involve time-consuming changes of aircraft en route.

However, if the change only involves a change of aircraft belonging to the same operator, time and trouble may be saved. This interlining within the same operator's network is important and is accomplished only where the link is at the same hub; a distinct advantage for the hub-spoke network favoured by a competitive model. Recently this system has received a massive boost by the introduction of comprehensive computer reservation systems (CRS). To have your services immediately accessible to the prospective customer, and linked to a far greater domestic and international airline network, is a powerful market tool. Competition internationally has been intense and a small number of large systems dominate the market, paralleling a similar development in the airlines themselves.

A competitive model, therefore, aims at a more rational, economic distribution of airline resources, with prices matched to the supply and demand for those services and subject to rules that ensure safety standards and social obligations. The model does not imply an overall reduction in fares; the fare levels are matched to the market and not determined by some arbitrary, regulatory assumption.

The airline record

Overall world development may be discerned by reference to Figure 13.1 which demonstrates the prime fact of continuing strong growth. In 1988, world airlines celebrated the 'billion mark' when over a billion passenger journeys were recorded for the first time (1.7 billion passenger km.). According to the International Civil Aviation Organization (ICAO) this figure is likely to top 2.2 billion passenger km. by the year 2000. The growth rate over the years since 1978 has been six per cent per year, despite dips in the early 1980s, with a similar 'likely' trend continuing to the end of the century. However, the recent Gulf war in Iraq has led to increased fuel prices and reduced international passenger traffic so that the confidence expressed by ICAO may prove to have been premature.

Regional patterns (Tables 13.1 and 13.4) show the dominance of the USA, which with the help of its enormous domestic network accounts for 42 per cent of the traffic followed by Europe with 30 per cent, making these two areas responsible for over 70 per cent of the total. More interesting is the fact that while Europe's share has fallen some four per cent since 1978, the regions of the Asia-Pacific rim have shown a rapid increase and now provide 18 per cent of world traffic. Elsewhere, the position is more variable, depending to some extent on the economic fortunes of the regions concerned, but also upon changes in bilateral and internal agreements. Countries like Australia and Canada have followed the deregulation course, whereas most African and Latin-American airlines remain regulated. On a world scale, the sensitivity of airline fortunes to economic events can be seen by comparing airline traffic with gross domestic product (GDP) as indicated in Figure 13.1.

To demonstrate the effects of deregulation we shall look at some examples of regions and airlines. Pride of place should go to the USA, which not only dominates the airline world, but was also the first to deregulate its industry under the Act of 1978. Doganis (1985) and Shaw (1982) analyse world airlines in detail, while a more topical account is provided by Naveau (1990).

Airlines in the USA

In 1978 there were ten trunk carriers certificated by the Civil Aeronautics Board (CAB) responsible for 88 per cent of the traffic, supported by some 16 local service carriers operating regional services. These terms were superseded in 1981, when the CAB recognized 'major airlines' as those with annual revenues greater than $1 billion and 'national' carriers with a revenue range of $100 million to $1 billion, the remaining airlines being regional or commuter carriers. Deregulation produced a varied response.

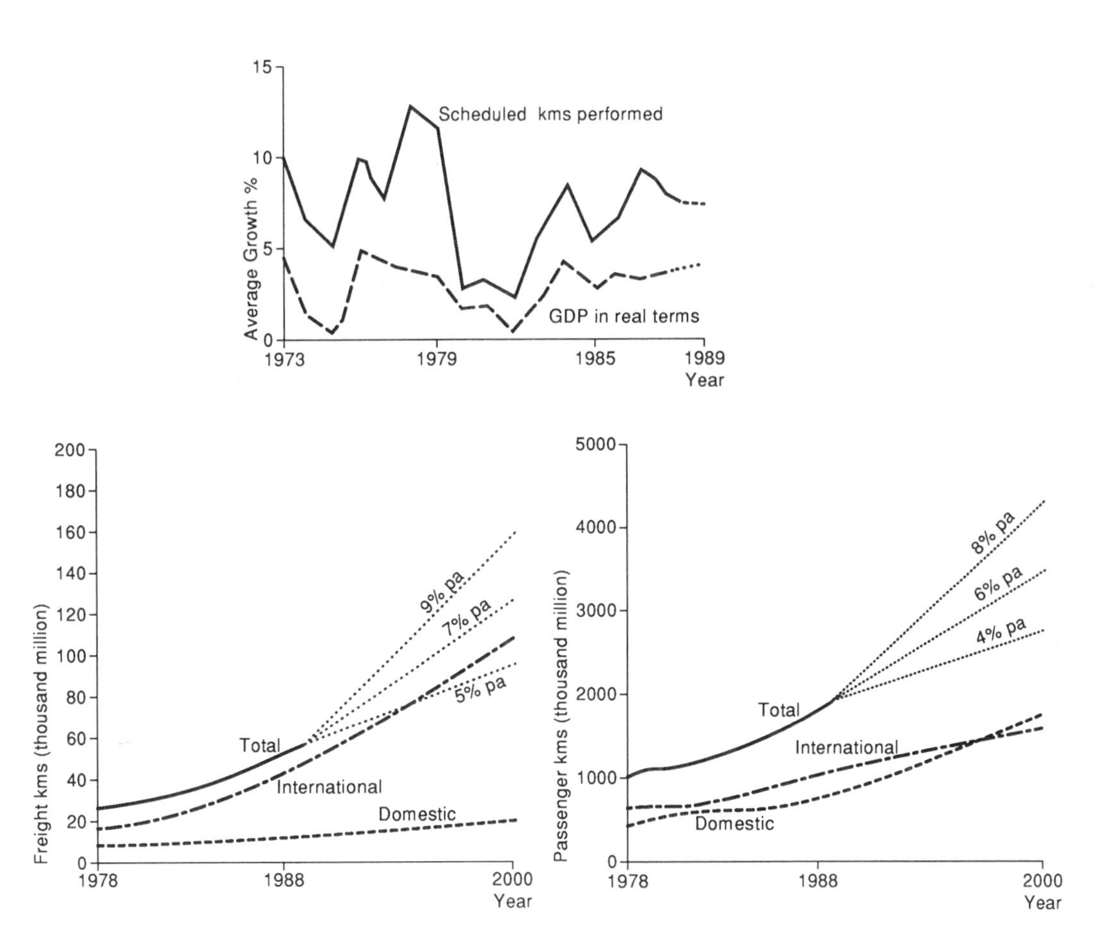

Figure 13.1 Trends in world air passenger and freight traffic, and a comparison between GDP and air traffic.

Source: ICAO Traffic Forecast 1989.

Braniff, a former trunk, expanded into 'city-pairs' it deemed potentially profitable, with no real network philosophy visible. It went bankrupt in 1982; after a brief reappearance around a hub at Kansas City, went bankrupt again in 1989. Other trunks, hampered by excessive fleets and over-stretched networks, took time to recover. Indeed, Pan American has accumulated losses of over $1 billion since 1980, so perhaps it was not surprising that in October 1990, United Airlines made a bid to take over Pan Am's London hub and associated routes from four US gateways, as well as proposing other market and asset agreements. As Figure 13.8 shows, this merger, finally agreed in March 1991, has made United Airlines one of the leading US international carriers, particularly over the Atlantic, as Pan Am, in 1989, accounted for 14 per cent of the total Atlantic traffic alone. Furthermore, United Airlines will inherit further access to European routes and they have also obtained agreement to use Pan Am's operating slots at Heathrow, a decision the airline termed essential as neither Gatwick nor Stansted were considered acceptable. The decision to allow United Airlines to operate from Heathrow, agreed in April 1991, has set a precedent that could well lead to the further dominance of the US megacarriers. Thus, if the deal involving American Airlines' bid for TWA's London

Table 13.1 World's leading airlines 1987: scheduled passengers carried (thousands)

(1) *Total traffic (domestic and international)* *Airline*		*Traffic*	*Rank*
American Airlines		59,091	1
United Airlines		55,184	2
Eastern Airlines		44,674	3
Continental Airlines		39,380	4
Trans World Airlines (TWA)		24,795	5
British Airways (BA)		19,100	6
Japan Air Lines (JAL)		17,877	7
Deutsche Lufthansa A.G.		16,865	8
Pan American World Airways (Pan Am)		14,888	9
(2) *International traffic*			
British Airways		14,482	1
Air France		10,516	2
Deutsche Lufthansa		9,888	3
Pan American World Airways		9,376	4
Japan Air Lines		6,639	5
Iberia		6,426	6
Scandinavian Airlines System (SAS)		6,384	7
Swissair		6,119	8
KLM		5,875	9
Alitalia		5,330	10
American Airlines		5,021	11
Eastern Airlines		4,214	12
(3) *Domestic traffic*			
USA:	American Airlines	54,670	1
	United Airlines	51,768	2
	Eastern Airlines	46,460	3
	Continental Airlines	36,604	4
	Trans World Airlines	21,561	5
Europe:	Alitalia	8,923	
	Iberia	7,624	
	Deutsche Lufthansa	6,977	
	Scandinavian Airlines Systems	6,249	
	British Airways	4,618	
	Air France	2,843	
	British Midland Airway	1,614	

Source: IATA World Air Transport Statistics, 1987.

routes goes through, the two carriers would command 30 per cent of the North Atlantic traffic. British Airways' share of the market would be approximately ten per cent; what price competition?

New carriers and former local service airlines have had mixed fortunes. The former local carrier, Trans Texas Airways, expanded into the national and later international market as Texas International Airlines. Later they absorbed the former trunks Continental and Eastern, among other smaller airlines, to become the largest organization in the industry as the Texas Air Corporation. Among the 'new' post-1978 carriers, only one, America West, has so far survived to become a major carrier in 1990, bringing the total for majors to ten. Apart from commuters, the only other national survivor, Midway Airlines, may be bought out by American Airlines (*Flight International*, 22–28 August 1990). In January

Table 13.2 US airlines percentage of total revenue passenger traffic

1978		*1988*		*Average annual % change 1984–88 (rev.pass.miles)*
United	18.1	United	17.6	+10.3
American	12.8	American	16.5	+15.3
TWA	12.3	Delta	13.2	+5.2
Eastern	11.0	Continental	10.4	−0.4
Delta	10.4	Northwest	10.4	+17.6
Pan Am	10.3	TWA	8.9	+5.2
Western	4.6	US Air	8.0	–
		Eastern	7.3	−0.4

Table 13.3 UK leading airlines 1990

Airline	*ton km. capacity*	*ton km. used*	*% of total load ton km. used*
British Airways	11,949	8,200	60.3
Britannia	1,290	1,141	8.4
Virgin Atlantic	901	509	3.7
Air Europe*	876	583	4.3
Dan Air	857	720	5.3
Monarch	731	555	4.1
Air 2000	418	370	2.7
Caledonian	359	284	2.1
Novair International**	303	230	1.7
British Midland	273	119	0.9
British Island	178	120	0.9
Air UK	138	71	0.5
47 remaining airlines			less 0.5

* Ceased operations March 1991.
** Ceased operations May 1990.

Source: CAA: UK airlines monthly operations and traffic statistics.

1991 Eastern filed for Chapter 11 Bankruptcy following fare discounting through their Philadelphia hub which subsequently helped to bankrupt Midway Airlines in April 1991.

The survivors all show a strong orientation to a hub-spoke network as the model predicted. Figure 13.3 shows typical patterns, but it is more difficult to show how each net meshes into the whole picture. Most majors have progressed to multiple hubs: for example, Delta Airlines has four main hubs at Atlanta, Dallas, Salt Lake City and Cincinnati with four more supplementary hubs, as shown in Figure 13.3. Feeder services through the hubs are largely controlled or owned by Delta under the 'Delta Connection', while this massive domestic network is extended by a considerable international net. Figure 13.3b shows how the acquisition of Western Air Lines in 1987 balanced up the network by adding a previously missing Western coverage. In conclusion, American experience confirmed the expectations of the competitive model up to 1985 but since then some of the fears of the opposing school have become manifest. The growth of the megacarriers, nearly all of whom are ex-trunks and most controlled by large corporations, have raised fears that a tight oligopoly of airlines will control the US system as surely as the old CAB regulations ever did. There is talk of a need for 'regulation to ensure competition', but at the time of writing, little evidence of any return to a controlled system.

Table 13.4 World airports 1988: leading airports by traffic volume total passengers handled (thousands)

	No. (000's)	*% change from 1987*	*Rank 1988*	*Rank 1987*
Chicago: O'Hare	56,773	0.9	1	1
Atlanta: Hartsfield	45,900	−3.7	2	2
Los Angeles International	44,399	−1.1	3	3
Dallas: Ft Worth	44,271	5.7	4	4
London: Heathrow	37,510	8.0	5	5
Tokyo: Hareda	32,177	7.5	6	8
Denver: Stapleton	31,798	−6.7	7	6
New York: Kennedy	31,166	3.2	8	7
San Francisco Internat.	30,504	2.3	9	9
Miami International	24,525	2.1	10	11

Remaining rank order 1988

Frankfurt	11	Osaka	19
New York: La Guardia	12	Detroit	20
Boston	13	Toronto	21
New York: Newark	14	Pittsburgh	22
Paris: Orly	15	Paris: de Gaulle	23
London: Gatwick	16	Minneapolis	24
St Louis	17	Orlando	25
Honolulu	18		

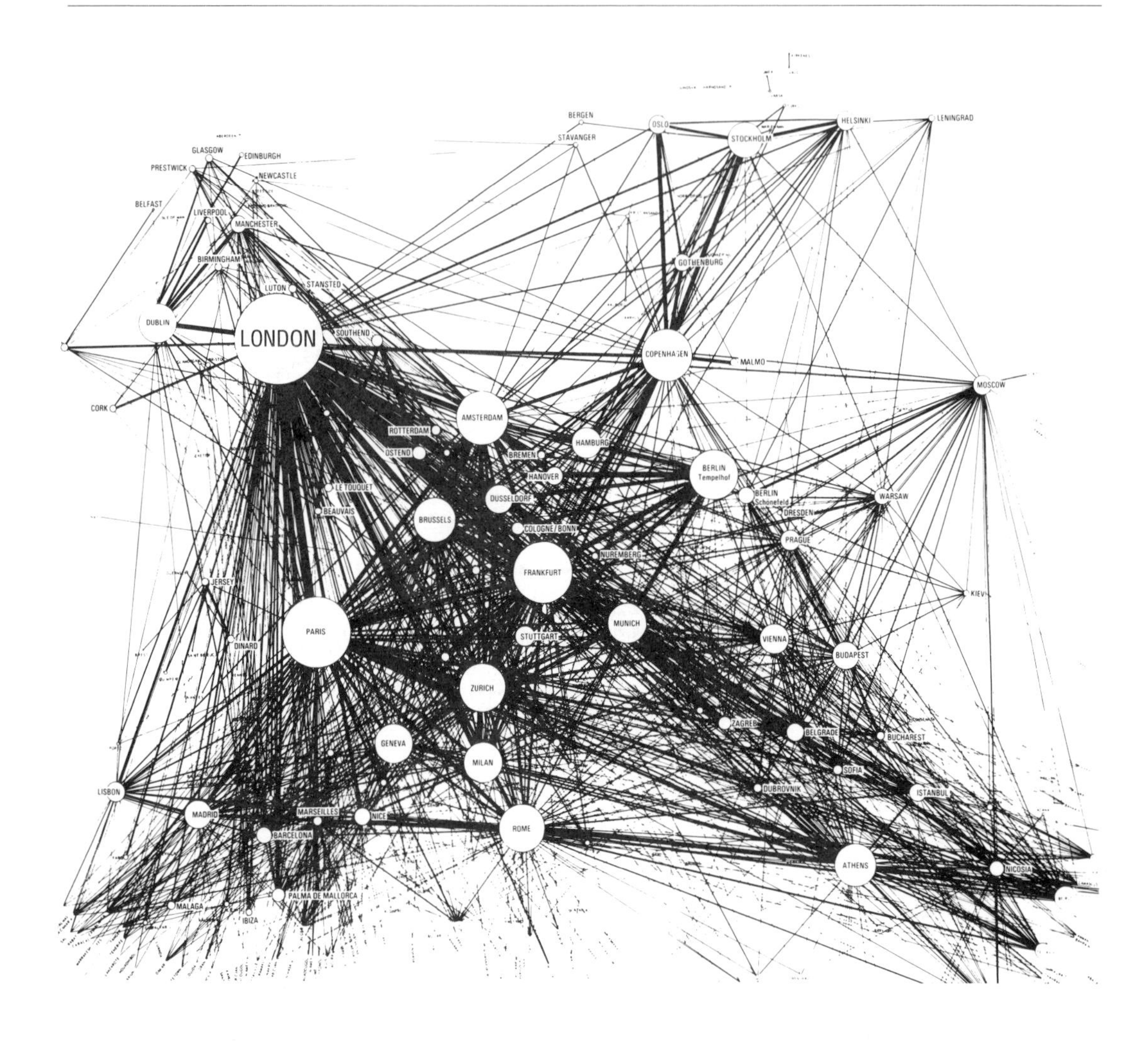

Figure 13.2 The West European air network: a regulated system.

Source: Based on ICAO data.

European airlines

American experience has not been lost on the European states. Some like the UK have deregulated the domestic system; other airlines, as in France, have sought to control the national network by mergers. The European Community in the shape of the European Commission have sought to make some concessions to liberalization, most notably the Regional Services Agreement 1984 (for a discussion, see EC, 1980). Significantly, the scramble for mergers has accelerated in 1990; apart from service agreements between European carriers and between European and US carriers, merging interests between national carriers has appeared: something not seen before. Figure 13.5 gives some impression of the growing complexity of the European case. Fears are growing that an oligopoly of European and US airlines will dominate any fully deregulated European market, as the big carriers are doing in the USA.

The European network is different from that of the USA (Figure 13.2) in that it is

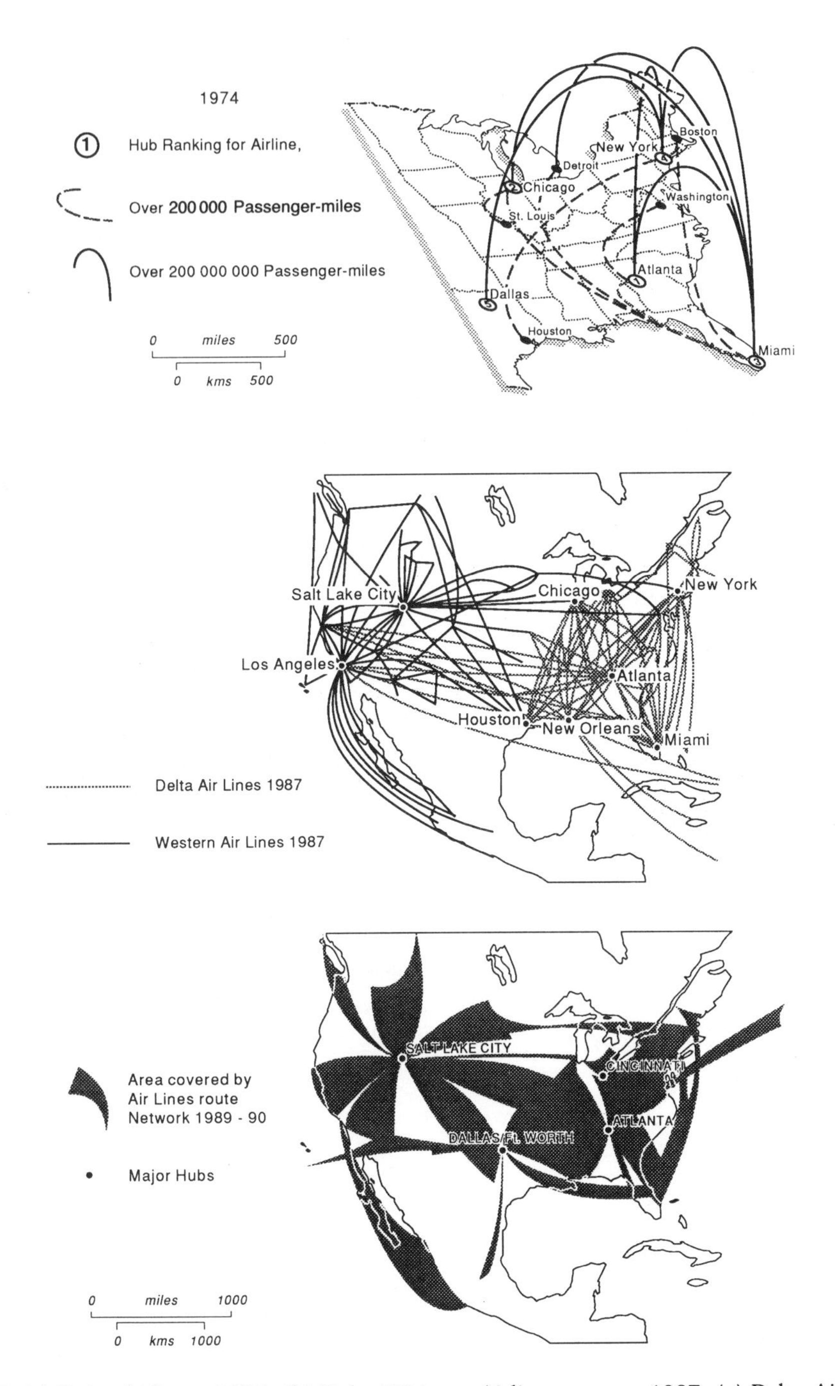

Figure 13.3 (a) Delta Airlines, 1974. (b) Delta/Western Airlines merger, 1987. (c) Delta Airlines, 1989–90.

Source: Delta Airlines and *Flight International*.

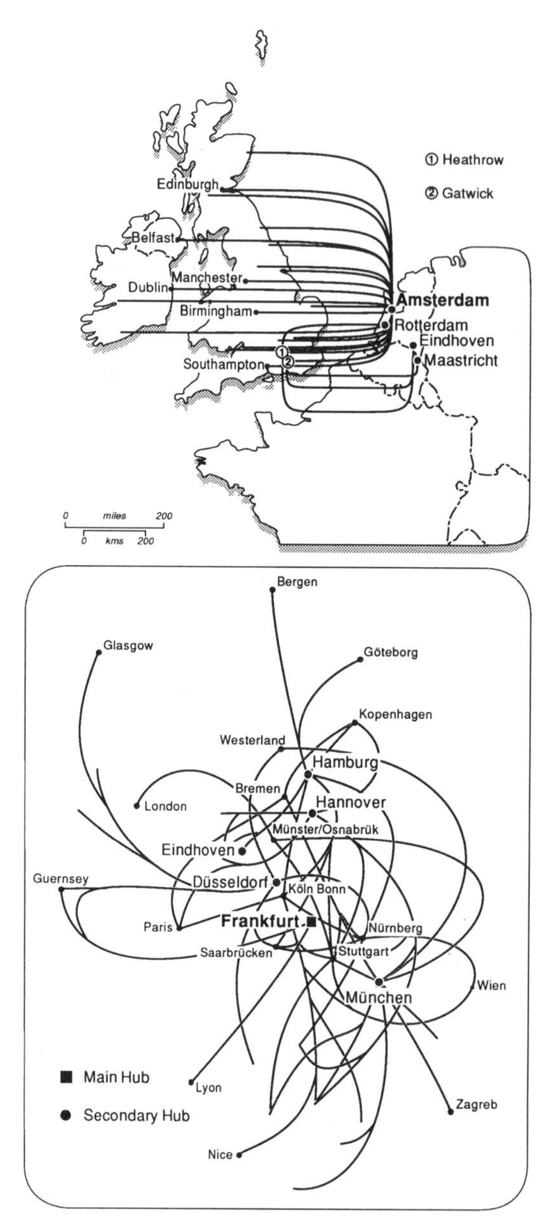

Figure 13.4 European regional air carriers. (a) UK–Amsterdam air services. (b) The DLT domestic network, Western Germany.

Sources: KLM and *Flight International.*

primarily an international network, as the traffic figures suggest (Tables 13.1, 13.4 and 13.5), with domestic operations on a small scale by US standards. Thus the network is already a hub-spoke system, only this time based on national hubs connected by direct bilaterally agreed direct links with each other, well-shown in Figure 13.2. The underlying domestic systems are not conspicuous, mainly because the catchments are smaller and open to severe surface transport competition. More pertinent up to the present has been the lack of a true 'secondary system' of international routes connecting medium-sized cities, as developed between US cities of comparable size. One of the only really useful liberalizing edicts agreed by the EC so far has been the Regional Services Agreement which has encouraged the growth of medium-size city links operated by the smaller airlines. Some of these have been hit by the decline in charter tour traffic in the later 1980s (Table 13.6) and find the possibility of obtaining stable schedules opportunities at least a start towards stability in the future. However, the currently evolving nets do not all follow the US pattern. The early fares agreement between the UK and The Netherlands signed in 1984 provides a longer-term glimpse of possible developments (Figure 13.4a). Figure 13.4b shows the international and domestic pattern of the West German regional carrier, DLT: a network rather different and perhaps best described as a grid system. How this pattern will develop under the newly created united German state is a matter of conjecture. Whatever its final form, the European system will supply geographers with some fascinating network problems for many years ahead.

As far as the EC is concerned, progress towards civil aviation's implementation of the Treaty of Rome may be termed a process of 'gradualization' towards liberalization. Documentation of the process must not only take account of the statements, suggestions and actions of the EC Council, the Commission and the Council of Ministers, but also of the representations of numerous national and regional bodies. Most recently, the Council of Ministers, meeting in June 1990, managed to agree on only two modifications dealing with the liberalizing of fares and 'confirmed'

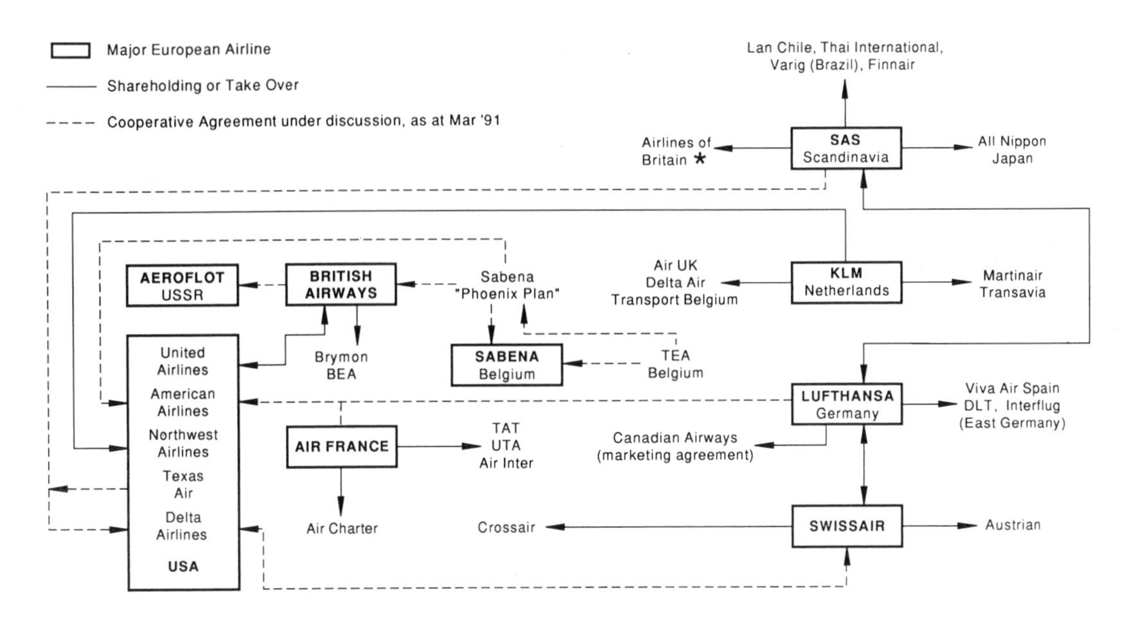

Figure 13.5 Agreements and mergers between European airlines, showing both confirmed and proposed agreements to April 1991.

Table 13.5 Leading airports by traffic volume: international passengers handled (thousands)

	No. (000's)	*% change from 1987*	*Rank 1988*	*Rank 1987*
London: Heathrow	30,659	7.1	1	1
Frankfurt	24,344	9.7	2	4
London: Gatwick	19,619	7.5	3	2
New York: Kennedy	18,021	3.5	4	3
Paris: de Gaulle	16,172	11.9	5	5
Hong Kong	15,277	20.6	6	7
Amsterdam	14,397	9.0	7	6
Tokyo	13,893	21.9	8	8
Singapore	11,381	13.7	9	9
Zurich	10,194	6.2	10	10

Remaining rank order 1988

Toronto	11	Taipei	22
Miami	12	Athens	23
Palma: Mallorca	13	Madrid	24
Copenhagen	14	Munich	25
Paris: Orly	15		
Bangkok	16		
Los Angeles International	17		
Manchester	18		
Düsseldorf	19		
Rome	20		
Brussels	21		

Source: ICAO Airport Traffic Digest of Statistics, 1988.

Table 13.6 UK inclusive tour traffic and margins

Year	*Total ITs (millions)*	*By air (millions)*	*Profit margin %*
1979	5.0	4.2	5.4
1980	6.2	4.9	4.8
1981	6.8	5.2	5.1
1982	7.7	5.8	2.1
1983	8.0	5.75	3.9
1984	9.0	7.0	3.0
1985	8.5	6.4	3.3
1986	10.66	8.4	1.7
1987	11.98	9.8	-0.9
1988	12.6	10.3	0.5
1989*	11.5	10.0	0.5

Source: International Passenger Survey by the UK Office of Population Censuses and Surveys. Profit margins from the CAA. (* = estimate)

the 'objectives' for 1993. Thus many hoped that they would have revoked the current 'double disapproval' principle governing fares, whereby fare changes would be automatically approved unless both countries on the route concerned disapproved. In the event they did not change it.

The problem with the gradual approach is that it gives time for the big, entrenched carriers to consolidate their position, holding up progress by endless arguments over whether actions are stifling competition. National differences arise. Thus the UK domestic network is deregulated and, despite skirmishes with British Airways, other airlines have been able to exploit the network. In France, the national carrier Air France has agreed terms to take control of its rival UTA, and the domestic carriers TAT and Air Inter, thus virtually controlling the whole of the French domestic net – and, incidentally, making it the second largest European airline after British Airways. This action has been condemned by the EC Commission in a running battle since autumn 1989 (*Flight International*, 24 January 1990). This is by no means the only reorganization that is taking place. It is quite impossible to map the changes at present, but Figure 13.5 gives the major moves in this game of European chess. It is interesting to note that a study of the gradual implementation of deregulation in Canada seems to have confirmed the dangers of the slow process. The idea was to smooth the transition, but in fact it enabled incumbent airlines to protect their positions (Button, 1989). There are already signs that this may occur in Europe where Air France's recent acquisition of the two main domestic carriers gives them control over much of French air traffic. Nevertheless, increasing freedom should eventually ensure that European airlines will be able to operate a far wider network than the restricted opportunities available to them up to now will allow (see, for example, Williams, 1990).

The rest of the world

The variety of situations in the rest of the world defy detailed treatment here. Basic principles still apply and where systems are closely controlled, networks will reflect that control. On the main international scene, bilateral agreements still dominate and often only give direct city-pair connections between national capitals, or a limited number of 'gateway' airports. Domestically, tight control may result in a simple spoke network radiating from the capital, or where regional cities are important, grids or multiple hub-spoke patterns will evolve. Thus, for example, Nigerian Airways, owned by the Federal government, virtually controls the whole aviation network serving all 19 states of the Federation as well as the major international routes. Despite its strength it has still been in financial difficulties. These have resulted in rationalization measures to reduce the costs arising from an oversized workforce and unprofitable services. By contrast, a number of airlines, apparently without a national catchment but situated in an expanding region and with operational freedom, have become truly world airlines. The best examples come from the Asian-Pacific rim, the fastest growing air market in the world. Apart from the three major Japanese airlines led by Japan Airlines, the growth of Singapore Airlines (SA) and the Hong Kong carrier Cathay Pacific Airlines (CPA) each

Table 13.7 European meeting points: members of organization

	ECAC	Community Community member	Eurocontrol
Albania			
Austria	Yes		
Belgium	Yes	Yes	Yes (founder)
Bulgaria			
Cyprus	Yes		
Czechoslovakia			
Denmark	Yes	Yes	
Finland	Yes		
France	Yes	Yes	Yes (founder)
Germany (East)			
Germany (West)	Yes	Yes	Yes (founder)
Greece	Yes	Yes	Yes
Hungary			
Iceland	Yes		
Ireland	Yes	Yes	Yes
Italy	Yes	Yes	
Luxembourg	Yes	Yes	Yes (founder)
Malta	Yes		Joining
Netherlands	Yes	Yes	Yes (founder)
Norway	Yes		
Poland			
Portugal	Yes	Yes	Yes
Romania			
Russia			
Spain	Yes	Yes	
Sweden	Yes		
Turkey	Yes		Joining
UK	Yes	Yes	Yes (founder)
Yugoslavia	Yes		
Total number	22	12	9 (+2)

Notes: A blank space means *no* membership or application for membership.

ECAC = European Civil Aviation Conference.

serving over 30 countries on a worldwide network is astonishing. Whereas neither SA nor CPA appear to have a 'natural' catchment since they directly serve only the small areas of Hong Kong and Singapore, appearances are deceptive. In fact, the industries and commerce of these areas are more than sufficient to provide a basis for a passenger and, just as important, an air-cargo trade. Add to this other commercial development in Korea, Taiwan, Indonesia and, more recently, in the mainland of China, and the basis of their success is apparent. Indeed, some forecasts suggest that this area will surpass Europe by the end of the century (Kelley, 1988). Not only is business traffic increasing from this commercial beehive, but tourism is too. For example, the number of Japanese travelling abroad jumped from seven per cent of the population in 1985 to 22 per cent in the first six months of 1987. Air cargo is particularly significant, since not only are many of the industrial products of Hong Kong, Taiwan and Korea eminently suitable for air freight (for example, optical products and clothing), but markets are worldwide. Thus although air cargo represents only one per cent of Hong Kong's traffic by volume, its value is far greater and growth is also rapid. In 1983, cargo volume reached just under 200,000 metric tons, while in 1988 the figure had reached 580,000 m. tons. The area has witnessed its share of fare wars, mergers and bankruptcies, especially among small airlines, but at the same time it has seen the growth of its own megacarriers. Mainland China has recently entered the international field, and the future will almost certainly see a rise in stature of the national airline, CAAC, and perhaps some successors.

Airports

The ICAO lists 1,012 airports as essential to world aviation, but numbers are reduced to 25 airports if we list only those sites catering for more than 15 million passengers per year, of which 16 are in the USA and five in Europe (Tables 13.4 and 13.5). The airport pattern on this scale manifestly reflects the dominance of the USA (especially the domestic sector) and Europe in the airline world. The rest of the world hardly appears apart from the airports serving the Asia-Pacific region.

Airport problems

The biggest problem facing the major airports is congestion; taking airport traffic figures and relating them to airport capacity shows that eight US and a similar number of European airports are currently constrained

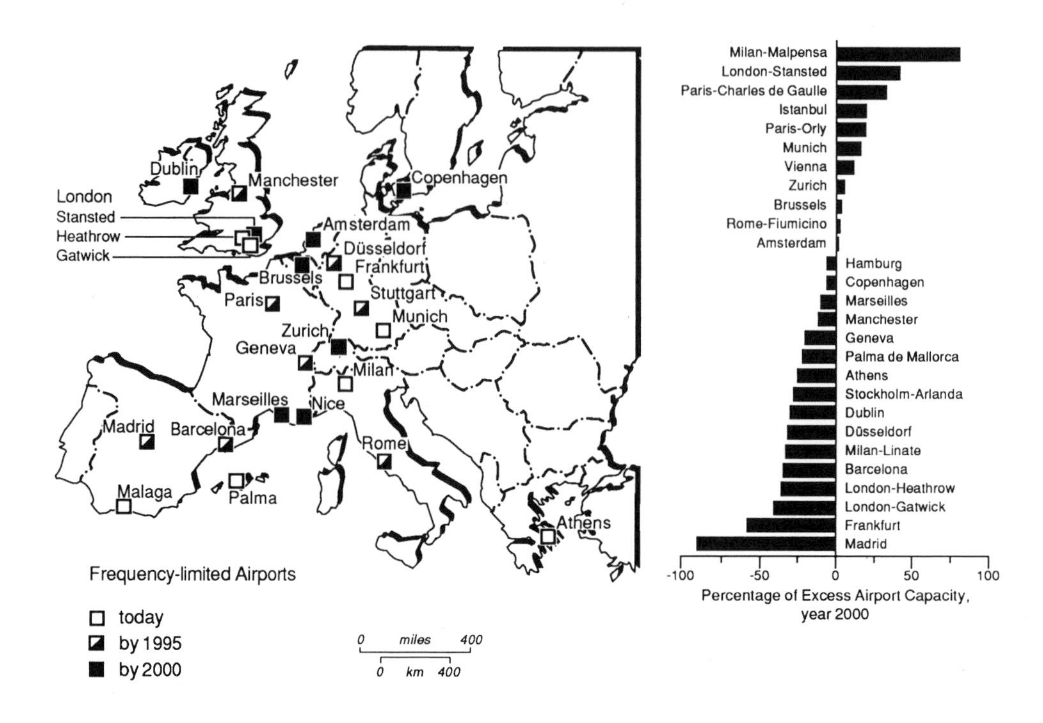

Figure 13.6a Airport congestion in Europe.
Source: Airbus Industries.

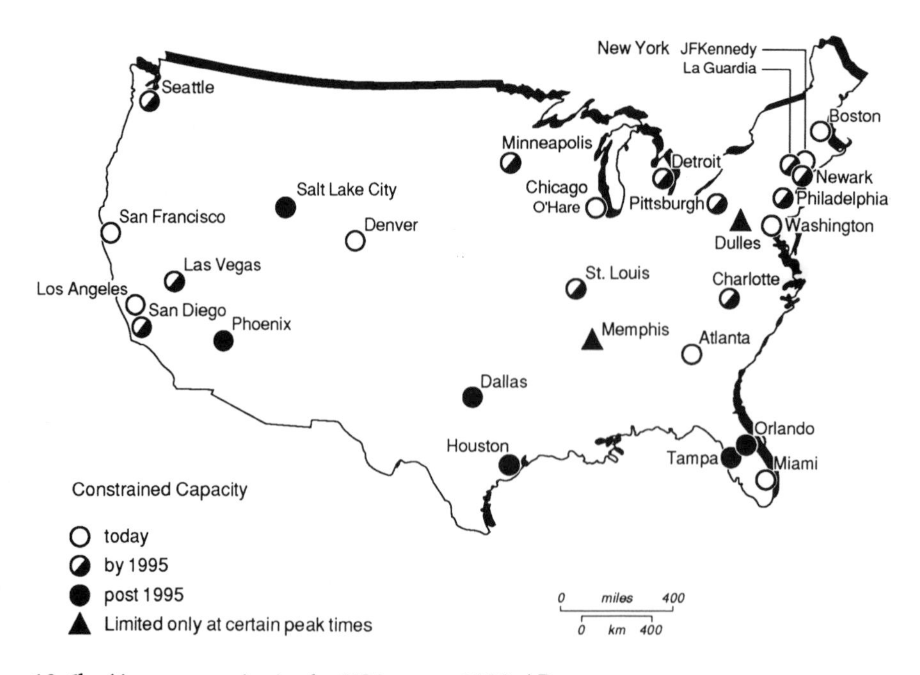

Figure 13.6b Airport capacity in the USA up to 2000 AD.
Source: SRI International 1990.

and suffer from congestion (Figure 13.6). A recent study by SRI International (1990) puts it more strongly, indicating that in all the major Western economies, air and ground congestion could strangle civil aviation within the next decade. A recent meeting of IATA in May 1990 voiced concern and called for more concerted action to improve the infrastructure.

Given the growth rate of the industry, it is odd to find that no new major airport has been completed since 1974. Only now are one or two sites emerging from prolonged wrangling towards completion, the most notable being the second airport at Denver, which with a projected acreage of 53 square miles and a capacity of 110 million passengers, will be one of the world's largest airports. A second airport for Munich is due to open in 1992. There have been many extensions to existing sites: for example, London Heathrow has added a fourth terminal, together with progressive upgrading of the other terminals in the 1980s. Even these developments have had to run the gauntlet of protest and enquiry on environmental and economic issues; and many schemes have either been abandoned, delayed or greeted with violence, as at Tokyo's Hareda airport.

However, while some major airports suffer congestion, many smaller regional and city airports have unused capacity. Whereas on the international front, bilateral agreements in the past have militated against the smaller airport, home policies have not helped either. Thus in the UK, the 1960s saw optimistic developments at many regional airports where traffic prospects depended heavily on charter IT traffic, as scheduled traffic opportunities were poor under the regulated climate of the time. Only recently have opportunities begun to favour the regional airport and there is still a long way to go.

Airport regulation

Under a regulated system airport traffic follows airline allocations which can accentuate the importance of favoured sites. Internationally, bilateral agreements usually name 'gateway' airports that foreign airlines may use, most often the capital or at best the major cities. In the original US-UK bilateral agreement, only New York and Washington were agreed US gateways, with London at the UK end. Even the most recent proposals (April 1991) provoked airline attack for failing to liberalize airport availability for UK carriers in the USA while giving the USA the use of Manchester together with two more routes to any of the UK regional airports. In return, one more UK carrier would get London-Boston while other airlines could operate between any main US gateway from any UK regional airport. Internal regulations favour direct city-pairs connections and it would seem that a wide spectrum of airports would be involved. Indeed, many airports might serve multiple destinations, but dense traffic would be confined to those serving major regions. Nevertheless it could be argued that airport authorities do know which carriers have been licensed to operate into their airports and can forecast likely levels of demand.

Opponents of regulation argue that such a system restricts the development of the market. It induces over-capacity at smaller airports, which serve relatively large numbers of aircraft flying on thin traffic routes, and under-capacity at the favoured major airports leading to heavy congestion. Furthermore, since air operations rarely provide more than 50 per cent of the revenue required to cover costs, smaller airports would either run at a loss and require a subsidy, or have to develop non-aviation activities such as shops, restaurants or even industrial parks. UK regional airports all operated at a loss under regulation and many still do; non-aviation income opportunities are not great at many smaller airports (Doganis and Nuutinen, 1983). The market approach, as with the airlines, argues that airports should be left to market forces and ownership should be in private hands rather than public corporations or municipal authorities. Freedom of entry to a route may benefit the airline but it could bring uncertainty to the airport, since airport managers do not know who will need their facilities in the future, for example, and economic recession might lead to route withdrawals and a loss of income. The situation seems to be exacerbated by the fact that free airline markets favour a hub-spoke system, making for hub congestion quite the equal of any regulated system at major airports, while smaller airports could lose traffic and be worse off than under regulation. As with the airlines, market supporters recognize that there will be a period of adjustment following the introduction of a liberal system, but that over time, supply will adjust to the trends in demand (Starkie and Thompson, 1985). Where social needs warrant support, a subsidy allocated through public tender would be necessary. Where opportunities exist, airport authorities should seek non-aviation income through concessions, agencies, hotels and support for local industrial parks and tourist developments. In other words, instead of everyone fighting to get on some vast 'Airport Plan', local initiative ought to be encouraged to fit airports into the mainstream of industrial and commercial growth. Theoretically this suggests more attention to the integration of airport systems rather than time spent on wider planning matters. In fact, such an approach raises further issues and must now be considered.

Air traffic control

The ability to coordinate the air and ground systems of an airport is a severe test of either the regulated or the market models. An airport can be conceived as an open system which for its successful operation must not only be accordant with the region it serves, but also rationalize its own sub-systems. In the air an ATC network must control movements within the pattern of air routes and airport-control areas; the airport ground system must marshall aircraft arriving and departing so that exchange between air and ground transport is as efficient as possible, while access to ground transport that links the airport with its markets is necessary. Each of these major nets has sub-systems within them. Thus, for example, terminal systems must ensure that passengers, freight and aircraft all meet at the departure gate at the right time. This in turn needs coordination with aircraft turn-round and maintenance systems, immigration and customs, freight and baggage handling, to name the most important examples. At first sight this would seem to favour a single controlling body with jurisdiction over all the main systems. Furthermore, since aircraft are by nature peripatetic, some international coordination of at least the ATC services would seem necessary. On the other hand, it is possible to distinguish 'control' as opposed to operational functions, including economic ones, so that it could be argued that market forces should deal with the latter, while the former are the preserve of national governments and international bodies. Figure 13.7 gives an impression of the complexity and propensity for congestion of the UK control system (see also Ogilvy, 1989).

British airports

The European Context — We may start by picking up the comments of the SRI report already noted (SRI, op. cit., 1990). The study bluntly says that the present national system will bring European airspace to a standstill in the late 1990s, yet it also noted that if present capacity were efficiently reorganized to serve Europe as a whole rather than individual countries, such a dilemma could be avoided. The price for inaction is high. Investment in enhancing airport systems could be paid back by one year of avoided loss. Thus, for example, constrained airspace costs $2,500 million a year in quantifiable losses; whereas the cost of rerouteing and resectoring the system would be $500 million, Trans European ATC harmonization would cost less than $2,000 million. There have been numerous discussions aimed at finding a unified

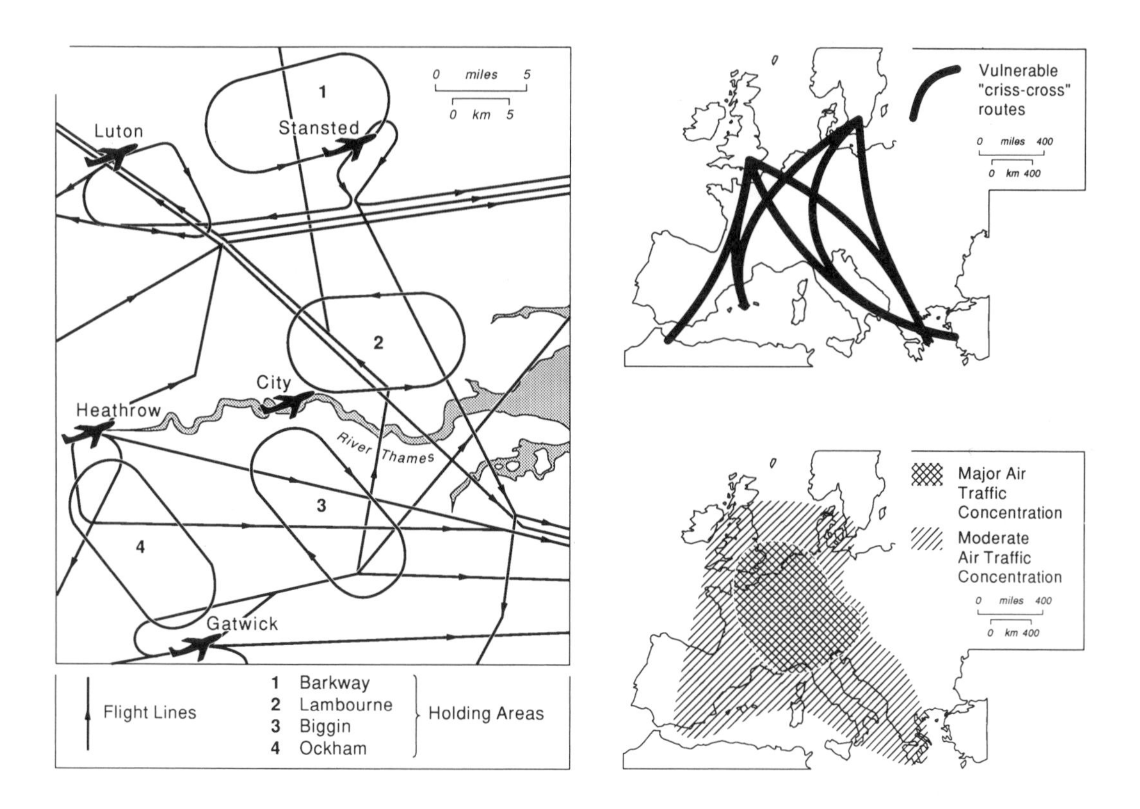

Figure 13.7 (a) The London Airports and the basic air traffic control structure. (b) Air traffic control problems in Europe: possible areas of conflict.

Sources: Flight International and *Aviation Week*.

solution to the problems of European ATC, most based on extending the powers of the currently restricted Eurocontrol organization, but with little success to date. European ATC remains a patchwork of 22 national systems operating over 40 control centres; needless to say, the weakest link in this chain sets the pace. The SRI study considers that a unified European system would enable ATC to handle twice the traffic it currently handles. Change may be spurred by technical development: for example, the introduction of micro landing-systems which give greater flexibility in the spacing and handling of aircraft. But even here there is a plethora of jargon and national plans. What is urgently needed is central flow management control presumably under the aegis of the basic Eurocontrol structure which would need to expand from its present membership of nine countries to the whole of Europe. It hardly needs emphasizing that this would make geographical sense of the issue (Table 13.7).

London airports

Closer ties with the EC have served to fuel the continued expansion of south-east England and with it the importance of the London airports which include Heathrow, Gatwick, Stansted and Luton with London City, Battersea Heliport and smaller airfields like Biggin Hill contributing to business travel. The two major airports, Heathrow and Gatwick, face increasing congestion, while Stansted stands on the threshold of a major expansion now that its new £90

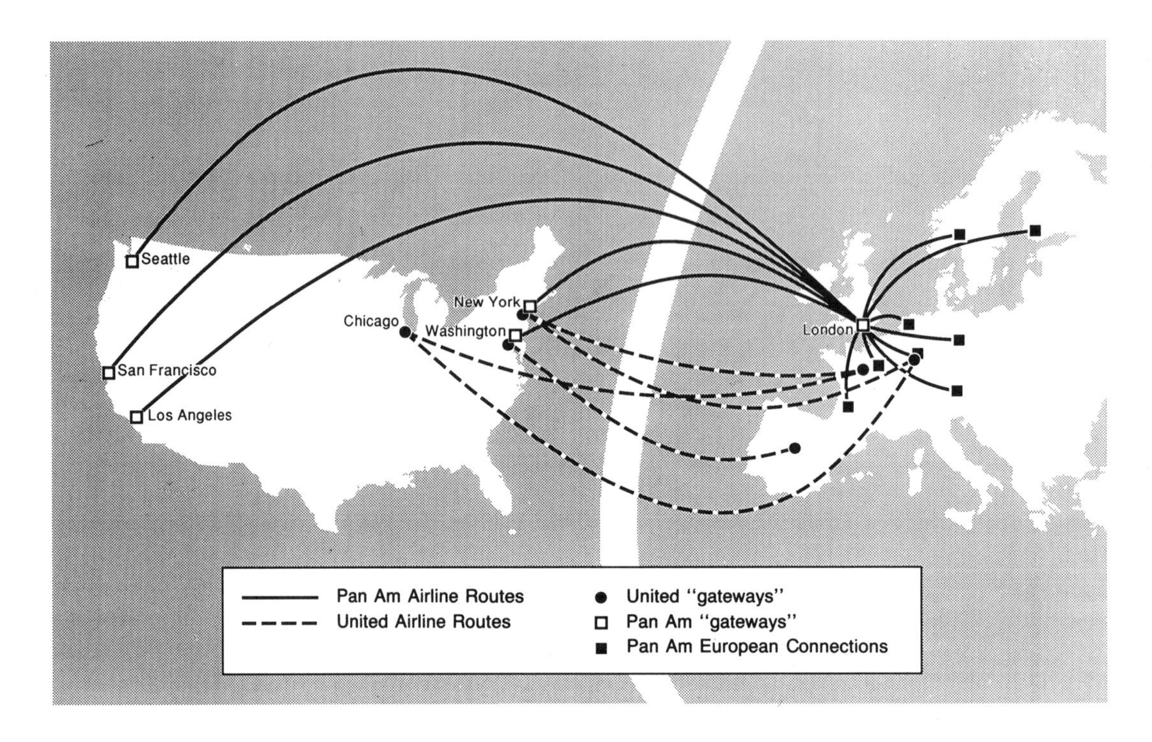

Figure 13.8 The United Airlines–Pan American Airlines agreement and its implications for the North Atlantic system (as at March 1991).

million terminal is coming into operation. It will be capable of expanding to serve eight million passengers a year as further elements of the overall £400 million development proceed. In 1989 it handled 1.3 million passengers. This development represents the only immediate prospect for expansion; the eventual completion of the Stage I plan, taking the capacity to 15 million passengers a year, looks like facing major battles from environmental groups. Luton, currently municipally owned and responsible for a large proportion of the charter-holiday traffic, faces congestion unless expansion is undertaken. Such a development is clouded by current financial constraints and the possibility of its sale to the private sector. London City, a useful development for business travel, needs to be upgraded to take a wider, more economic range of services, while a second helicopter pad at Cannon Street faces a public inquiry at the time of writing. To support a useful network of services covering European as well as UK destinations, both developments seem necessary. There is no overall ownership of this system. Heathrow and Gatwick are constituent companies under the privatized British Airports Authority (BAA) holding company; Luton and Biggin Hill are municipal airports; London City is owned by the Mowlem Construction Group.

There have been many battles over airport ownership. In many ways, as far as coordination is concerned, ownership need not be important. It all depends upon a willingness to work within an agreed frame. Indeed, maximizing the use of the heavy investment involved may only be possible with such cooperation. Whereas none of the airports consciously specialize, the technical capabilities of the site and the geographical pattern of markets has produced some degree of specialization. Thus Heathrow is the prime international airport – indeed, a world leader – Gatwick has increasingly displayed a similar international spectrum, but both airports reflect their geographical positions north and south of the Thames with some bias towards this division in their traffic origins. This is more noticeable in the case of Gatwick which has a strong 'southern' element in its catchment. Luton's position north of London favoured its growth as a national tourist airport and currently handles not only charter but also an increasing flow of scheduled traffic. London City has its own niche as a short-haul business strip within the growing commercial redevelopment of London's Docklands, complemented by the broader business capacity available to fixed-wing, and increasingly, helicopter operators at Biggin Hill. Battersea Heliport is the helicopter equivalent of the City airstrip, but its imminent saturation and its offset city site create problems hopefully to be resolved by a new development near Cannon Street railway station (Figure 13.7).

In the air, coordination of this complex group rests with the London Air Traffic Control Centre (LATCC). The advent of London City highlighted the congested nature of the skies over south-east England, the first services from City having to run the gauntlet to the south coast before entering the main ATC system. The development of Stansted was strongly favoured because of its north-easterly situation where ATC problems in the LATCC area were least. Any future expansion of Luton must take account of the development of its close neighbour at Stansted. On the ground, similar problems of congestion exist. Heathrow's future here must rest with rail access, the newly confirmed development of a second rail link from Paddington is long overdue. Rail services to Gatwick have been upgraded and a rail link to Stansted is part of its overall development. London City faces road congestion from its city catchment and would benefit from extensions to the existing underground system and the Docklands Light Railway. Inter-airport access is still poor and congested roads and the need to pass through central London on rail links increases the journey time even for direct services like the Thameslink BR service which connects Luton and Gatwick.

Increasing congestion in south-east England, both in the air and on the ground, forces a reappraisal for the future. If Stansted reaches its current limits by the year 2000, then future capacity in the south-east will demand either an increase in Stansted's role, or the development of new runway space elsewhere, or possibly both. Alternatively, some reorienting of traffic towards regional airports may be considered, together with the greater use of high-speed ground transport, notably the rail system linked to the Channel Tunnel. All these possibilities have been discussed, but here only air developments will be briefly outlined.

The possibilities for new runway space have been studied by the Civil Aviation Authority (CAA) over a number of years, culminating most recently in the study on airport capacity, published in July 1990 (CAA, 1990). This report sees a need for new capacity in the early years of the new century, basing its assertions on the UK's dominating role in European air transport. Briefly, five UK carriers are in the top ten European operators, while three top the EC charter market, which taken together accounts for more than 50 per cent of European international passengers; London airports serve 18 of the 30 busiest EC routes (CAA, op. cit., 1990). Largely on the basis of this extensive market, the study sees little likelihood of being able to divert traffic to regional airports. Having established the need, the study outlines the possibilities, including extensions and new runways at Heathrow, Gatwick and Luton, together with 'outside' possibilities at five other sites. None of these suggestions are trouble free. Most, except possibly Lydd, involve ATC problems, while extensions at the majors, although possible, have so far not been accepted by the government.

It has often been suggested that some London and south-eastern traffic could be diverted to regional airports, as some 20 per cent of London traffic originates from the regions. However, traffic numbers are no indication of the viability of such a move. All hubs generate traffic from the surrounding regions, attracted by the vastly superior international interlining system at the hub. The creation of a sufficiently attractive regional hub is dependent upon the normal growth of the region providing most of the incentive. Recent events, including the battles with the US airlines over Heathrow operations, have shown how important they rate Heathrow's facilities; even Gatwick has been termed a 'second class' airport for major trunk systems. The power hub of the developing EC lies eastward from London and the feeling is that any forced diversion of traffic will lead to a further drift in that direction. Schipol (Amsterdam) and Charles de Gaulle (Paris) airports currently have spare capacity and are pushing for increased shares of European international traffic. In any case, in a truly liberalized civil-aviation system you can only 'divert' traffic by reverting to regulation; otherwise the market will decide whether such a move is viable. Current indications suggest that it is not.

Regional airports

As far as the regional airports are concerned, the CAA study sees their expansion continuing as liberalization in the EC proceeds and regional services expand. Of the UK regionals, Manchester and Birmingham have shown the greatest development, with Manchester now the UK's 'third gateway' after Heathrow and Gatwick. Birmingham illustrates a trend visible in many EC cities: the growth of a local market as a result of industrial and commercial development. In Birmingham's case, it not only serves the West Midlands region, but this airport is closely associated with an international conference and exhibition centre. It is precisely this kind of international focus that drives airline demand, especially when linked with a growing local catchment. The link between airport and local catchment can be paralleled in other EC countries, notably in Germany (*Flight International*, 14 January 1989). The growth of local markets also depends on good surface transport links; far from destroying air transport, such links can be vital for its growth. The most successful air routes are those that bring together communities where journey time favours air transport and this includes many of the growing regional centres of the EC. Recent development in Eastern Europe and the USSR will almost certainly reinforce the trend, to the benefit of many regional airports.

In common with many of their peers in other countries, the UK airports pose some fascinating geographical problems. Airports are not only airline hubs, but are also the centres of ATC and ground networks. Success depends upon the coordination of each of these functions. Under a regulated system, ownership and control could be centralized within the airport system and 'integrated' with ground transport. Critics see this as a recipe for a bureaucracy unable to cope with such a geographical and economic spread. On the other hand, a free market would seemingly divorce control of the three main networks; coordination would need some 'management'. In practical terms, a regulated system would need to decide which airports should be developed; under the old UK system, this meant 'essential for the public service'. Seemingly it would be problematic whether you would encourage the development of both Manchester and Birmingham, since they are so close together and well-served by rail and road. Indeed, many arguments did take place in the 1960–1970s, notably on the provision of northern and Midlands airports (Department of Trade, 1976). As noted in earlier sections, route systems were also strictly controlled, so that few opportunities existed for regional expansion. Under the market model, no such restrictions exist and certainly such a system has already provided an impetus for EC regional airport expansion. In the case of Manchester and Birmingham, what matters is not their close geographical proximity, but their relationships to the regional markets in their areas on the one hand and their relationships to the rest of liberalized Europe on the other. Given large enough markets, or market differentiation, then both can survive and grow. The arguments for survival are similar for even such close rivals as Birmingham and the East Midlands airports, some 52 km. apart. Perhaps the point is made by noting that Heathrow and Gatwick are 40 km. apart.

The future outlook

Taken to its ultimate point, a European Community expanded to include the eastern bloc provides some interesting possibilities for the future. Thus one possibility to offset traffic congestion in London and the south-east would be to extend Stansted to the stature of a truly major airport handling 25 million passengers a year. Alternatively or in addition, two new runways might be built at Heathrow, giving a four-runway, double-parallel system as at Los Angeles International, or to extend runways at another south-eastern site, such as Manston in Kent. On the other hand, these may seem parochial. We should seek the creation of an international 'Western Europe multihub' using airports at London, Amsterdam, Brussels and Paris as one coordinated system connected by high-speed rail and the Channel Tunnel and by a commuter interline system from city airports like London City.

Finally, environmental aspects will continue to loom large in whatever solutions are made to ease congestion: a subject in itself. One immediate problem for the future will centre around the extension of Stansted as the next phase of its development is required. In the longer-term, environmental issues may finally lead to a more rational use of European airport capacity.

References

Association of European Airlines (1977), updated 1982, *Air fares in Europe* (Brussels).

Barrett, S.D. (1987), *Flying high: airline prices and European regulation* (Aldershot: Avebury).

Button, K. (1989), *Liberalising the Canadian scheduled aviation market* (London: Institute of Fiscal Studies).

Civil Aviation Authority (1990), *Traffic distribution policy and airport and airspace capacity: the next 15 years* (London: CAA).

Department of Trade (1976), *Airport strategy for Great Britain*, Part I: *The London airports*; Part II: *The regional airports* (London: HMSO).

Doganis, R. S. and Nuutinen, H. (1983), *The Economics of European airports* (London: Polytechnic of Central London).

Doganis, R. S. (1985), *Flying off course, the economics of international airlines* (London: Allen & Unwin).

European Civil Aviation Conference (1982), Report on competition in intra-European services (Paris).

European Community (1980), Report on regional air services (Brussels).

Flight International (14 January 1989), 'Moving Out: regional airports in Europe'; (29 November 1989), 'France defies EC on liberalisation'; (24 January 1990), 'UTA takeover by Air France'; (22 August 1990), 'American considers buying Midway'.

Kelley, A. (1988), 'Pacific Potential', *Flight International*, 23 January, 28–30.

Naveau, J. (1990), *International air transport in a changing world* (Dordrecht: Martinus Nijhoff).

Ogilvy, D. (1989), *UK airspace: is it safe?* (Haynes: Yeovil).

Pryke, R. (1987), *Competition amongst international airlines* (Aldershot: Gower).

Rosenberg, A. (1970), *Air travel within Europe* (Stockholm: The National Swedish Consumer Council).

Shaw, W. (1982), *Air transport, a marketing perspective* (London: Pitman).

SRI International (1990), *A European planning strategy for air traffic to the year 2010*, Vol. I; *Analysis and recommendations*, Vol. II, *Supporting data*, Report for IATA, Project 7474.

Starkie, D. and Thompson, D. (1985), *Privatising London's airports* (London: Institute for Fiscal Studies).

Wheatcroft, S. F. (1964), *Air transport policy* (London: Michael Joseph).

Williams, G. (1990), Achieving a competitive environment for Europe's airline industry. National Westminster Bank, *Quarterly Review*, May 1990, 2–14.

Additional references

Ashworth, M. H. and Forsyth, P. J. (1984), *Civil aviation and the privatisation of British Airways* (London: Institute of Fiscal Studies).

Association of European Airlines (1984), *Comparison of air transport in Europe and the USA* (Brussels).

Civil Aviation Authority (1979), CAP 420 *Domestic services – a review of regulatory policy*; (1984), Paper 84009 *Deregulation of air transport*; (1989), CAP 548 *Traffic distribution policy for the London area and strategic options for the long term*; (1989), CAP 559 *Traffic distribution policy for airports serving the London area.*

Department of Transport (1984), *Airline competition policy* Cmd. 9366 (London: HMSO).

European Community (1981), *Progress towards the development of a common air transport policy*, Civil Aviation Memo 2 (Brussels).

14

Transport and the Future

Alan Williams

Successful transport is subject to innovation and spread, playing a developing role until challenged by something new which can perform better. Today, the world of transport is characterized by overlapping, competing technologies which operate over a patchwork of dominant transport zones. To the end of the century, most change will come from modifications to tried and tested systems. Thereafter, major transport developments will be influenced not only by technology, economy and public policy, but also by environment and energy. Early in the twenty-first century, research and development will need to focus upon fuels as well as vehicles. The most developed countries will edge towards the 'technetronic' society quite quickly because of the need to provide for still-rising human mobility. A century on, burgeoning world population increases may suggest an 'interconnected global city', but with what kind of mobility and access?

Introduction

Mechanical transport has existed for only a brief part of the time in which the world has had some elements of civilization and in our own era it may seem that technological developments in transport have never been arriving more swiftly. This final chapter examines some aspects of our transport future considered in the short-term (to the end of the twentieth century), in the mid-term (2000–2040 AD) and on towards the end of the twenty-first century.

Trying to look into the future is a notoriously dangerous activity. Necessarily, probabilities give way to possibilities, possibilities to speculations. Only the years immediately ahead can be viewed with a reasonable certainty of what will happen and predictions tend to be based on known technology and current fashions. So Napoleon's engineers thought that a Channel tunnel could be operated by horse-drawn coaches, Victorian artists drew pictures of the skies of the twentieth century darkened by steam-driven balloons manned by gentlemen in top hats, while the 1930s ideal of the aviation of the future was an enormous multi-propeller aeroplane with promenade decks, cane chairs and cocktail lounges. Before looking forward it is useful to look back, for this helps to keep a sense of perspective. A sometime President of the CIT reported the imaginative forecast of the editor of *The Scientific American* when he wrote, in 1906, an impressive view of the future: 'The improvement in every city's conditions by the general adoption of the motor-car can hardly be overestimated. Streets clean, dustless and odourless without horses; light, rubber-tyred vehicles moving swiftly and noiselessly over their smooth expanses. All this will eliminate a greater part of the nervousness, distraction, pollution and strain so characteristic of modern metropolitan life.' Fifty years later the same magazine had to write: 'Horse-drawn carriages used to travel smoothly and quietly, with just that audible "clip-clop", at an average speed of 11½ miles an hour in New York's mid-town traffic. Today the average speed of the much noisier automobile in the same streets is six miles an hour' (Masefield, 1976).

History suggests that radical improvements in transport have been obtained through the application of new technologies. 'Technology' being 'the science of the application of knowledge to practical purposes' (Webster, 1988), transport technology may be defined as 'the application of inventions which convey people or transfer freight to practical purposes'. The first mechanically propelled vehicle was John Fitch's river boat *Experiment* which sailed between Philadelphia and Trenton, New Jersey, in 1790. On the roads at that time the stage coach was still dominant. In 1829 urban transport made a significant advance with Shillibeer's three-horse 20-passenger omnibus which ran from Paddington Green to the City of London. Then, just over 100 years ago (1885) Gottlieb Daimler put his 'horseless carriage' on the road, developed in only 13 years as a public transport motor vehicle. By 1913 more than 2,600 B-type motor buses were in use in Great Britain and the horse-bus gave way to the electric tram. In the meantime, since their beginning in 1825, steam railways became the major form of land transport between towns in many parts of the developed world and even aviation was on its way. Commercial flying began before the First World War in Florida only ten years after the Wright brothers' first powered flight, and the world's first scheduled daily international air-passenger service began in 1919 between Hounslow Heath and Le Bourget, Paris. Only 22 years after this came Whittle's invention of the gas-turbine engine. Following its military use, this became the mainspring of a leap forward in commercial flying, leading directly to today's continent-hopping Jumbo Jets.

By such examples we see that, after research and development, each successful transport invention undergoes trial, acceptance and dispersal, perhaps to become part of a new commercially viable transport system. Subject to its own limitations by land, sea or air, it may become widespread or even dominant unless challenged by some other invention that can do the job better (which almost always means easier, faster, cheaper, safer and more reliable). The transport world of today is characterized by overlapping, competing technologies which operate over a patchwork of 'dominant transport' zones (Boyce and Williams, 1979) (Figure 14.1) each of which is continuously modified. Having conquered the developed world, the petroleum-based road and off-road vehicle makes further incursions, especially within the developing countries of Latin America, Africa and South-east Asia. Air transport, too, has made significant advances into the waterway zones of South America and into the otherwise 'traffic-dead' zones of the other continents.

Five of the most significant transport innovations of the last 30 years are indicated in Table 14.1. Awareness of the successes, however, may cause us to forget the failures. The first practical bicycle was built in 1839, but who remembers the BSA company's Otto Dicycle of 1880? Or Ernest Bazin's screw-propelled cross-Channel 'roller vessel' of 1896? Bazin incorporated a raised hull to minimize water resistance (the principle to be successfully applied in modern craft such as the hydrofoil and hovercraft). The tray-like passenger platform was mounted on three giant axles, each rotating a pair of huge wheels. Bazin's vessel was seaworthy and stable, but since a swift crossing was the principal boast and it failed to reach the promised speed, it also failed to find commercial backing (Healey, 1983).

It may also be difficult to distinguish between 'blind-alley' technologies and those whose time has not yet come. A quarter of a

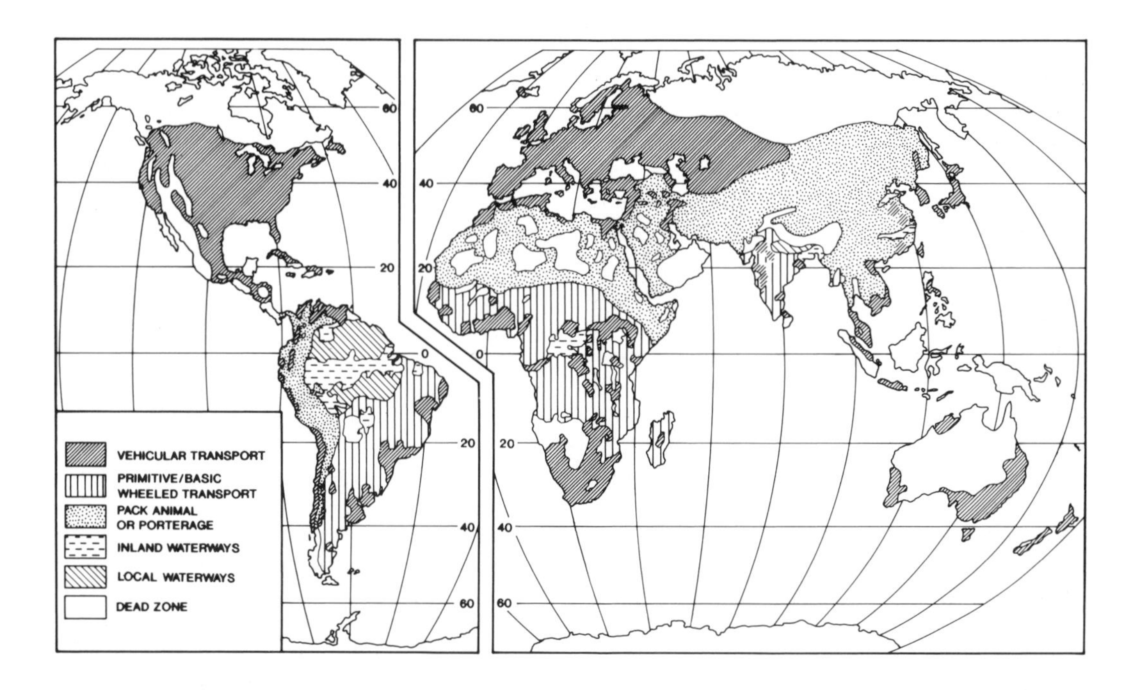

Figure 14.1 Major world transport regions.
Source: Williams, 1979.

Table 14.1 Thirty years of transport technology

*Innovation**	*Characteristics*	*Major role*
'Superships' (1960–)	Bulk and tank vessels, up to half a million deadweight tonnes	Reduce the cost of transporting natural resources between continents, promoting continued concentration of world manufacturing in OECD countries
Container vessels (1968–)	Vessels transporting up to 4,000 20-foot containers (TEUs)	Carry primarily manufactured goods, increasingly from NIEs (newly industrialized economies), such as Taiwan and Korea
Jet aircraft (1958–)	Capable of providing non-stop services between an increasing number of world cities	Create major stimulus for service industries, such as banking, consulting and tourism
Fuel-efficient cars and trucks (1970–)	Significant reduction in fuel consumption due to lower weight and more efficient engines	Enable highway transport to continue to increase its share of urban and intercity transport
High-speed trains (1964–)	Shinkansen ('bullet train', Japan) TGV (train à grande vitesse, France)	Effective competition for intercity air and car transport in high-density areas

* Year indicated is an approximate date of entry into service.

Source: Sletmo, 1989.

century ago the hovercraft was supposed to change the face of travel; repeatedly, by Californian inventors, we are promised 'the car-plane to soar over traffic jams'. In the same period, vehicles run by rechargeable batteries conjure visions of milk floats and golf-carts. Models tested in the 1960s could do no more than 30–40 mph and 'could not mix safely with freeway traffic' (Gunston, 1972). Electric cars are so far constrained by a lack of storage batteries yielding high energy and power from a given bulk and weight which are both safe to use and inexpensive. GM's 'concept electric car' of the early 1990s, the *Impact*, can outperform many petrol-driven production sports cars (0–60 in eight seconds; top speed 100 mph) but only has a 120-mile range. As for the hydrogen-powered vehicle, many experimental types have been displayed. So in 1977 a national daily reported: 'THE CAR THAT RUNS ON WATER: It's here. The car that runs on water at the fantastic rate of 100 miles per gallon and could make petrol out of date' (*The Daily Mail*, 19 August 1977). This was a Mini that used hydrogen in a near-normal engine. Fourteen years later the same newspaper reported anew: 'THE CAR THAT RUNS ON WATER' (17 June 1991), this time a Fiesta driven by an electric motor, making its hydrogen overnight. Perhaps, like Fitch's steamboat, it should have been called *Experiment*, for *The Daily Telegraph* claimed that the first production car could appear by the end of the year and *Today* said we could soon be off to the holiday sun in planes flying on water-based fuel!

It may seem unlikely that we shall gravitate quickly from our dependence on oil to a twenty-first-century dependence on hydrogen. But people who lived before the first paved roads, or before the first four-wheeled vehicle, first steam engine, first petrol engine or first aeroplane could neither predict their uses nor wish to treat them seriously. In particular, when steam power was supreme, no one could foresee our switch to oil.

> Thanks to steam, man could roam the seas at will, or reach the furthermost corner of the country from London in a matter of hours. Someone aged 50 in 1872 was born three years before the first railway, in an age when man was still dependent on animal- and wind-power. But if that same Victorian had lived on for another 50 years he would have seen changes that would have made 1872 look like the Dark Ages. For one thing in 1922, *everyone* could now travel, not just the wealthy and privileged. Underground electric trains carried commuters to distant suburbs. The whole of Greater London was covered by a network of tramways and motorbus routes. Private cars had superseded the horse. . . . most overseas travellers voyaged on ocean liners in unexcelled comfort. Those people of 1922 would have found it hard to believe that those same liners would be made obsolete by jet liners flying at more than 600 mph. They would have regarded it as incomprehensible that the fast, efficient British trams would become as extinct as dinosaurs. They would have been astonished that car ownership would come within reach of the major part of the population, with fearful consequences. They would have been amazed that ordinary people would spend their holidays in Spain, Italy and Yugoslavia and think little of it. And they certainly would have had difficulty in accepting that men actually would go to the moon. (Perry, 1972)

Moving through the 1990s

In Chapter 1 it was pointed out that, necessarily, our inherited transport systems and superseded decision-making processes provide part of the framework within which present-day decisions are taken and future developments planned. During the years to the end of the present century, the greater part of change in transport will come from modifications and improvements to these inherited systems. Once decided, large-scale investments in transport take time to achieve: the M40 in southern central England took 19 years from strategy-planning documents (1973) to full opening (1991); the Channel Tunnel (with go/stop/go and design changes) will have taken 30 years and the CrossRail passenger line, announced in 1989 as an answer to traffic congestion in London north of the Thames, is not scheduled for completion until 1999. This long lead time suggests years of traffic disruption: congestion to get rid of congestion, leading to 'gridlock' (grinding to a halt). These fears in London come about because the CrossRail scheme joins a large number of other transport-related works-in-progress: the reformation of Kings Cross and Waterloo Stations to accommodate Channel Tunnel passenger trains (four years); the Jubilee line extension to Docklands (seven years), Canary Wharf office tower and satellites (60,000 new commuters by completion in 1996), road-improvement schemes in Westminster, the completion of a water ring-main under the capital, the laying of 1,000 miles of cable for television and telephone by 1995 and a programme to prepare 600 road bridges for the introduction of heavier lorries by the end of the 1990s.

Investment in such schemes (as explained in Chapter 2), is a matter for political negotiation, economic calculation and environmental consideration. In the near future, among the many related influences (Figure 14.2) it is likely that the energy, policy and environmental aspects will loom large. With the 1990s decade immediately in view, the influences can be explored separately (although we must remember that in practice they interact).

In relation to energy, as with vehicle technology and our inherited transport infrastructures, it must be supposed that the next ten years is too short a time to bring about great change. With an open international oil market the oil-based transport technologies of the developed and developing world may expect to flourish, despite under-investment within the Soviet Union (the world's largest oil producer) and the devastations caused by war in Kuwait and Iraq. But this scenario requires the recovery of the Soviet and Kuwaiti oil industries to avoid inflationary price surges. A most important fact is that in 1991, petrol prices in the USA, the biggest consumer of oil in the Western world, are lower in real terms than in 1972 before the first Arab oil embargo. The present balance of world supply and demand is in part a function of the US and EEC recession which has kept petroleum consumption at its lowest since 1983. Western economic recovery before the recovery in oil output in the Soviet Union and the Middle East can bring price rises, but the unrestricted production and international movement of oil most certainly stabilizes crude oil prices close enough to present prices to be acceptable to consumers and just about what OPEC needs for new investment in its oil fields.

While cheap oil will delay the search for alternative fuel and vehicle technologies, strong Western economies will promote even higher levels of road, rail and air travel. International air trips, a measure of international travel and tourism, shows much variation between major countries (Table 14.2) and is indicative of opportunities for more as living standards rise. On the ground it is the added congestion and greater pollution accompanying rising vehicle numbers that may bring about most change. In the early 1990s, 130,000 new cars are made in the world *each day*. By the year 2000 the number of cars on Britain's roads will have increased from 19.2 million in 1989 to between 23.5 and 25.5 million. Car traffic is expected to rise by at least 28 per cent but this could be as much as 49 per cent, while lorry traffic is expected to rise by up to 27 per cent (British Road Federation, 1991). Road vehicles are, on average, travelling greater distances each year which means that the amount of traffic is growing faster than the number of vehicles.

If all the current evidence suggests that traffic congestion in our towns and cities will continue to increase at a disturbing rate, it is also seen that present solutions are largely aimed at accommodating personal mobility through the construction of new highways and traffic-management measures, or encouraging people to use public transport. In the latter case, we may be persuaded that

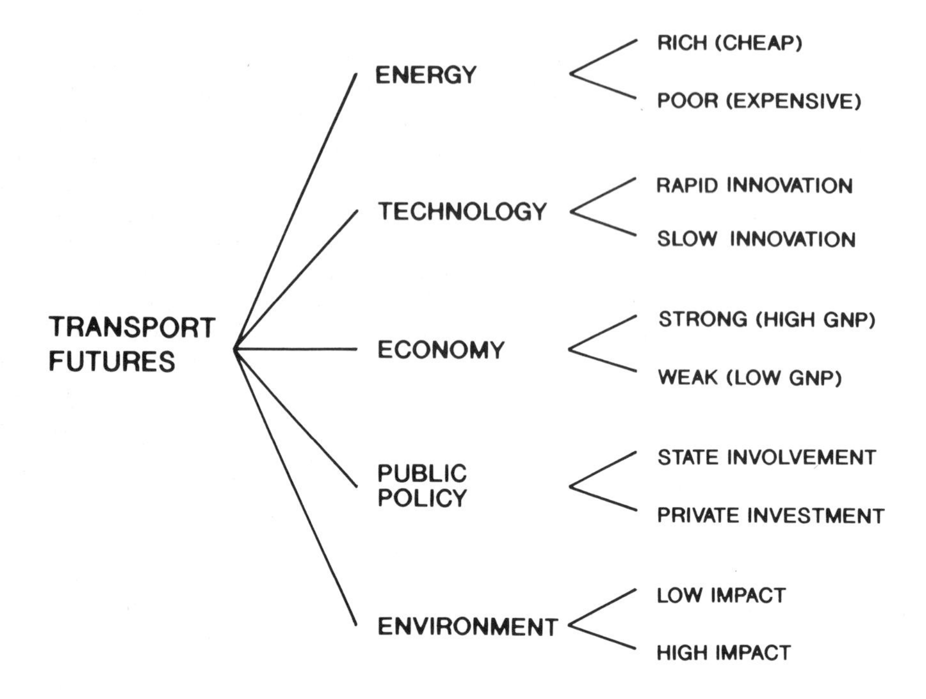

Figure 14.2 Factors influencing transport futures.

Table 14.2 Air travel and tourism, selected countries

1985	*Air travel (number of trips)*				
	Domestic		*International*		*Foreign visitors*
	millions	per 1000 inhab.	millions	per 1000 inhab.	millions
Canada	14.8	578	4.9	191	35.9
USA	328.2	1,358	29.4	122	52.7
France	13.5	224	11.0	198	36.7
UK	8.7	154	20.0	348	14.4
Japan	45.0	370	6.3	52	2.3
China	6.0	0.6	1.3	0.1	1.4

Sources: UN, UNCTAD, Statistics Canada. Numbers refer to 1985 whenever possible, but in some cases may be slightly older.

new LRT schemes projected for many cities in the UK and elsewhere will make a significant contribution. But we must again be mindful of lead times: if a single LRT 'show-off' route takes 15 years to grow into a fully-fledged network, it is unlikely to keep pace with either passenger demand or the general rate of growth in traffic congestion.

A recent research study exploring the UK's transport problems sets its own agenda for the 1990s. It recognizes the intolerable imbalance between expected trends in mobility and the capacity of the land-based transport system, causing problems to industry and to the environment, and also to people's ability to lead fulfilling lives. The growth in reliance on car use will not succeed in realizing its own objectives, the study asserts, and it confirms that it is not possible to provide sufficient road capacity to meet unrestrained demands for movement in and around cities (Road Transport Forum, 1991). In the 1990s decade, therefore, it sees it necessary to devise systems of managing demand which are economically efficient, provide attractive possibilities for travel for car- and non-car owners and give priority to essential traffic (such as emergency services, freight, buses and limited categories of need). National and local policies to accomplish these tasks are, the argument goes, 'technically possible'. These will include much greater attention to land-use planning, road improvements (including some, but not much extra capacity, such as is represented by essential new motorways and town bypasses); extensive use of traffic management (especially priority systems); substantial public-transport improvements; traffic calming both at the local and the strategic level, and consistent charging and financing of all modes, preferably by road pricing.

All this is expected to make radical changes in the patterns of car use and attitudes to car ownership, perhaps modifying the forecasts given above. But the enactment of these policies demands the political will to see them through, with the necessary and appropriate proportions of state and private investment and with each transport mode given its rightful opportunities. At present in the UK, road transport is in a special category, since virtually all investment in 'track' is met from public funds. If it were costed on the same basis as rail transport, it would become more expensive and there would be more investment in the railways. As long as road transport is subsidized, the argument goes, it will be used even where it is inappropriate, because it appears cheaper to the user. Should the Department of Transport take over direct responsibility for the infrastructure of rail lines, leaving BR the job of running trains? Should BR networks be privatized, or merely its train services, so that competition between train-operating companies offering different styles of service brings benefit to the customer? With the M25 Dartford crossing of the Thames being built with private finance and operated by tolls, it may be that the present decade will see the next UK motorways built and operated as turnpikes. The first of these, the 30-mile Birmingham Northern Relief Road, with the adjoining Western Orbital Road near the Black Country, will breach the long-established principle in Britain that roads should be free at the point of use. In urging the government to accelerate the trunk road programme, the British Road Federation suggests that the projected Home Counties Orbital, to run 30 to 40 miles from London to relieve the M25 and congested London ends of the radial motorways, should also be built as a toll-earning expressway. But – lead times again – the Birmingham Northern Relief Road is not scheduled for completion until 1997 while the Home Counties Orbital seems more remote than the purpose-built, high-speed, limited-access railway between London and the Channel Tunnel, not now expected before 2001.

Following the completion of the railway electrification schemes between London, Leeds and Edinburgh, further electrification will proceed on BR through the 1990s. But, overall, we see most clearly that our travel and transport will still be served in this period by the oil-based modes. On land, as in the air, we have good cause to remind ourselves just how polluting these modes are. Air pollution is most obvious in such Mediterranean cities as Athens and Marseille where temperature inversion holds down a yellow photochemical smog caused by vehicle exhaust gases. But the problem is present in all the major cities of the developed world. Motor vehicles create a fifth of the carbon dioxide that makes up the greenhouse effect, three-quarters of the world's carbon monoxide and half of the smog-forming hydrocarbons. In the UK the car is responsible for 17 per cent of carbon emissions, a figure that will rise through the 1990s even allowing for the expected improvements in engine efficiency and emission abatement technology. These last (as noted in Chapter 4) are not necessarily mutually reinforcing technologies.

Throughout Europe it seems that we are unable to curb the gases of combustion to better than 1990 levels by the year 2000 or 2005; but well before this many more countries are likely to discourage pollution from road vehicles by making the owners of 'gas-guzzlers' pay a higher road tax than those with smaller, fuel-efficient models. Pollution checks on Californian cars have been imposed through the 1980s and in the 1990s Los Angeles will be subject to the strictest measures. Cars will have to be 80 per cent 'cleaner' in the next ten years than now and by 2003, cars sold there must produce only a tenth of the hydrocarbons they do today. Such is the size of the Los Angeles market for cars (eight million vehicles for 12 million people driving 100 million miles every day) that the city's administration can push the manufacturers towards the appropriate technology (Horizon Science Series, 1991).

Towards the end of the decade, a series of new information and communication technologies presently on trial may be in common use with a variety of surface vehicles, edging the most developed countries and regions towards the 'technetronic' (technological/electronic) society (Figure 14.3). An electronic system of road-pricing, for example, has been tested successfully in Hong Kong and is to be tried in Cambridge. Cars have 'black boxes' which automatically operate a metering system each time the vehicle crosses over coils in the road surface at the city centre. There are other systems under development to provide the sort of on-board electronic tracking James Bond enjoyed chasing Goldfinger in his Aston Martin. To ensure pinpoint accuracy an expensive combination of roadside beacons and satellites is required, but the digital mapping needed is available today and, through it, a computer-generated route-finding system was tested in Berlin in 1991 (Institute of Electrical Engineers, 1991). By 1993 a system that warns drivers of traffic jams (again by using fixed road sensors relaying information to the driver's dashboard) will be available for the UK motorway network. The humble bus, in the meantime, together with its more capacious fixed-track companion the LRT street-based 'supertram', will be able to change traffic lights and avoid bunching by electronic means. Progress between stops will be relayed via post-based transmitters to a control computer which signals back to stopping-place display boards giving next vehicle arrival time, available capacity and route details.

Lastly, by the later 1990s electronics may well bring 'intelligent cruise control' to road vehicles whereby they can brake as well as accelerate automatically. But according to one authority (Wootton, 1991) we shall have to wait until well beyond the turn of the century for the kind of vehicle-to-vehicle communication which can 'offer the possibility of doubling the capacity of existing roads'. A more pollution-conscious and environmentally aware electorate may be more receptive than now to policies which address problems of mobility and access seriously. For long it has been widely supposed that stronger measures will need to be taken to discourage urban car use (Price, 1971; Preston, 1991). High parking charges and selective road-pricing schemes may begin to ration roadspace in towns, thereby curbing the greatest excesses of the private car. Yet every motorist is a voter who will not willingly give up his much-prized asset unless an attractive alternative is available; it will be necessary to maintain and improve collective transport systems that are compatible with personal mobility. There are problems of equity. Who would stand to gain the most from the elimination of congestion? Central zone vehicle charges need to be high enough to deter sufficient car commuters, and this bears unjustly on car-users with lower incomes. As Ralph Nader observed, 'There is nothing more democratic than a good traffic jam.'

Into the twenty-first century

There have to be fundamental changes in transport in the next 50 years because almost all mechanical transport today is dependent on oil and 50 years from now oil may be getting scarce. Beyond 2010–2020 AD there must be doubt about the ability of fossil-based oil and natural gas to energize a still-developing economic and social system. The long-term energy alternatives have to be considered a priority in the early years of the twenty-first century because of the research and development needed. In Western

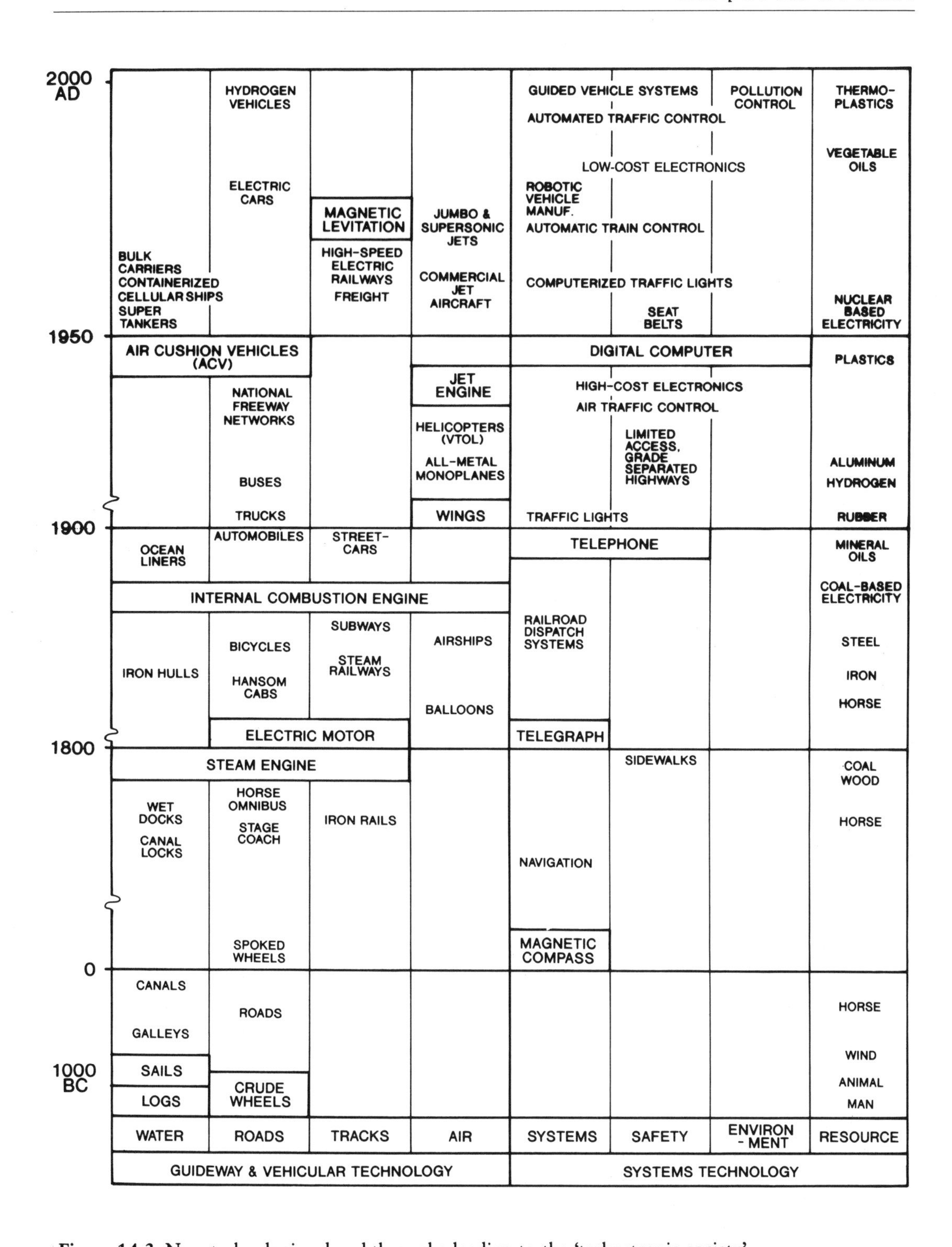

Figure 14.3 New technologies: breakthroughs leading to the 'technetronic society'.

Europe, given large remaining coal reserves and their potential for cleaner combustion, gasification and liquidification, it is possible to sustain a hydrocarbon-based economy into the second quarter of the twenty-first century, 'some 100 years on from when the first European coal age started to disintegrate under competition from oil' (Odell, 1981). With worldwide oil extraction beginning to decline, leading to net increases in price, synthetic fuels will replace oil. Such is the enormous investment in our inherited transport systems that it is far cheaper to alter the fuels than to make major changes in vehicle technology. Liquid fuels will minimize alteration to the distribution chain leading to the user pumps.

Oil from coal is not new (it was produced in Germany during the Second World War), but it may be more expensive than oil from the great tar sands of the world (the best-known are in Alberta, Canada). Both may be more expensive than producing oil from plants. In the 1970s Brazil started diluting its petrol with sugar-cane-derived 20 per cent alcohol (the maximum possible without altering car engines). Maize, sugar-beet, soya, peanuts, sunflower and rape oils have all been tested as additive or alternative fuels. These open up the possibility in areas like the US Mid-West and the North European Plain that set-aside acres may be brought back to produce gasohol. This 'harvesting' approach might also be directed to euphorbia plants: these yield a latex which can be processed into a hydrocarbon. In the meantime, methanol produced from natural gas is available from fuel pumps in California for use in standard engines. Methanol is cleaner than leaded or unleaded petrol but it is toxic and highly corrosive, so the vehicle requires non-corroding fuel lines. Usable alone in warm climates, it requires to be mixed with petrol for cold-starting elsewhere, such as in Alberta where Ford experimented with it in the late 1980s.

Perhaps the main point about the arrival of such oil substitutes is that, as far as transport is concerned, the overall production stabilizes. Rising fuel prices may act as a deterrent to private motoring, leading to less car travel and to the greater use of public transport. But with the utility of the private car so high and the future land-use system still so dependent upon road vehicles, we shall take considerable steps to mitigate fuel-price rises. So cars will be built with even lighter materials than today (more aluminium than steel with lightweight ceramic-aluminium composites used in moving parts) and ever-lower fuel consumption. By 2010–20 the electric town car will have made its long-awaited breakthrough to join a range of quiet, economic electric vehicles, both public and private. In sunny climates such cars will plug in to solar panels, converting sunlight to electric power.

Towards the middle of the century, with congestion, pollution and environmental concerns well to the fore, the balance between public and private transport may be changed slowly in favour of the former. Strenuous attempts will have been made to reduce the dependency of the land-use system on the car and additional taxes will have been imposed related to the distance travelled per year. But, propelled by one or a combination of fuels, the private car will survive strongly. These and other possible developments are depicted in Figure 14.4. Non-polluting electric modes will predominate among all modern public transport systems, drawing energy directly or indirectly from a national grid which derives it from many different sources. The oil-based railway fades fast beyond 2000, having been replaced by the international, purpose-built, high-speed operations whose foundations were laid in the late twentieth century by TGV and NBS-style technology (see Chapter 12). This technology has by this time been exported to Canada, the US, China and Australia. Within Europe and later in the US, the high-speed rail renaissance gradually squeezes out internal air travel, made almost redundant by city centre to city centre speeds of up to 250 mph (although of little significance before 2010, this replacement of air by rail had been initiated as early as the 1980s by Lufthansa in Germany). All other

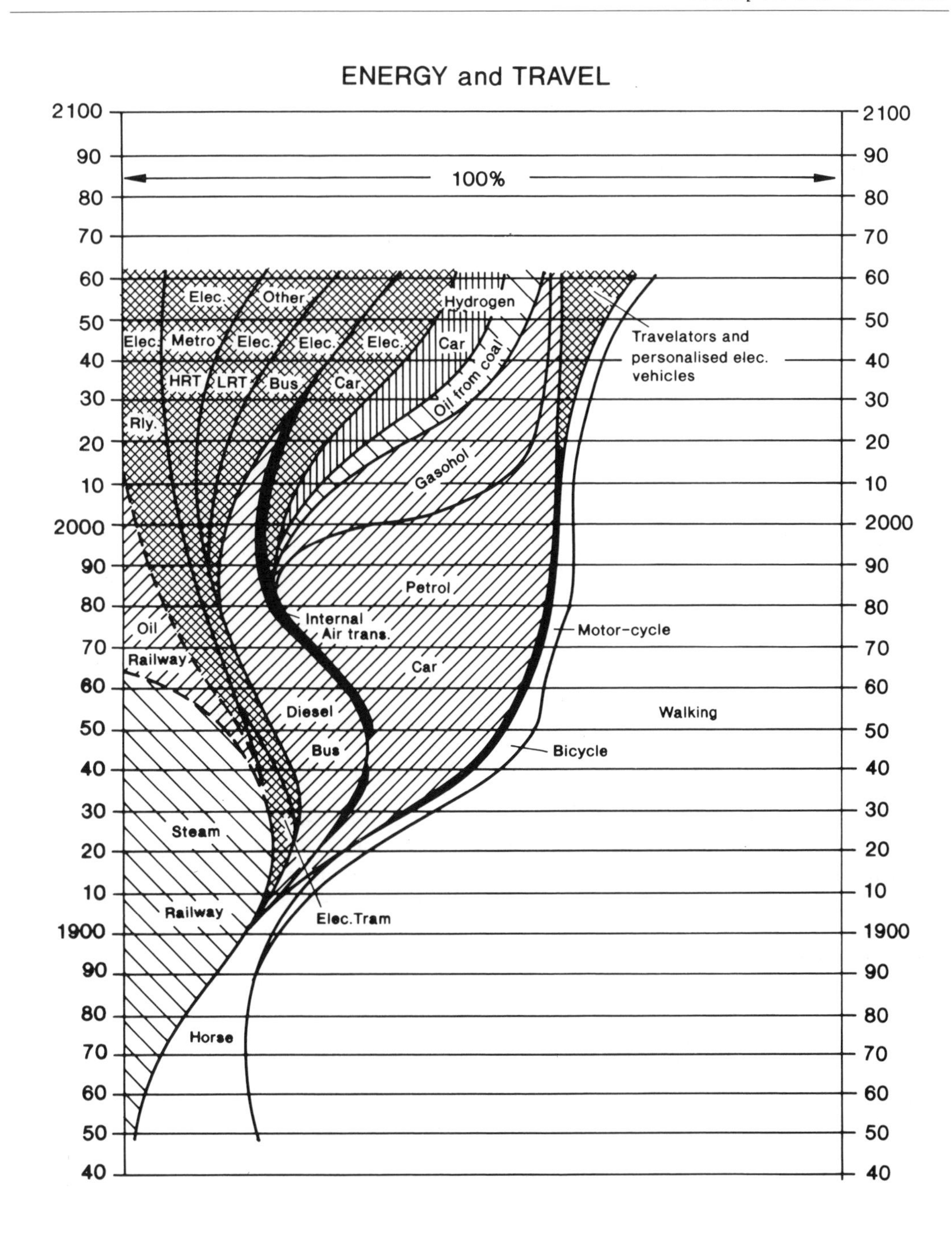

Figure 14.4 Energy and travel: a mid-term future.

Source: Williams, 1982.

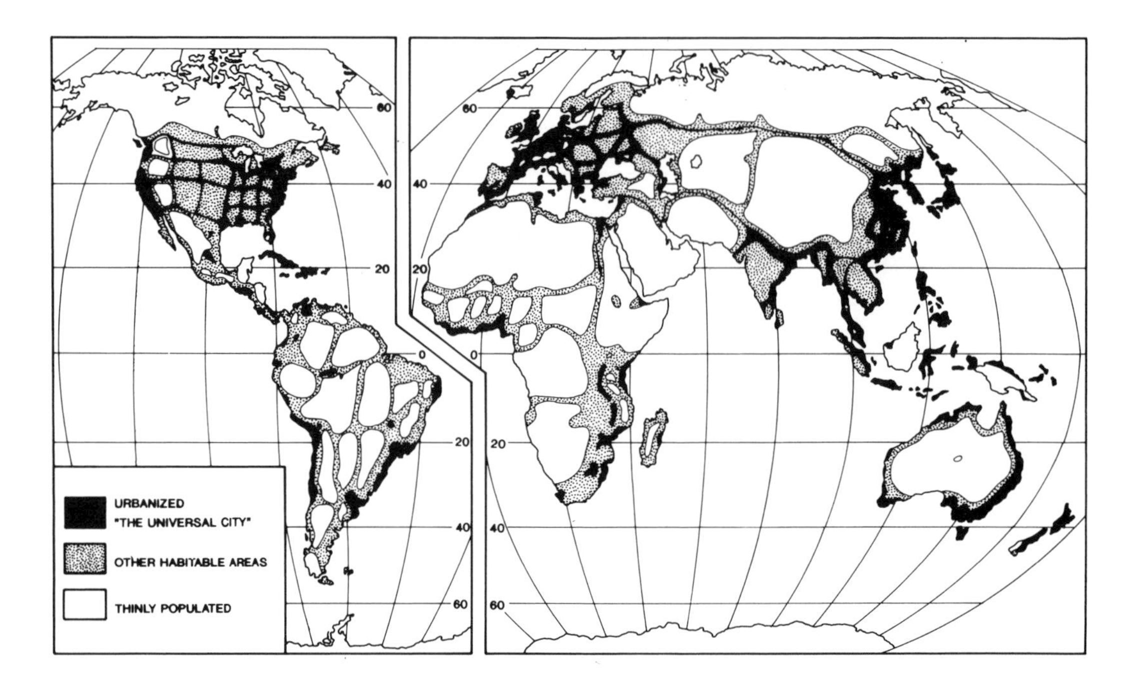

Figure 14.5 'Ecumenopolis' at the end of the 21st century?
Source: Doxiadis, 1970.

intercity railways throughout the world derive benefit from standardization and automated control, with their freight (and passenger) operations benefiting greatly from the development of the landbridge principle (Chapter 11). Tunnelling technology has advanced as far beyond the Channel Tunnel of the 1990s as this itself had advanced since the 1880s Severn Tunnel. In consequence, there are railways under straits such as Bering and Gibraltar and railway-carrying bridges such as over the Oresund.

As for international air transport, by 2020 it will have reached maximum route length as every major city around the globe is linked with almost every other, and it soon after reaches 'maturity' in passenger-miles flown. A two-tier aviation technology, 'old' and 'new', will clearly emerge. The bulk of airline passengers will be leisure-bound and carried over distances ranging from 400 miles to 14,000 miles in 1,000-seat subsonic air-buses. The much smaller business market will be served by thermo-plastic built SSTs, the 'gas-guzzlers' of their age; their ancestor being a NASA-linked SST on the planning board in 1991, to be built larger than the largest passenger jet currently flying (the Boeing 747–400), first flown in 2005, and put into service in 2011. As Concorde had once doubled speeds at one bound, so the newer generation SSTs will do likewise:

> once people (especially people in the business world) have experienced the ability to get to New York in three hours and Australia in 14 hours there will be no going back . . . The sort of journey times we can expect in the first decades of the next century are probably London to New York in 40 minutes, London to Sydney in one and a half hours . . . the passenger rocket will certainly evolve . . . our airports will, perhaps 50 years from now, provide the launch platforms for automatic, vertical landing of long-range passenger rockets which will bring New York to within half an hour and Australia to under one hour of London. (Masefield, 1976, 218–19)

The chief development to have reduced the need for international and other business travel will be a spectacular growth in telecommunications systems, universally available and affordable. There will consequently be more working and shopping from home and much use will be made of information technology for remote business meetings. The Industrial Society has estimated that in the UK as many as 40 per cent of the total workforce may become 'telecommuters' by the middle of the century.

So 60 years from now, can we envisage John Bull and family living in their weatherproof, half-buried pod house in the now well-established leafy *Waterton* built along the Thames from the heavily-populated Docklands to Kent (a linear city first suggested by an environment minister in the 1990s)? Bull's wife is an international airline pilot on the Tokyo run, so there are few days when they do not see each other either morning or evening. Bull commutes to Brussels where he is an EEC archivist. This morning he takes ten whole minutes to reach the East London River Crossings in his community-owned electric runabout, which he plugs in (a fine if he forgets) before boarding the high-speed electric bus to Ashford. He knows he will reach his office in one hour from here, even if his train is using the old Eurotunnel rather than the new. At work, he tests a new programme which accesses the world's holdings in 42 languages on the subject of twentieth-century transport policies ('AUTOTRANS'). He can do this at home, of course, but he is needed in Amsterdam that afternoon. On the screen, his eye is caught by: 'KEEP LONDON MOVING. Massive developments like Canary Wharf were planned in docklands with a totally inadequate public transport network and with far more commuter car spaces than the roads can cope with. Great efforts have been made to help motorists, including building new roads and by giving tax incentives for the company car perk . . .' The date is 1989, the report cost only £5. Bull calls up fellow specialists in Stockholm, Helsinki, Belgrade and Madrid to arrange their home-based teleconference for the next day and his wife, videophoning from Osaka, tells him not to forget to collect John Jr. from the PITS (People's Information Technology School) at Newham. Due to address historians after lunch, he reaches Amsterdam by VTOL company car-plane in 20 minutes. Later, he indulges himself by flying the Busyjet STOL shuttle into the old London City Airport. Now, where had he left his community car that morning? He takes a self-guided taxi to Newham and home. He and John Jr. are due to meet Mary at Canary Wharf by 5.30 pm. They take the 30-year-old Waterton MAGLEV and catch a glimpse of an old hovercraft on the river. Quaint, that 1989 report, thinks John: back in the old days, didn't they understand that, to solve all their daily needs of moving in and between cities to reach all the good things in life, they needed to dovetail their transport policies with their land-development policies?

We have not been conspicuously successful at any time in the past when we have tried to look forward. It is unlikely that we shall reach the middle of the twenty-first century with only those transport technologies we know today. In the developed world, at least, perhaps we surrender too completely to modern technology, believing that our possession of it represents a sharp break with the past and a preparation for a future of great expectations. But we should neither be too euphoric nor too pessimistic. Will the global realities be shaped by too many people, conspicuous consumption, depleting resources and the millstones of poverty and hunger? How may we be affected by great events (wars; the fall of communism) and by environmental deterioration and global warming (Intergovernmental Panel on Climate Change, 1990)? Can we ease our way forward after setting ourselves goals and objectives, in transport as in other areas of human endeavour? Shall we proceed towards the realization of C. A. Doxiadis's dream (Figure 14.5) of the late twenty-first century, interconnected global city, Ecumenopolis, where movement is conducted along intensive corridors and led upwards by mile-high elevators (Doxiadis, 1970; Danforth, 1970)?

Or do we have some other vision of the

way the world of transport should be? We remind ourselves that technology creates new possibilities for human advice and action, but leaves their disposition uncertain (Mesthene, 1968). What its effects will be and what ends it will serve are not inherent in the transport technology, but depend on what people will do with it. What kind of mobility, what kind of access?

References

Boyce, R. R. and Williams, A. F. (1979), *The Bases of Economic Geography* (British edition), (London: Holt, Rinehart & Winston).

British Road Federation (1991), *Basic Road Statistics* (London: The Federation).

Danforth, Paul M. (1970), *Transport Control: A Technology On The Move* (London: Aldus).

Doxiadis, Constantinos A. (1970), 'Role of transportation in the cities of the future', *Traffic Engineering and Control*, 18–29.

Gunston, B. (1972), *Transport: Problems and Prospects* (London: Thames & Hudson).

Healey, Tim (1983), *Extraordinary Inventions* (London: Readers Digest).

Horizon Science Series (1991), *California Dreaming* (London: BBC Publications).

Institute of Electrical Engineers with Royal Institute of Navigation (1991), *Colloquium: Technology For Road Pricing and Route Guidance*, May 22 (London: IEE).

Intergovernmental Panel on Climate Change (1990), *Potential Impacts Working Group 2 Report: Summary of Likely Impacts of Climate Change on Transport*, Sect.5, 25–32 (Geneva: WMO/UNEP).

Masefield, Sir Peter G. (1976), 'The challenge of change in transport', *Geography* Vol. 61, Pt. 4, 206–20.

Mesthene, E. G. (1968), 'How technology will shape the future', *Science* CLXI July, 135–43.

Odell, P. R. (1981), 'The energy economy of Western Europe', *Geography* Vol. 66, Pt. 1, 1–14.

Perry, George (1972), 'Future passive: looking into the 21st century', *Sunday Times*, December 3.

Preston, Barbara (1991), *The Impact of the Motor Car* (Tregaron, Dyfed: Brefi Press).

Price, B. T. (1971), 'Transport in the eighties', *The Advancement of Science* 27, No. 133, 216–26.

Road Transport Forum (1991), *Transport And Society, Final Report* (London: Institute of Highways and Transportation).

Sletmo, Gunnar K. (1989), 'The quest for mobility', *Forces*, No. 87, 4–13.

Webster (1988), *The New World Dictionary* (New York: Prentice Hall).

Williams, A. F. (1982), 'Transport and energy' in J. Wreford Watson, *The United States: Habitation of Hope* (London: Longman).

Wootton, J. (1991), 'Road transport solutions with 2020 visions' (Lecture to the Road Transport Forum) (London: The Fellowship of Engineering).

Index